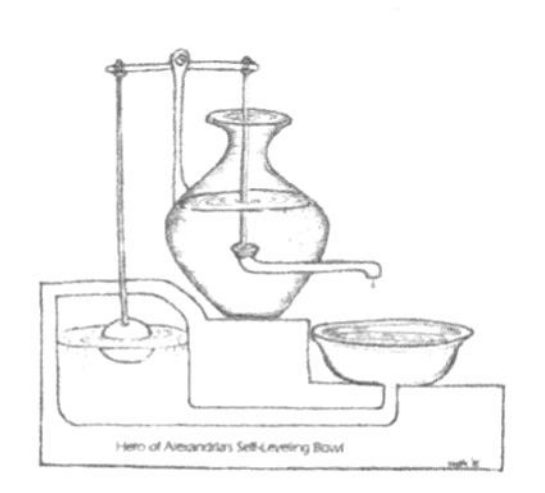

A Mathematical Introduction to Control Theory

Second Edition

SERIES IN ELECTRICAL AND COMPUTER ENGINEERING

Editor: Wai-Kai Chen *(University of Illinois, Chicago, USA)*

Published:

Vol. 1: Net Theory and Its Applications
Flows in Networks
by W. K. Chen

Vol. 2: A Mathematical Introduction to Control Theory
by S. Engelberg

Vol. 3: Intrusion Detection
A Machine Learning Approach
by Zhenwei Yu and Jeffrey J. P. Tsai

Vol. 4: A Mathematical Introduction to Control Theory (2nd Edition)
by S. Engelberg

Vol. 5: Security Modeling and Analysis of Mobile Agent Systems
by Lu Ma and Jeffrey J. P. Tsai

Vol. 6: Multi-Objective Group Decision Making: Methods Software and Applications with Fuzzy Set Techniques
by Jie Lu, Guangquan Zhang, Da Ruan and Fengjie Wu

SERIES IN ELECTRICAL AND COMPUTER ENGINEERING VOL. 4

Second Edition

A Mathematical Introduction to Control Theory

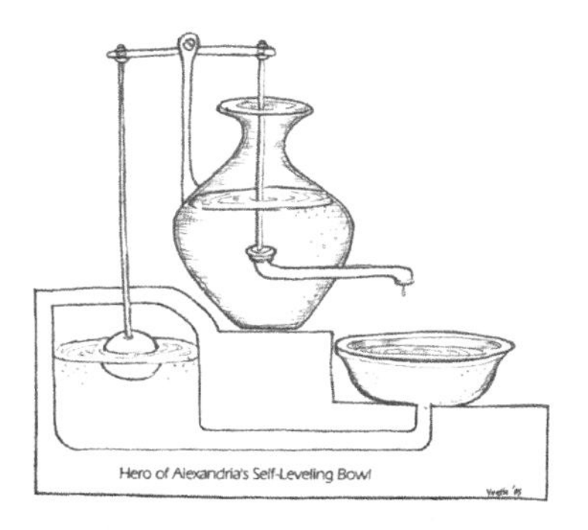
Hero of Alexandria's Self-Leveling Bowl

Shlomo Engelberg
Jerusalem College of Technology, Israel

ICP
Imperial College Press

Published by

Imperial College Press
57 Shelton Street
Covent Garden
London WC2H 9HE

Distributed by

World Scientific Publishing Co. Pte. Ltd.
5 Toh Tuck Link, Singapore 596224
USA office: 27 Warren Street, Suite 401-402, Hackensack, NJ 07601
UK office: 57 Shelton Street, Covent Garden, London WC2H 9HE

British Library Cataloguing-in-Publication Data
A catalogue record for this book is available from the British Library.

For MATLAB and Simulink product information, please contact:
The MathWorks, Inc.
3 Apple Hill Drive
Natick, MA, 01760-2098 USA
Tel: 508-647-7000
Fax: 508-647-7001
E-mail: info@mathworks.com
Web: mathworks.com
How to buy: www.mathworks.com/store

A MATHEMATICAL INTRODUCTION TO CONTROL THEORY
Second Edition

ISBN 978-1-78326-779-8

Printed in Singapore

Dedication

This book is dedicated to the memory of my beloved uncle
Stephen Aaron Engelberg (1940–2005)
who helped teach me how a *mensch* behaves
and how a person can love and appreciate learning.
May his memory be a blessing.

Preface to the Second Edition

Nearly ten years after the first edition of *A Mathematical Introduction to Control Theory* appeared, Imperial College Press in the person of Steven Patt asked if I would like to write a second edition. I agreed with alacrity.

Over the past ten years I have taught a control theory course of one sort or another many times, and there were whole subjects I wanted to add to the book. There were also errors to correct, explanations to improve, and examples and many exercises to add.

This edition treats several new subjects. In §2.6, we discuss systems with zeros in the right half-plane and show that they generally suffer from undershoot. We develop the conditions under which a system with feedback tracks a function of the form $t^n u(t)$ in §3.6. In §3.7, we define and determine conditions for the internal stability of a system with feedback. A brief history of formulas for calculating the roots of polynomials is given in §4.1. In §7.8, we provide an introduction to the Ziegler-Nichols rules for tuning a PID controller, and in §7.9, we discuss and prove Bode's sensitivity integrals. Finally, in §9.11, we consider observer design.

Over forty-five exercises have been added. The complete solutions to the new odd numbered exercises are given in the book's final chapter.

As always, there are many people to thank. I would like to thank all of my students at the Jerusalem College of Technology and at Bar Ilan University—without them there would neither have been a first nor a second edition of this book. It is my pleasure to thank Professor Aryeh Weiss of Bar Ilan University; he brought me to the Jerusalem College of Technology and to Bar Ilan University. Without him this book would never have been written.

I would like to thank my children, Chananel, Nediva, and Oriya for their patience and their willingness to allow me to spend time working on this book. Finally, I would like to thank my wife, Yvette, for her patience and for her encouragement to engage in activities that I love—like teaching and writing.

Shlomo Engelberg
Jerusalem, Israel

Preface to the First Edition

Control theory is largely an application of the theory of complex variables, modern algebra, and linear algebra to engineering. The main question that control theory answers is "given reasonable inputs, will my system give reasonable outputs?" Much of the answer to this question is given in the following pages. There are many books that cover control theory. What distinguishes this book is that it provides a complete introduction to control theory without sacrificing either the intuitive side of the subject or mathematical rigor. This book shows how control theory fits into the worlds of mathematics and engineering.

This book was written for students who have had at least one semester of complex analysis and some acquaintance with ordinary differential equations. Theorems from modern algebra are quoted before use—a course in modern algebra *is not* a prerequisite for this book; a single course in complex analysis is. Additionally, to properly understand the material on modern control a first course in linear algebra is necessary. Finally, sections 5.3 and 6.4 are a bit technical in nature; they can be skipped without affecting the flow of the chapters in which they are contained.

In order to make this book as accessible as possible many footnotes have been added in places where the reader's background—either in mathematics or in engineering—may not be sufficient to understand some concept or follow some chain of reasoning. The footnotes generally add some detail that is not directly related to the argument being made. Additionally, there are several footnotes that give biographical information about the people whose names appear in these pages—often as part of the name of some technique. We hope that these footnotes will give the reader something of a feel for the history of control theory.

In the first seven chapters of this book classical control theory is developed. The next three chapters constitute an introduction to three important areas of control theory: nonlinear control, modern control, and the control of hybrid systems. The final chapter contains solutions to some of the exercises. The first seven chapters can be covered in a reasonably paced one semester course. To cover the whole book will probably take most students and instructors two semesters.

The first chapter of this book is an introduction to the Laplace transform, a brief introduction to the notion of stability, and a short introduction to MATLAB®. MATLAB is used throughout this book as a very fancy calculator. MATLAB allows students to avoid some of the work that would once have had to be done by hand but which cannot be done by a person with either the speed or the accuracy with which a computer can do the same work.

The second chapter bridges the gap between the world of mathematics and of engineering. In it we present transfer functions, and we discuss how to use and manipulate block diagrams. The discussion is in sufficient depth for the non-engineer, and is hopefully not too long for the engineering student who may have been exposed to some of the material previously.

Next we introduce feedback systems. We describe how one calculates the transfer function of a feedback system. We provide a number of examples of how the overall transfer function of a system is calculated. We also discuss the sensitivity of feedback systems to their components. We discuss the conditions under which feedback control systems track their input. Finally we consider the effect of the feedback connection on the way the system deals with noise.

The next chapter is devoted to the Routh-Hurwitz Criterion. We state and prove the Routh-Hurwitz theorem—a theorem which gives a necessary and sufficient condition for the zeros of a real polynomial to be in the left half plane. We provide a number of applications of the theorem to the design of control systems.

In the fifth chapter, we cover the principle of the argument and its consequences. We start the chapter by discussing and proving the principle of the argument. We show how it leads to a graphical method—the Nyquist plot—for determining the stability of a system. We discuss low-pass systems, and we introduce the Bode plots and show how one can use them to determine the stability of such systems. We discuss the gain and phase margins and some of their limitations.

In the sixth chapter, we discuss the root locus diagram. Having covered a large portion of the classical frequency domain techniques for analyzing and designing feedback systems, we turn our attention to time-domain based approaches. We describe how one plots a root locus diagram. We explain the mathematics behind this plot—how the properties of the plot are simply properties of quotients of polynomials with real coefficients. We explain how one uses a root locus plot to analyze and design feedback systems.

In the seventh chapter we describe how one designs compensators for linear systems. Having devoted five chapters largely to the analysis of systems, in this chapter we concentrate on how to design systems. We discuss how one can use various types of compensators to improve the performance of a given system. In particular, we discuss phase-lag, phase-lead, lag-lead and PID (position integral derivative) controllers and how to use them.

In the eighth chapter we discuss nonlinear systems, limit cycles, the describing function technique, and Tsypkin's method. We show how the describing function is a very natural, albeit not always a very good, way of analyzing nonlinear circuits. We describe how one uses it to predict the existence and stability of limit cycles. We point out some of the limitations of the technique. Then we present Tsypkin's method which is an exact method but which is only useful for predicting the existence of limit cycles in a rather limited class of systems.

In the ninth chapter we consider modern control theory. We review the necessary background from linear algebra, and we carefully explain controllability and observability. Then we give necessary and sufficient conditions for controllability and observability of single-input single-output system. We also discuss the pole placement problem.

In the tenth chapter we consider discrete-time control theory and the control of hybrid systems. We start with the necessary background about the z-transform. Then we show how to analyze discrete-time system. The role of the unit circle is described, and the bilinear transform is carefully explained. We describe how to design compensators for discrete-time systems, and we give a brief introduction to the modified z-transform.

In the final chapter we provide solutions to selected exercises. The solutions are generally done at sufficient length that the student will not have to struggle too much to understand them. It is hoped that these solutions will be used instead of going to a friend or teacher to check one's answer. They should not be used to avoid thinking about how to go about solving the exercise or to avoid the real work of calculating the solution. In order to develop a good grasp of control theory, one must do problems. It is not enough to "understand" the material that has been presented; one must *experience* it.

Having spent many years preparing this book and having been helped by many people with this book, I have many people to thank. I am particularly grateful to Professors Richard G. Costello, Jonathan Goodman, Steven Schochet, and Aryeh Weiss who each read this work, critiqued it, and helped me improve it. I also grateful to the many anonymous referees whose comments helped me to improve my presentation of the beautiful results herein described.

I am happy to acknowledge Professor George Anastassiou's support. Professor Anastassiou has both encouraged me in my efforts to have this work published and has helped me in my search for a suitable publisher. My officemate, Aharon Naiman, has earned my thanks many, many times; he has helped me become more proficient in my use of LaTeX, put up with my enthusiasms, and helped me clarify my thoughts on many points.

My wife, Yvette, and my children, Chananel, Nediva, and Oriya, have always been supportive of my efforts; without Yvette's support this book would not have been written. My students been kind enough to put up with my penchant for handing out notes in English without complaining too bitterly; their comments have helped improve this book in many ways. My parents have, as always, been pillars of support. Without my father's love and appreciation of mathematics and science and my mother's love of good writing I would neither have desired to nor been suited to write a book of this nature. Because of the support of my parents, wife, children, colleagues, and students, writing this book has been a pleasant and meaningful as well as an interesting and challenging experience.

Though all of the many people who have helped and supported me over the years have made their mark on this work I, stubborn as ever, made the final decisions as to what material to include and how to present that material. The nicely turned phrase may well have been provided by a friend or mentor, by a parent or colleague; the mistakes are my own.

Shlomo Engelberg
Jerusalem, Israel

Contents

Preface to the Second Edition vii

Preface to the First Edition ix

1. Mathematical Preliminaries 1
   - 1.1 An Introduction to the Laplace Transform 1
   - 1.2 Properties of the Laplace Transform 2
   - 1.3 A Second Proof of the Initial Value Theorem 12
     - 1.3.1 The Intuition Behind the Proof 13
     - 1.3.2 The Proof 13
   - 1.4 Finding the Inverse Laplace Transform 15
     - 1.4.1 Some Simple Inverse Transforms 16
     - 1.4.2 The Quadratic Denominator 18
   - 1.5 Integro-Differential Equations 19
   - 1.6 An Introduction to Stability 24
     - 1.6.1 Some Preliminary Manipulations 24
     - 1.6.2 Stability 26
     - 1.6.3 Why We Obsess about Stability 28
     - 1.6.4 The Tacoma Narrows Bridge—a Brief Case History 29
   - 1.7 MATLAB® 29
     - 1.7.1 Assignments 29
     - 1.7.2 Commands 30
   - 1.8 Exercises 32

2. Transfer Functions 37

2.1 Transfer Functions 37
2.2 The Frequency Response of a System 39
2.3 Bode Plots 42
2.4 The Time Response of Certain "Typical" Systems 45
2.4.1 First Order Systems 45
2.4.2 Second Order Systems 46
2.5 Three Important Devices and Their Transfer Functions . 48
2.5.1 The Operational Amplifier (op amp) 48
2.5.2 The DC Motor 51
2.5.3 The "Simple Satellite" 52
2.6 Zeros in the Right Half-Plane 54
2.7 Block Diagrams and How to Manipulate Them 55
2.8 A Final Example 58
2.9 Exercises 61

3. Feedback—An Introduction 67

3.1 Why Feedback—A First View 67
3.2 Sensitivity 68
3.3 More about Sensitivity 70
3.4 A Simple Example 71
3.5 System Behavior at DC 72
3.6 Tracking Inputs of the Form $t^n u(t)$ 76
3.7 Internal Stability 78
3.8 Disturbance Rejection 80
3.9 Exercises 81

4. The Routh-Hurwitz Criterion 87

4.1 Roots of Polynomials—A Little History 87
4.2 Theorem, Proof, and Applications 88
4.3 A Design Example 97
4.4 Exercises 101

5. The Principle of the Argument and Its Consequences 105

5.1 More about Poles in the Right Half Plane 105
5.2 The Principle of the Argument 106
5.3 The Proof of the Principle of the Argument 107
5.4 How are Encirclements Measured? 109
5.5 First Applications to Control Theory 112

5.6 Systems with Low-Pass Open-Loop Transfer Functions . 114
5.7 MATLAB® and Nyquist Plots 120
5.8 The Nyquist Plot and Delays 121
5.9 Delays and the Routh-Hurwitz Criterion 125
5.10 Relative Stability 127
5.11 The Bode Plots 132
5.12 An (Approximate) Connection between Frequency Specifications and Time Specification 133
5.13 Some More Examples 136
5.14 Exercises 143

6. The Root Locus Diagram 151

6.1 The Root Locus—An Introduction 151
6.2 Rules for Plotting the Root Locus 153
6.2.1 The Symmetry of the Root Locus 153
6.2.2 Branches on the Real Axis 154
6.2.3 The Asymptotic Behavior of the Branches . . . 155
6.2.4 Departure of Branches from the Real Axis . . . 158
6.2.5 A "Conservation Law" 163
6.2.6 The Behavior of Branches as They Leave Finite Poles or Enter Finite Zeros 164
6.2.7 A Group of Poles and Zeros Near the Origin . . 166
6.3 Some (Semi-)Practical Examples 168
6.3.1 The Effect of Zeros in the Right Half-Plane . . . 168
6.3.2 The Effect of Three Poles at the Origin 169
6.3.3 The Effect of Two Poles at the Origin 169
6.3.4 A System that Tracks $t\sin(t)$ 171
6.3.5 Variations on Our Theme 174
6.3.6 The Effect of a Delay on the Root Locus Plot . 176
6.3.7 The Phase-lock Loop 179
6.3.8 Sounding a Cautionary Note—Pole-Zero Cancellation 182
6.4 More on the Behavior of the Roots of $Q(s)/K + P(s) = 0$ 184
6.5 Exercises 186

7. Compensation 193

7.1 Compensation—An Introduction 193

7.2 The Attenuator . . . 193
7.3 Phase-Lag Compensation . . . 194
7.4 Phase-Lead Compensation . . . 201
7.5 Lag-Lead Compensation . . . 206
7.6 The PID Controller . . . 207
7.7 An Extended Example . . . 214
7.7.1 The Attenuator . . . 215
7.7.2 The Phase-Lag Compensator . . . 215
7.7.3 The Phase-Lead Compensator . . . 217
7.7.4 The Lag-Lead Compensator . . . 219
7.7.5 The PD Controller . . . 221
7.8 The Ziegler-Nichols Rules . . . 222
7.8.1 Introduction . . . 222
7.8.2 The Assumption . . . 223
7.8.3 Characterizing the Plant . . . 224
7.8.4 A Partial Verification . . . 226
7.8.5 Estimating the System's Phase Margin . . . 228
7.8.6 Estimating the System's Gain Margin . . . 228
7.8.7 Estimating the System's Settling Time . . . 229
7.9 Bode's Sensitivity Integrals . . . 230
7.9.1 Introduction . . . 230
7.9.2 The Sensitivity Function . . . 230
7.9.3 Bode's Sensitivity Integrals . . . 230
7.9.4 Case I . . . 231
7.9.5 Case II . . . 233
7.9.6 Case III . . . 234
7.9.7 Case IV . . . 236
7.9.8 Available Bandwidth . . . 239
7.9.9 Open-Loop Unstable Systems with Limited Available Bandwidth . . . 240
7.10 Exercises . . . 241

8. Some Nonlinear Control Theory 247

8.1 Introduction . . . 247
8.2 The Describing Function Technique . . . 248
8.2.1 The Describing Function Concept . . . 248
8.2.2 Predicting Limit Cycles . . . 251
8.2.3 The Stability of Limit Cycles . . . 252
8.2.4 More Examples . . . 255

8.2.4.1 A Nonlinear Oscillator 255
8.2.4.2 A Comparator with a Dead Zone 256
8.2.4.3 A Simple Quantizer 257
8.2.5 Graphical Method 258
8.3 Tsypkin's Method 260
8.4 The Tsypkin Locus and the Describing Function Technique 265
8.5 Exercises 267

9. An Introduction to Modern Control 271

9.1 Introduction 271
9.2 The State Variables Formalism 271
9.3 Solving Matrix Differential Equations 273
9.4 The Significance of the Eigenvalues of the Matrix 274
9.5 Poles and Eigenvalues 276
9.6 Inhomogeneous Matrix Differential Equations 277
9.7 The Cayley-Hamilton Theorem 278
9.8 Controllability 279
9.9 Pole Placement 280
9.10 Observability 281
9.11 Observer Design 283
9.12 Choosing States When a System is Characterized by an ODE 285
9.13 Converting Transfer Functions to State Equations 286
9.14 Examples 289
9.14.1 Pole Placement 289
9.14.2 Adding an Integrator 290
9.14.3 Modern Control Using MATLAB® 292
9.14.4 A System that is not Observable 293
9.14.5 A Second System that is not Observable 294
9.14.6 A System that is neither Observable nor Controllable 295
9.14.7 Designing an Observer 296
9.15 Some Technical Results about Series of Matrices 297
9.16 Exercises 299

10. Control of Hybrid Systems 303

10.1 Introduction 303

10.2 The Definition of the Z-Transform 303
10.3 Some Examples 304
10.4 Properties of the Z-Transform 305
10.5 Sampled-data Systems 309
10.6 The Sample-and-Hold Element 310
10.7 The Delta Function and its Laplace Transform 312
10.8 The Ideal Sampler 313
10.9 The Zero-Order Hold 313
10.10 Calculating the Pulse Transfer Function 314
10.11 Using MATLAB® to Perform the Calculations 318
10.12 The Transfer Function of a Discrete-Time System 320
10.13 Adding a Digital Compensator 321
10.14 Stability of Discrete-Time Systems 323
10.15 A Condition for Stability 325
10.16 The Frequency Response 328
10.17 A Bit about Aliasing 330
10.18 The Behavior of the System in the Steady-State 330
10.19 The Bilinear Transform 331
10.20 The Behavior of the Bilinear Transform as $T \to 0$. 336
10.21 Digital Compensators 337
10.22 When Is There No Pulse Transfer Function? 340
10.23 An Introduction to The Modified Z-Transform 341
10.24 Exercises 343

11. Answers to Selected Exercises 347

11.1 Chapter 1 347
11.1.1 Exercise 1 347
11.1.2 Exercise 3 348
11.1.3 Exercise 5 349
11.1.4 Exercise 7 350
11.1.5 Exercise 9 350
11.1.6 Exercise 11 351
11.2 Chapter 2 352
11.2.1 Exercise 1 352
11.2.2 Exercise 3 352
11.2.3 Exercise 5 354
11.2.4 Exercise 7 355
11.2.5 Exercise 9 356
11.2.6 Exercise 11 356

11.3 Chapter 3 357
11.3.1 Exercise 1 357
11.3.2 Exercise 3 358
11.3.3 Exercise 5 359
11.3.4 Exercise 7 359
11.3.5 Exercise 9 360
11.3.6 Exercise 11 360
11.3.7 Exercise 13 360
11.4 Chapter 4 361
11.4.1 Exercise 1 361
11.4.2 Exercise 3 362
11.4.3 Exercise 5 363
11.4.4 Exercise 7 363
11.4.5 Exercise 9 364
11.4.6 Exercise 11 365
11.4.7 Exercise 13 366
11.5 Chapter 5 366
11.5.1 Exercise 1 366
11.5.2 Exercise 3 369
11.5.3 Exercise 5 370
11.5.4 Exercise 7 371
11.5.5 Exercise 9 371
11.5.6 Exercise 11 372
11.5.7 Exercise 13 373
11.5.8 Exercise 15 374
11.5.9 Exercise 17 375
11.5.10 Exercise 19 379
11.5.11 Exercise 21 381
11.5.12 Exercise 23 383
11.6 Chapter 6 385
11.6.1 Exercise 1 385
11.6.2 Exercise 3 386
11.6.3 Exercise 5 387
11.6.4 Exercise 7 389
11.6.5 Exercise 9 390
11.6.6 Exercise 11 392
11.6.7 Exercise 13 395
11.6.8 Exercise 15 395
11.6.9 Exercise 17 397

11.7 Chapter 7 398
11.7.1 Exercise 1 398
11.7.2 Exercise 3 400
11.7.3 Exercise 5 402
11.7.4 Exercise 7 403
11.7.5 Exercise 9 406
11.7.6 Exercise 11 407
11.7.7 Exercise 13 408
11.7.8 Exercise 15 409
11.7.9 Exercise 17 410
11.8 Chapter 8 411
11.8.1 Exercise 1 411
11.8.2 Exercise 3 414
11.8.3 Exercise 5 415
11.8.4 Exercise 7 415
11.9 Chapter 9 416
11.9.1 Exercise 1 416
11.9.2 Exercise 3 416
11.9.3 Exercise 4 417
11.9.4 Exercise 5 417
11.9.5 Exercise 6 418
11.9.6 Exercise 7 419
11.9.7 Exercise 11 419
11.9.8 Exercise 13 420
11.9.9 Exercise 15 420
11.10 Chapter 10 421
11.10.1 Exercise 4 421
11.10.2 Exercise 10 421
11.10.3 Exercise 13 422
11.10.4 Exercise 16 423
11.10.5 Exercise 17 424
11.10.6 Exercise 19 425

Bibliography 427

Index 429

Chapter 1

Mathematical Preliminaries

1.1 An Introduction to the Laplace Transform

Much of this chapter is devoted to describing and deriving some of the properties of the one-sided Laplace transform. The Laplace transform is the engineer's most important tool for analyzing the stability of linear, time-invariant, continuous-time systems. The Laplace transform is defined as:

$$\mathcal{L}(f(t))(s) \equiv \int_0^\infty e^{-st} f(t)\, dt.$$

We often write $F(s)$ for the Laplace transform of $f(t)$. It is customary to use lower-case letters for functions of time, t, and to use the same letter—but in its upper-case form—for the Laplace transform of the function; throughout this book, we follow this practice.

We assume that the functions $f(t)$ are of *exponential type*—that they satisfy an inequality of the form $|f(t)| \leq Ce^{\alpha t}, C \in \mathcal{R}$. If the real part of s, $\Re(s)$, satisfies $\Re(s) > \alpha$, then the integral that defines the Laplace transform converges. The Laplace transform's usefulness comes largely from the fact that it allows us to convert differential and integro-differential equations into algebraic equations.

We now calculate the Laplace transform of some functions. We start with the unit step function (also known as the Heaviside [1] function):

$$u(t) = \begin{cases} 0 & t < 0 \\ 1 & t \geq 0 \end{cases}.$$

[1] After Oliver Heaviside (1850-1925) who between 1880 and 1887 invented the "operational calculus" [OR]. His operational calculus was widely used in its time. The Laplace transform that is used today is a "cousin" of Heaviside's operational calculus [Dea97].

From the definition of the Laplace transform, we find that:

$$\begin{aligned}
U(s) &= \mathcal{L}(u(t))(s) \\
&= \int_0^\infty e^{-st} \cdot 1\, dt \\
&= \left.\frac{e^{-st}}{-s}\right|_0^\infty \\
&= \lim_{t\to\infty} \frac{e^{-st}}{-s} - \frac{1}{-s}.
\end{aligned}$$

Denote the real part of s by α and its imaginary part by β. Continuing our calculation, we find that:

$$\begin{aligned}
U(s) &= \lim_{t\to\infty} e^{-\alpha t} \frac{e^{-j\beta t}}{-s} + \frac{1}{s} \\
&= 0 + \frac{1}{s} = \frac{1}{s}.
\end{aligned}$$

This holds as long as $\alpha > 0$. In this case, the first term in the limit:

$$\lim_{t\to\infty} e^{-\alpha t} \frac{e^{-j\beta t}}{-s}$$

is approaching zero while the second term—though oscillatory—is bounded. In general, we assume that s is chosen so that integrals and limits that must converge do. For our purposes, the region of convergence (in terms of s) of the integral is not terribly important.

Next we consider $\mathcal{L}(e^{at})(s)$. We find that:

$$\begin{aligned}
\mathcal{L}(e^{at})(s) &= \int_0^\infty e^{-st} e^{at}\, dt \\
&= \left.\frac{e^{(a-s)t}}{a-s}\right|_0^\infty \\
&= \frac{1}{s-a}.
\end{aligned}$$

1.2 Properties of the Laplace Transform

The first property of the Laplace transform is its *linearity*.

Theorem 1

$$\mathcal{L}\left(\alpha f(t) + \beta g(t)\right)(s) = \alpha F(s) + \beta G(s).$$

Simply put, "the Laplace transform of a linear combination is the linear combination of the Laplace transforms."

PROOF: Making use of the properties of the integral, we find that:

$$\begin{aligned}\mathcal{L}\left(\alpha f(t) + \beta g(t)\right)(s) &= \int_0^\infty e^{-st}\left(\alpha f(t) + \beta g(t)\right)\,dt \\ &= \alpha \int_0^\infty e^{-st} f(t)\,dt + \beta \int_0^\infty e^{-st} g(t)\,dt \\ &= \alpha F(s) + \beta G(s).\end{aligned}$$

We see that the linearity of the Laplace transform is part of its "inheritance" from the integral which defines it.

The Laplace Transform of $\sin(t)$ I—An Example

Following the engineering convention that $j \equiv \sqrt{-1}$, we write:

$$\sin(t) = \frac{e^{jt} - e^{-jt}}{2j}.$$

By linearity we find that:

$$\mathcal{L}(\sin(t))(s) = \frac{1}{2j}\left(\mathcal{L}(e^{jt})(s) - \mathcal{L}(e^{-jt})(s)\right).$$

Making use of the fact that we know what the Laplace transform of an exponential is, we find that:

$$\mathcal{L}(\sin(t))(s) = \frac{1}{2j}\left(\frac{1}{s-j} - \frac{1}{s+j}\right) = \frac{1}{s^2+1}.$$

The next property we consider is the property that makes the Laplace transform transform so useful. As we shall see, it is possible to calculate the Laplace transform of the solution of a constant-coefficient ordinary differential equation (ODE) *without solving the ODE.*

Theorem 2 *Assume that $f(t)$ is has a well defined limit as t approaches zero from the right*[2]*. Then we find that:*

$$\mathcal{L}(f'(t))(s) = sF(s) - f(0^+).$$

PROOF: This result is proved by making use of integration by parts. We see that:

$$\mathcal{L}(f'(t))(s) = \int_0^\infty e^{-st} f'(t)\, dt.$$

Let $u = e^{-st}$ and $dv = f'(t)dt$. Then $du = -se^{-st}$ and $v = f(t)$. Assuming that $\alpha = \mathcal{R}(s) > 0$, we find that:

$$\begin{aligned}\int_0^\infty e^{-st} f'(t)\, dt &= -\int_0^\infty \frac{d}{dt} e^{-st} f(t)\, dt + e^{-st} f(t)|_{0+}^\infty \\ &= s\int_0^\infty e^{-st} f(t)\, dt + \lim_{t\to\infty} e^{-st} f(t) - f(0^+) \\ &= sF(s) + 0 - f(0^+) \\ &= sF(s) - f(0^+).\end{aligned}$$

We take the limit of $f(t)$ as $t \to 0^+$ because the integral itself deals only with positive values of t. Often we dispense with the added generality that the limit from the right gives us, and we write $f(0)$.

We can use this theorem to find the Laplace transform of the second (or higher) derivative of a function. To find the Laplace transform of the second derivative of a function, one applies the theorem twice. I.e.:

$$\begin{aligned}\mathcal{L}(f''(t))(s) &= s\mathcal{L}(f'(t))(s) - f'(0) \\ &= s(sF(s) - f(0)) - f'(0) \\ &= s^2F(s) - sf'(0) - f(0).\end{aligned}$$

The Laplace Transform of $\sin(t)$ II—An Example

[2]The limit of $f(t)$ as t tends to zero from the right is the value to which $f(t)$ tends as t approaches zero through the positive numbers. In many cases, we assume that $f(t) = 0$ for $t \leq 0$. Sometimes there is a jump in the value of the function at $t = 0$. As the zero value for $t \leq 0$ is often something we do not want to relate to, we sometimes consider only the limit from the right. The limit as one approaches a number, a, from the right is denoted by a^+. By convention $f(0^+) \equiv \lim_{t\to 0^+} f(t)$. Of course, if $f(t)$ is continuous at 0, then $f(0^+) = f(0)$.

We now calculate the Laplace transform of $\sin(t)$ a second way. Let $f(t) = \sin(t)$. Note that $f''(t) = -f(t)$ and that $f(0) = 0, f'(0) = 1$. We find that:

$$\begin{aligned}
\mathcal{L}(-\sin(t))(s) &= s^2\mathcal{L}(\sin(t))(s) - s \cdot 0 - 1 \Leftrightarrow \\
-\mathcal{L}(\sin(t))(s) &= s^2\mathcal{L}(\sin(t))(s) - 1 \Leftrightarrow \\
(s^2+1)\mathcal{L}(\sin(t))(s) &= 1 \Leftrightarrow \\
\mathcal{L}(\sin(t))(s) &= \frac{1}{s^2+1}.
\end{aligned}$$

The Laplace Transform of $\cos(t)$—An Example

From the fact that $\cos(t) = (\sin(t))'$ and that $\sin(0) = 0$, we see that:

$$\mathcal{L}(\cos(t))(s) = s\mathcal{L}(\sin(t))(s) - 0 = \frac{s}{s^2+1}.$$

An easy corollary of Theorem 2 is:

Corollary 3 $\mathcal{L}\left(\int_0^t f(y)\,dy\right)(s) = \frac{F(s)}{s}$.

PROOF: Let $g(t) = \int_0^t f(y)\,dy$. Clearly, $g(0) = 0$, and $g'(t) = f(t)$. From Theorem 2, we see that $\mathcal{L}(g'(t))(s) = s\mathcal{L}(g(t))(s) - 0 = \mathcal{L}(f(t))(s)$. We find that $\mathcal{L}(\int_0^t f(y)\,dy) = F(s)/s$.

We have seen how to calculate the transform of the derivative of a function; the transform of the derivative is s times the transform of the original function less a constant. We now show that the derivative of the transform of a function is the transform of $-t$ times the original function. By linearity this is identical to:

Theorem 4

$$\mathcal{L}(tf(t))(s) = -\frac{d}{ds}F(s)$$

PROOF:

$$\begin{aligned}
-\frac{d}{ds}F(s) &= -\frac{d}{ds}\int_0^\infty e^{-st}f(t)\,dt \\
&= -\int_0^\infty \frac{d}{ds}e^{-st}f(t)\,dt \\
&= -\int_0^\infty e^{-st}\left(-tf(t)\right)dt \\
&= \mathcal{L}(tf(t)).
\end{aligned}$$

The Transforms of $t\sin(t)$ and te^{-t}—An Example
Using Theorem 4, we find that:

$$\mathcal{L}(t\sin(t)) = -\frac{d}{ds}\left(\frac{1}{s^2+1}\right) = \frac{2s}{(s^2+1)^2}.$$

Similarly, we find that:

$$\mathcal{L}(te^{-t}) = -\frac{d}{ds}\left(\frac{1}{s+1}\right) = \frac{1}{(s+1)^2}.$$

We see that there is a connection between transforms whose denominators have repeated roots and functions that have been multiplied by powers of t.

As many equations have solutions of the form $y(t) = e^{-at}f(t)$, it will prove useful to know how to calculate $\mathcal{L}\left(e^{-at}f(t)\right)$. We find that:

Theorem 5

$$\mathcal{L}\left(e^{-at}f(t)\right)(s) = F(s+a).$$

PROOF:

$$\begin{aligned}
\mathcal{L}\left(e^{-at}f(t)\right)(s) &= \int_0^\infty e^{-st}e^{-at}f(t)\,dt \\
&= \int_0^\infty e^{-(s+a)t}f(t)\,dt \\
&= F(s+a).
\end{aligned}$$

The Laplace Transform of $te^{-t}\sin(t)$—An Example

Consider $\mathcal{L}\left(e^{-t}\sin(t)\right)(s)$. As we know that $\mathcal{L}\left(\sin(t)\right)(s) = \frac{1}{s^2+1}$, using our theorem we find that:

$$\mathcal{L}\left(e^{-t}\sin(t)\right)(s) = \frac{1}{(s+1)^2+1}.$$

As we know that multiplication in the time domain by t is equivalent to taking minus the derivative of the Laplace transform we find that:

$$\mathcal{L}\left(te^{-t}\sin(t)\right)(s) = -\frac{d}{ds}\left(\frac{1}{s^2+2s+2}\right).$$

Using the chain rule[3] we find that:

$$(1/f(s))' = (-1/f^2(s))f'(s).$$

Thus we find that $\mathcal{L}(te^{-t}\sin(t))(s)$ is equal to:

$$\begin{aligned} -\frac{d}{ds}\left(\frac{1}{s^2+2s+2}\right) &= -\frac{-1}{(s^2+2s+2)^2}\frac{d}{ds}\left(s^2+2s+2\right) \\ &= \frac{2s+2}{(s^2+2s+2)^2}. \end{aligned}$$

Often we need to calculate the Laplace transform of a function $g(t) = f(at), a > 0$. It is important to understand the effect that this "dilation" of the time variable has on the Laplace transform. We find that:

Theorem 6

$$\mathcal{L}(f(at))(s) = \frac{1}{a}F(s/a), a > 0.$$

PROOF:

$$\begin{aligned} \mathcal{L}(f(at))(s) &= \int_0^\infty e^{-st} f(at)\, dt \\ &\overset{u=at}{=} \frac{1}{a}\int_0^\infty e^{-(s/a)u} f(u)\, du \\ &= \frac{1}{a}F(s/a). \end{aligned}$$

---

[3]The chain rule states that:

$$\frac{d}{dt}f(g(t)) = \left.\frac{df(y)}{dy}\right|_{y=g(t)} g'(t).$$

$\sin(\omega t)$—An Example

Suppose $f(t) = \sin(\omega t)$. What is $\mathcal{L}(f(t))(s)$? Using Theorem 6, we find that:

$$\mathcal{L}(\sin(\omega t))(s) = \frac{1}{\omega}\frac{1}{(s/\omega)^2+1} = \frac{\omega}{s^2+\omega^2}.$$

Because we often need to model the effects of a delay on a system, it is important for us to understand how the addition of a delay to a function affects the function's Laplace transform. We find that:

Theorem 7

$$\mathcal{L}(f(t-T)u(t-T))(s) = e^{-Ts}F(s), T \geq 0.$$

PROOF:

$$\begin{aligned}
\mathcal{L}(f(t-T)u(t-T))(s) &= \int_0^\infty e^{-st}f(t-T)u(t-T)\,dt \\
&= \int_T^\infty e^{-st}f(t-T)\,dt \\
&\stackrel{u=t-T}{=} \int_0^\infty e^{-s(u+T)}f(u)\,du \\
&= e^{-sT}\int_0^\infty e^{-su}f(u)\,du \\
&= e^{-sT}F(s).
\end{aligned}$$

It is important to be able to determine the steady-state output of a system—the output of the system after all the transients have died down. For this purpose one can sometimes make use of the final value theorem:

Theorem 8 *If $f(t)$ approaches a limit as $t \to \infty$ then:*

$$\lim_{t\to\infty} f(t) = \lim_{s\to 0^+} sF(s).$$

PROOF: In order to prove this result, we make use of the triangle inequality:

$$|a+b| \leq |a| + |b|$$

which states that the absolute value of the sum of two numbers is less than or equal to the sum of the absolute value of the numbers considered separately. We also make use of the generalized triangle inequality:

$$\left| \int_a^b f(t)\, dt \right| \leq \int_a^b |f(t)|\, dt, \quad b \geq a$$

which states that the absolute value of the integral of a function is less than or equal to the integral of the absolute value of the function.

We start the proof of the theorem by considering functions, $f(t)$, that tend to zero at infinity. For such $f(t)$, for any $\epsilon > 0$ there exists a value of T for which $|f(t)| < \epsilon$ for all $t \geq T$.

Consider $F(s)$ for $s = a > 0$. We find that:

$$\begin{aligned} |F(a)| &= \left| \int_0^T e^{-at} f(t)\, dt + \int_T^\infty e^{-at} f(t)\, dt \right| \\ &\leq \int_0^T e^{-at} |f(t)|\, dt + \int_T^\infty e^{-at} |f(t)|\, dt \\ &\leq C + \int_T^\infty e^{-at} \epsilon\, dt \\ &= C + e^{-aT} \epsilon / a \\ &\leq C + \epsilon / a. \end{aligned}$$

That is, for $s = a > 0$, $|aF(a)| \leq Ca + \epsilon$. As $a \to 0^+$, we find that this expression is less than or equal to ϵ. As this is true for all $\epsilon > 0$, we find that:

$$\lim_{s \to 0^+} |sF(s)| = 0.$$

Consequently, when $f(t) \to 0$ as $t \to \infty$, we find that:

$$\lim_{s \to 0^+} sF(s) = 0 = \lim_{t \to \infty} f(t).$$

Next, consider the case that $f(t)$ tends to a non-zero limit, D. Let $g(t) \equiv f(t) - Du(t)$. Clearly,

$$\lim_{s \to 0^+} sG(s) = 0.$$

From the definition of the Laplace transform, we find that $G(s) = F(s) - D/s$. Thus, we find that:

$$\lim_{s \to 0+} s(F(s) - D/s) = 0.$$

This is equivalent to:

$$\lim_{s \to 0+} sF(s) = D = \lim_{t \to \infty} f(t)$$

which is the result we set out to prove.

A *Mis*application of the Final Value Theorem—An Example

Let us find the final value of $\sin(t)$. Using Theorem 8, we find that:

$$\lim_{t \to \infty} \sin(t) = \lim_{s \to 0} s \frac{1}{s^2 + 1} = 0.$$

This is ridiculous; the sine function has no final value. What happened? One of the conditions of Theorem 8 is that the function under consideration have a limit as $t \to \infty$. If the function does not have a limit, then the theorem does not apply, and when the theorem does not apply one can get nonsensical results if one uses the theorem (but see exercise 9).

The Delayed Unit Step Function—An Example

Consider the function $f(t) = u(t - a), a > 0$. Clearly, here a final value does exist; the final value of a delayed unit step function is 1. The Laplace transform of the delayed unit step function is (according to Theorem 7):

$$\mathcal{L}(f(t))(s) = \frac{e^{-as}}{s}.$$

From the final value theorem, we see that:

$$\begin{aligned}
\lim_{t \to \infty} u(t - a) &= \lim_{s \to 0+} s\mathcal{L}(u(t - a))(s) \\
&= \lim_{s \to 0+} s \frac{e^{-as}}{s} \\
&= \lim_{s \to 0+} e^{-as} \\
&= 1.
\end{aligned}$$

This is just as it should be.

We will also find use for the initial value theorem:

Theorem 9

$$f(0^+) = \lim_{t \to 0+} f(t) = \lim_{\Re(s) \to +\infty} sF(s).$$

PROOF: In the proof of Theorem 8, we found that:

$$\begin{aligned} sF(s) &= s \int_0^\infty e^{-st} f(t)\, dt \\ &= f(0^+) + \int_0^\infty e^{-st} f'(t)\, dt. \end{aligned}$$

As long as $f'(t)$ is of exponential type, we know that:

$$|f'(t)| \le \mathrm{C}e^{at}.$$

If this is so, then from the generalized triangle inequality we find that:

$$\begin{aligned} \left| \int_0^\infty e^{-st} f'(t)\, dt \right| &\le \int_0^\infty |e^{-st} f'(t)|\, dt \\ &= \int_0^\infty |e^{-st}||f'(t)|\, dt \\ &\le \int_0^\infty |e^{-st}|\mathrm{C}e^{at}\, dt \\ &= \mathrm{C} \int_0^\infty e^{-\mathcal{R}(s)t} e^{at}\, dt \\ &= \mathrm{C} \int_0^\infty e^{-(\mathcal{R}(s)-a)t}\, dt. \end{aligned}$$

It is clear that as $\mathcal{R}(s) \to \infty$ the integral above tends to zero. This completes the proof of the theorem. (For a second proof of the theorem, see §1.3.)

The Delayed Cosine—An Example

Let us make use of the initial value theorem to find the value of $\cos(t - \tau)u(t - \tau), \tau > 0$ when $t = 0$. Evaluating the expression we see that it is equal to $\cos(-\tau)u(-\tau) = \cos(\tau) \cdot 0 = 0$. We know that:

$$\mathcal{L}(\cos(t - \tau)u(t - \tau))(s) = e^{-\tau s} \frac{s}{s^2 + 1}.$$

Using the initial value theorem we find that:

$$\cos(0-\tau)u(0-\tau) = \lim_{\Re(s)\to\infty} se^{-\tau s}\frac{s}{s^2+1} = 0$$

as it must.

We present one final, important, theorem without proof.

Theorem 10 *If $f(t)$ and $g(t)$ are piecewise continuous*[4] *and if $\mathcal{L}(f(t))(s) = \mathcal{L}(g(t))(s)$, then $f(t) = g(t)$ (except, possibly, at the points of discontinuity). I.e., the Laplace transform is (to all intents and purposes) unique.*

Thus, if one recognizes $F(s)$ as the Laplace transform of $f(t)$ one can state that $F(s)$ does come from $f(t)$. It is not possible that there is a second piecewise continuous function $\tilde{f}(t)$ whose transform is also $F(s)$.

Solving Ordinary Differential Equations—An Example

Let us solve the equation $y'(t) = -2y(t)$, $y(0) = 4$. We start by finding the Laplace transform of both sides of the equation. We find that:

$$sY(s) - 4 = -2Y(s) \Leftrightarrow Y(s) = \frac{4}{s+2}.$$

We know that $\mathcal{L}(e^{-2t})(s) = \frac{1}{s+2}$. From the linearity of the Laplace transform we know that $\mathcal{L}(4e^{-2t})(s) = \frac{4}{s+2}$. From our uniqueness result, we find that $y(t) = 4e^{-2t}$.

1.3 A Second Proof of the Initial Value Theorem

We now present a second proof of the initial value theorem:

$$f(0^+) = \lim_{t\to 0+} f(t) = \lim_{s\to+\infty} sF(s).$$

We assume that s tends to infinity through the positive real numbers.

[4] Piecewise continuous functions are functions that are "pieced together" from continuous functions. An example is the delayed unit step function $u(t-2)$ which is 0 until 2, and is 1 afterwards. Both 0 and 1 are continuous functions. At the interface of the continuous pieces, at $t = 2$, $u(t-2)$ is not continuous.

1.3.1 *The Intuition Behind the Proof*

Before starting the "formal" part of the proof, we consider the function $se^{-st}, \quad t \geq 0$ for large value of s. We find that:

- When $t = 0$, the function is equal to s—a very large value.
- By the time $t = 1/\sqrt{s}$, after a very short period has elapsed, the function is "essentially" zero. That is, the function is very large near zero and tends to zero almost immediately.
- Additionally, the integral of the function is:

$$\int_0^\infty se^{-st}\,dt = -e^{-st}\Big|_0^\infty = 1.$$

That is, the function is very large near $t = 0$, is very small everywhere else, and its integral is one. The function looks very much like a one-sided delta function. If as $s \to \infty$ it indeed tended to a one-sided function, then it would be clear that:

$$\lim_{s\to\infty} \int_0^\infty se^{-st} f(t)\,dt = \int_0^\infty \delta_{\text{one-sided}} f(t)\,dt = f(0^+).$$

We now make these ideas precise without actually using a delta function.

1.3.2 *The Proof*

We start the proof by making use of the definition of the $f(0^+)$. For any $\epsilon > 0$, there exists a value $S > 0$ such that for all $s > S$ and $t \in [0, 1/\sqrt{s}]$ we have $|f(t) - f(0^+)| < \epsilon$. Let $s > S$. Then:

$$\int_0^\infty se^{-st} f(t)\,dt = \int_0^{1/\sqrt{s}} se^{-st} f(t)\,dt + \int_{1/\sqrt{s}}^\infty se^{-st} f(t)\,dt.$$

Our goal is to show that the first integral on the right tends to $f(0^+)$ and the second tends to zero.

Let us consider the first of the two integrals. We find that:

$$\int_0^{1/\sqrt{s}} se^{-st} f(t)\,dt = \int_0^{1/\sqrt{s}} se^{-st}(f(t) - f(0^+))\,dt + \int_0^{1/\sqrt{s}} se^{-st} f(0^+)\,dt.$$

The absolute value of the first integral is bounded by:

$$\begin{aligned}
\left|\int_0^{1/\sqrt{s}} se^{-st}(f(t)-f(0^+))\,dt\right| &\leq \int_0^{1/\sqrt{s}} se^{-st}|f(t)-f(0^+)|\,dt \\
&\leq \int_0^{1/\sqrt{s}} se^{-st}\epsilon\,dt \\
&= \epsilon\left(-e^{-st}\Big|_0^{1/\sqrt{s}}\right. \\
&\leq \epsilon.
\end{aligned}$$

As ϵ can be chosen to be arbitrarily small, this integral can be made arbitrarily small. Considering the second of the two integrals, we find that:

$$\int_0^{1/\sqrt{s}} se^{-st}f(0^+)\,dt = (1-e^{-\sqrt{s}})f(0^+).$$

Clearly, as $s \to \infty$, this integral tends to $f(0^+)$.

Finally, consider:

$$\int_{1/\sqrt{s}}^{\infty} se^{-st}f(t)\,dt.$$

Assuming that $f(t)$ is of exponential order, $|f(t)| < Ce^{dt}$. Let us bound the absolute value of the integral for any such function. We find that:

$$\begin{aligned}
\left|\int_{1/\sqrt{s}}^{\infty} se^{-st}f(t)\,dt\right| &\leq \int_{1/\sqrt{s}}^{\infty} se^{-st}|f(t)|\,dt \\
&\leq \int_{1/\sqrt{s}}^{\infty} se^{-st}Ce^{dt}\,dt \\
&\leq C\int_{1/\sqrt{s}}^{\infty} se^{-(s+d)t}\,dt \\
&\leq -C\frac{s}{s+d}\,e^{-(s+d)t}\Big|_{1/\sqrt{s}}^{\infty} \\
&= C\frac{s}{s+d}e^{-(s+d)/\sqrt{s}}.
\end{aligned}$$

As $s \to \infty$, the exponential drives the integral to zero.

We find that as $s \to \infty$, all the terms but one tend to zero, and the remaining integral tends to $f(0^+)$. Thus,

$$\lim_{s\to\infty} sY(s) = f(0^+).$$

1.4 Finding the Inverse Laplace Transform

We will not attempt to find a formula that gives us the inverse Laplace transform of a function. (Such a formula exists, but it is somewhat complicated.) In general, one calculates the inverse Laplace transform of a Laplace transform by inspection. That is, one "massages" the transform until one has it in a form that one recognizes. Note that with the exception of Theorem 7 all of the transforms that we have encountered and all of the theorems that we have seen lead to transforms that are rational functions with real coefficients—that are of the form:

$$\frac{P(s)}{Q(s)}, \quad P(s) = a_n s^n + \cdots + a_0, \quad Q(s) = b_m s^m + \cdots + b_0$$

where the coefficients are all real numbers.

From modern algebra we know that all such fractions can be written in the form:

$$\begin{aligned}\frac{P(s)}{Q(s)} &= R(s) + \frac{a_{00}}{s-p_0} + \frac{a_{01}}{(s-p_0)^2} + \cdots \frac{a_{0n_0}}{(s-p_0)^{n_0}} \\ &+ \cdots \\ &+ \frac{a_{l0}}{s-p_l} + \frac{a_{l1}}{(s-p_l)^2} + \cdots + \frac{a_{p_{n_l}}}{(s-p_l)^{n_l}} \\ &+ \frac{b_{00}s + c_{00}}{(s-pc_0)(s-\overline{pc_0})} + \cdots + \frac{b_{0nc_0}s + c_{0nc_0}}{((s-pc_0)(s-\overline{pc_0}))^{nc_0}} \\ &+ \cdots \\ &+ \frac{b_{L0}s + c_{L0}}{(s-pc_L)(s-\overline{pc_L})} + \cdots + \frac{b_{Lnc_L}s + c_{Lnc_L}}{((s-pc_L)(s-\overline{pc_L}))^{nc_L}}\end{aligned}$$

where $R(s)$ is a polynomial with real coefficients, a_{ij}, b_{ij}, and c_{ij} are real constants, p_i are real poles[5] of the fraction, pc_i and $\overline{pc_i}$ are complex poles of the fraction, l is the number of distinct real poles of the fraction, L is the number of distinct pairs of complex poles, n_i is the number of times the real pole p_i is repeated, and nc_i is the number of times the complex pole pair $pc_i, \overline{pc_i}$ is repeated. Note that $R(s) = 0$ as long as the order of the denominator is greater than the order of the numerator—the most common case in control theory applications. As $(s-a)(s-\overline{a}) = s^2 - 2\Re(a)s + |a|^2$ is a polynomial with real coefficients, we find that all of the terms in the expansion above are rational function with real coefficients. The expression

[5]The poles of a rational function are those points at which the function becomes unbounded.

above is called the *partial fraction expansion* of $\frac{P(s)}{Q(s)}$. (It is often used in calculus to evaluate integrals of rational functions [Tho68].)

1.4.1 Some Simple Inverse Transforms

(1) $P(s)/Q(s) = 1/(s+1)^3$. This fraction is already in partial fraction form. We must determine the function whose Laplace transform it "obviously" is. We know that $\mathcal{L}(e^{-t})(s) = 1/(s+1)$. We note that the second derivative of the Laplace transform is $2/(s+1)^3$. Taking a second derivative in the Laplace domain is, according to Theorem 4, the same as multiplying the time function by t^2. Thus, $\mathcal{L}(t^2 e^{-t}/2)(s) = 1/(s+1)^3$.

(2) $P(s)/Q(s) = 1/(s^2-1)$. We can factor the denominator into $(s-1)(s+1)$. We find that:

$$\frac{1}{s^2-1} = \frac{a}{s-1} + \frac{b}{s+1}.$$

One way of finding the coefficients a and b is to multiply both sides of the equation by $s^2 - 1$ and to equate the coefficients of the resulting polynomials. One finds that:

$$1 = a(s+1) + b(s-1) \Leftrightarrow 1 = (a+b)s + (a-b).$$

Equating coefficients of like powers of s, we find that $a + b = 0$, and $a - b = 1$. Adding the two equations we find that $a = 1/2$. Clearly $b = -1/2$. We find that:

$$\frac{1}{s^2-1} = \frac{1/2}{s-1} - \frac{1/2}{s+1}.$$

Recognizing that:

$$\mathcal{L}((1/2)e^{-t})(s) = \frac{1/2}{s-1}$$

and that:

$$\mathcal{L}((1/2)e^{t})(s) = \frac{1/2}{s+1},$$

we find that:

$$\mathcal{L}((1/2)(e^{-t} - e^{t}))(s) = \frac{1}{s^2-1}.$$

(3) $P(s)/Q(s) = 1/(s+1)^2$. This fraction is also in partial fraction form. We once again must determine its "obvious" inverse transform. One way to proceed is to note that $1/s^2$ is the Laplace transform of t and a shift by -1 in the s-plane is equivalent to multiplication by e^{-t} in time. The "obvious" inverse Laplace transform is te^{-t}. Another way to get to the same answer is to note that $\mathcal{L}(e^{-t})(s) = 1/(s+1)$ and that $-d/ds\{1/(s+1)\} = 1/(s+1)^2$. As the differentiation of the Laplace transform is equivalent to multiplication of the original function by $-t$, we find that the original function must have been te^{-t}.

(4)

$$P(s)/Q(s) = \frac{s+1}{s^3+4s^2+4s}.$$

We can factor the denominator as follows:

$$s^3+4s^2+4s = s(s^2+4s+4) = s(s+2)^2.$$

We see that:

$$\frac{s+1}{s^3+4s^2+4s} = \frac{A}{s} + \frac{B}{s+2} + \frac{C}{(s+2)^2}.$$

Multiplying through by s^3+4s^2+4s we find that:

$$s+1 = A(s^2+4s+4) + B(s^2+2s) + Cs.$$

Equating coefficients of like powers of s we find that:

$$\begin{aligned} A + B &= 0 \\ 4A + 2B + C &= 1 \\ 4A &= 1. \end{aligned}$$

We find that $A = 1/4$, $B = -1/4$, and $C = 1/2$. Following the logic of the previous example, we see that:

$$\mathcal{L}(te^{-2t})(s) = \frac{1}{(s+2)^2}.$$

Thus, the original function must have been:

$$\left(\frac{1}{4} - \frac{1}{4}e^{-2t} + \frac{1}{2}te^{-2t}\right) u(t).$$

1.4.2 *The Quadratic Denominator*

Consider a fraction of the form:

$$\frac{\alpha s + \beta}{as^2 + bs + c}$$

where all of the constants are real. The poles of the function are the zeros of the denominator. Using the quadratic formula we find that the poles are:

$$s = \frac{-b \pm \sqrt{b^2 - 4ac}}{2a}.$$

If:

$$b^2 - 4ac \geq 0,$$

then there are two real poles and we can find the inverse transform quite easily using the techniques demonstrated in the previous example. Suppose, however, that:

$$b^2 - 4ac < 0.$$

Then there are two complex conjugate poles:

$$s = \frac{-b \pm j\sqrt{4ac - b^2}}{2a}.$$

In this case, the simplest way to proceed is to complete the squares in the denominator and then manipulate the fractions into fractions whose inverse transforms we know. We find that:

$$\begin{aligned}\frac{\alpha s + \beta}{as^2 + bs + c} &= \frac{\alpha s + \beta}{a(s^2 + (b/a)s + (c/a))} \\ &= \frac{\alpha s + \beta}{a\left(\left(s + \frac{b}{2a}\right)^2 + \frac{4ac - b^2}{4a^2}\right)}\end{aligned}$$

$$= \frac{1}{a} \frac{\alpha\left(s + \frac{b}{2a}\right) - \frac{\alpha b - 2a\beta}{2a}}{\left(s + \frac{b}{2a}\right)^2 + \frac{4ac-b^2}{4a^2}}$$

$$= \frac{1}{a}\left(\frac{\alpha\left(s + \frac{b}{2a}\right)}{\left(s + \frac{b}{2a}\right)^2 + \frac{4ac-b^2}{4a^2}} - \frac{\frac{\alpha b - 2a\beta}{2a}}{\left(s + \frac{b}{2a}\right)^2 + \frac{4ac-b^2}{4a^2}}\right)$$

$$= \frac{1}{a}\left(\frac{\alpha\left(s + \frac{b}{2a}\right)}{\left(s + \frac{b}{2a}\right)^2 + \frac{4ac-b^2}{4a^2}} - \frac{\alpha b - 2a\beta}{2a\sqrt{\frac{4ac-b^2}{4a^2}}} \frac{\sqrt{\frac{4ac-b^2}{4a^2}}}{\left(s + \frac{b}{2a}\right)^2 + \frac{4ac-b^2}{4a^2}}\right)$$

$$= \frac{1}{a}\left(\frac{\alpha\left(s + \frac{b}{2a}\right)}{\left(s + \frac{b}{2a}\right)^2 + \frac{4ac-b^2}{4a^2}} - \frac{\alpha b - 2a\beta}{\sqrt{4ac - b^2}} \frac{\sqrt{\frac{4ac-b^2}{4a^2}}}{\left(s + \frac{b}{2a}\right)^2 + \frac{4ac-b^2}{4a^2}}\right)$$

After these simple (albeit tedious) operations, we find that "by inspection" the inverse transform is:

$$\frac{e^{-\frac{b}{2a}t}}{a}\left(\alpha \cos\left(\sqrt{\frac{4ac - ba^2}{4a^2}}t\right) - \frac{\alpha b - 2a\beta}{\sqrt{4ac - b^2}} \sin\left(\sqrt{\frac{4ac - ba^2}{4a^2}}t\right)\right) u(t).$$

Comparing the inverse Laplace transform and the poles of the fraction, we find that the rate of growth of the inverse Laplace transform is a function of the real part of the pole—$-b/(2a)$—and the frequency of the oscillations is controlled by the imaginary part of the pole:

$$\pm j\sqrt{\frac{4ac - ba^2}{4a^2}} = \pm j\frac{\sqrt{4ac - ba^2}}{2a}.$$

1.5 Integro-Differential Equations

An integro-differential equation is an equation in which integrals and derivatives appear. Many systems can be modeled using integro-differential equations. From Theorem 2 and Corollary 3, it is clear that both linear constant coefficient differential and integro-differential equation in $y(t)$ lead to expressions for $Y(s)$ that are rational functions of s. Using the techniques of the previous section, we should be able to find the inverse Laplace transform of such expressions. We consider a few examples.

The Laplace Transform of $\sinh(\omega t)$—An Example

If $y(t) = \sinh(\omega t)$, then $y'(t) = \omega\cosh(\omega t)$, and $y''(t) = \omega^2\sinh(\omega t)$. Clearly $y(0) = 0$, and $y'(0) = \omega$. Let us find the Laplace transform of the solution of the equation $y''(t) = \omega^2 y(t)$, $y(0) = 0, y'(0) = \omega$ in order to find the Laplace transform of $\sinh(\omega t)$.

$$\begin{aligned} y''(t) = \omega^2 y(t), \quad y(0) = 0, y'(0) = \omega \Leftrightarrow \\ s^2\mathcal{L}(y(t))(s) - y'(0) - sy(0) = \omega^2\mathcal{L}(y(t))(s) \Leftrightarrow \\ \left(s^2 - \omega^2\right)\mathcal{L}(y(t))(s) = \omega \Leftrightarrow \\ \mathcal{L}(y(t))(s) = \frac{\omega}{s^2 - \omega^2}. \end{aligned}$$

At this point we have found the Laplace transform of $\sinh(\omega t)$. We note that the Laplace transform is not in partial fraction form. As $s^2 - \omega^2 = (s - \omega)(s + \omega)$, find that:

$$\frac{\omega}{s^2 - \omega^2} = \frac{A}{s - \omega} + \frac{B}{s + \omega}$$

It is easy to see that $A = 1/2$ and $B = -1/2$. The inverse transform of $\mathcal{L}(y(t))$ is thus:

$$\sinh(\omega t) = \frac{e^{\omega t} - e^{-\omega t}}{2}$$

which is just as it should be.

A Simple Circuit—An Example

If one has a series circuit composed of a switch, a 5V battery, a 10Ω resistor[6], a 10 mH inductor[7] and a 625 μF capacitor[8] in series (see Figure 1.1) and there is initially no current flowing, the charge

[6]Ohm's law states that the voltage across a resistor, $V_R(t)$, is equal to the current flowing through the resistor, $i_R(t)$, times the resistance, R, of the resistor. I.e. $V_R(t) = i_R(t)R$.

[7]Recall that the voltage across an inductor, $V_L(t)$, is equal to the inductance, L, of the inductor times the time derivative of the current flowing through the inductor, $di_L(t)/dt$. That is, $V_L(t) = L di_L(t)/dt$.

[8]The charge stored by a capacitor, $Q_C(t)$, is equal to the capacitor's capacitance, C, times the voltage across the capacitor, $V_C(t)$. I.e. $Q_C(t) = CV_C(t)$. Recall further that charge, $Q_C(t)$, is the integral of current, $i_C(t)$. Thus, $Q_C(t) = \int_0^t i_C(t)\,dt + Q_C(0)$.

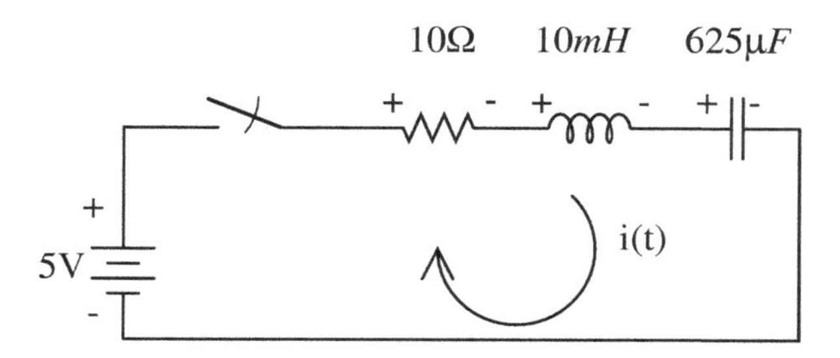

Fig. 1.1 A Simple R-L-C Circuit

on the capacitor is initially zero, and one closes the switch when $t = 0s$, then from Kirchoff's voltage law[9] (KVL) one finds that the resulting current flow, $i(t)$, is described by the equation:

$$\overbrace{10i}^{resistor} + \overbrace{0.01\frac{di}{dt}}^{inductor} + \overbrace{1600\int_0^t i(z)\,dz}^{capacitor} = \overbrace{5u(t)}^{battery}$$

where t it the time measured in seconds. Taking Laplace transforms, one finds that:

$$(10 + 0.01s + 1600/s)\,I(s) = 5/s.$$

Rearranging terms, we find that:

$$I(s) = \frac{500}{(s^2 + 1000s + 160000)}.$$

The denominator has two real roots:

$$p_{1,2} = -200, -800.$$

[9]Kirchoff's voltage law states that the sum of the voltage drops around any closed loop is equal to zero. In our example KVL says that:

$$-5u(t) + V_R(t) + V_L(t) + V_C(t) = 0.$$

Thus:

$$I(s) = \frac{500}{(s+200)(s+800)} = \frac{A}{s+200} + \frac{B}{s+800}.$$

Multiplying the fractions by $(s+200)(s+800)$, we find that:

$$500 = A(s+800) + B(s+200) \Rightarrow A = 5/6, B = -5/6.$$

We find that the solution of the equation is:

$$i(t) = \frac{5}{6}\left(e^{-200t} - e^{-800t}\right)u(t)$$

where the current is measured in amperes.

A Look at Resonance—An Example

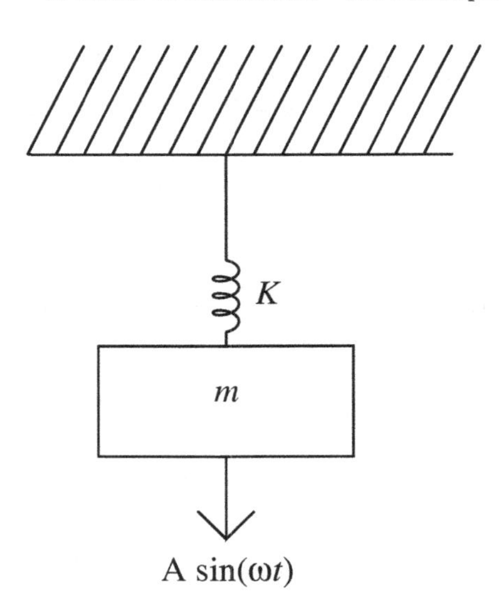

Fig. 1.2 The Free Body Diagram of the Spring-Mass System

Consider the equation satisfied by a spring-mass system with an m Kg mass and a spring whose spring constant is k N/M that is being excited by a sinusoidal force of amplitude A newtons and angular frequency Ω rad/sec. (See Figure 1.2.) Let $y(t)$ by the variation of the mass from it equilibrium position (measured in meters). From Newton's second law[10] we see that:

$$\overbrace{my''(t)}^{ma} = \overbrace{-ky(t) + A\sin(\Omega t)}^{\sum F}.$$

Assume that the mass starts from its equilibrium position, $y(0) = 0$, and that the mass starts from rest, $y'(0) = 0$. Taking the Laplace transform of the equation, we find that:

$$ms^2Y(s) = -kY(s) + \frac{A\Omega}{s^2+\Omega^2}.$$

That is:

$$Y(s) = \frac{A\Omega}{s^2+\Omega^2}\frac{1/m}{s^2+k/m}.$$

Assume that $k/m \neq \Omega^2$. Then the above fraction is not in partial fraction form. The partial fraction expansion of the fraction is:

$$\frac{A\Omega/m}{(s^2+\Omega^2)(s^2+k/m)} = \frac{As+B}{s^2+\Omega^2} + \frac{Cs+D}{s^2+k/m}.$$

Multiplying both sides of the equation by $(s^2+\Omega^2)(s^2+k/m)$ leaves us with the equation:

$$(As+B)(s^2+k/m) + (Cs+D)(s^2+\Omega^2) = A\Omega/m \Leftrightarrow$$
$$(A+C)s^3 + (B+D)s^2 + \left(\frac{Ak}{m} + C\Omega^2\right)s + \frac{Bk}{m} + D\Omega^2 = \frac{A\Omega}{m}.$$

Equating coefficients of powers of s in the last equation gives us:

$$A + C = 0 \tag{1.1}$$
$$B + D = 0 \tag{1.2}$$
$$C\Omega^2 + \frac{Ak}{m} = 0 \tag{1.3}$$
$$\frac{Bk}{m} + D\Omega^2 = \frac{A\Omega}{m}. \tag{1.4}$$

[10]Which states that:

$$\text{sum of forces} = \sum F = ma.$$

Combining (1.1) and (1.3), we find that $A = C = 0$. Combining (1.2) and (1.4), we find that $B = A\Omega/(k - m\Omega^2)$ and $D = -B$. Thus, we find that:

$$y(t) = \frac{A}{k - m\Omega^2}\left(\sin(\Omega t) - (\Omega/\sqrt{k/m})\sin(\sqrt{k/m}t)\right)$$

where $y(t)$ is measured in meters.

Now let us assume that $\Omega^2 = k/m$. Then the Laplace transform is already in partial fraction form. In this case, we must find the inverse transform of:

$$\frac{A\Omega/m}{(s^2 + \Omega^2)^2}.$$

Note that:

$$\mathcal{L}(t\sin(\Omega t))(s) = \frac{2s\Omega}{(s^2 + \Omega^2)^2}.$$

We are interested in $A\mathcal{L}(t\sin(\Omega t))(s)/(2ms)$. From Corollary 3 we know that this is:

$$\frac{A}{2m}\int_0^t z\sin(\Omega z)\,dz = \frac{A}{2m}\left(-\frac{t\cos(\Omega t)}{\Omega} + \frac{\sin(\Omega t)}{\Omega^2}\right).$$

We find that so long as the resonant frequency of the spring, $\sqrt{k/m}$, is not the same as the forcing frequency, Ω, the output of the system is bounded. If the resonant frequency is the same as the forcing frequency, then the solution grows "essentially linearly" in time.

1.6 An Introduction to Stability

It is important for us to know that a given system behaves in a "reasonable fashion." In this section we make the kind of reasonableness that we want precise, and we give a criterion that a system must satisfy in order to behave reasonably. First, however, we put our generic system in a form that makes reasonableness easy to check.

1.6.1 *Some Preliminary Manipulations*

Consider an arbitrary linear, constant-coefficient, inhomogeneous integro-differential equation in $y(t)$. After differentiating the equation a sufficient number of times to eliminate all of the integrals, one can convert the equation into a linear, constant-coefficient, differential equation.

A Simple R-C Circuit—An Example

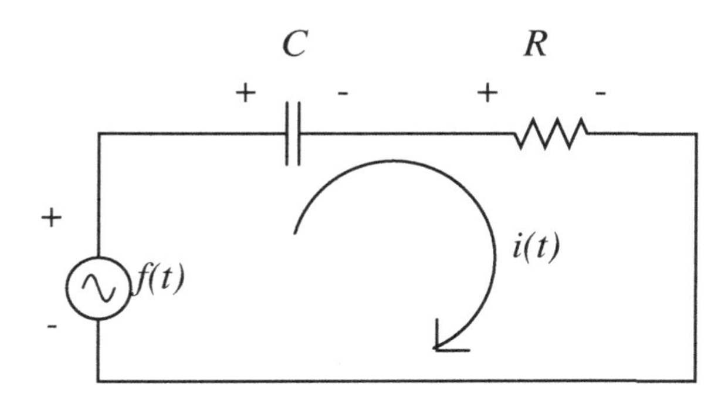

Fig. 1.3 A Simple R-C Circuit

Consider a circuit in which a capacitor of capacitance C farads, and a resistor of resistance R ohms are in series with a voltage source that sources the voltage $f(t)$ volts. Let $i(t)$ be the current (measured in amperes) traversing the circuit at time t (measured in seconds) and let $Q(0)$ be the charge on the capacitor at time $t = 0$. (See Figure 1.3.) The equation that describes the current in the circuit at time t is:

$$Ri(t) + \left(\int_0^t i(y)\,dy + Q(0)\right)/C = f(t) \tag{1.5}$$

We would like to manipulate this into an ordinary differential equation; we differentiate the equation once. We find that:

$$Ri'(t) + i(t)/C = f'(t).$$

We seem to be missing an initial condition—we must find $i(0)$. Looking at (1.5), we find that when $t = 0$, the equation reads

$Ri(0) + Q(0)/C = f(0)$. This equation gives us the initial value of the current—$i(0) = -Q(0)/C + f(0)$. This is the initial condition we need.

Let us assume that we have differentiated the integro-differential equation that describes our system sufficiently many times that we are left with an ODE of the form:

$$a_n y^{(n)}(t) + a_{n-1}y^{(n-1)}(t) + \cdots + a_0 y(t) = b_m f^{(m)}(t) + \cdots + b_0 f(t)$$
$$y^{(n-1)}(0) = c_{n-1}, \ldots, y(0) = c_0$$

where $f(t)$ is the input to the system and $y(t)$ is the output of the system.

Taking the Laplace transform of our equation, we find that:

$$\left(a_n s^n + a_{n-1}s^{n-1} + \cdots a_0\right) Y(s) = P(s) + \left(b_m s^m + b_{m-1}s^{m-1} + \cdots b_0\right) F(s)$$

where $P(s)$ is related to the initial value of $y(t)$ and its derivatives and of the initial value of $f(t)$ and its derivatives. Denoting $a_n s^n + a_{n-1}s^{n-1} + \cdots a_0$ by $Q(s)$, and denoting $b_m s^m + b_{m-1}s^{m-1} + \cdots b_0$ by $R(s)$, we find that:

$$Y(s) = P(s)/Q(s) + F(s)R(s)/Q(s)$$

where $P(s)$ is determined by our initial conditions, and $Q(s)$ and $R(s)$ are determined by the differential equation.

1.6.2 *Stability*

We would like to find a condition which guarantees that "if the input to the system is reasonable, then so is the output of the system." The first question one must ask is, how does one define reasonable? We take a simple definition—we say that *a system is stable if* for any *(reasonable) bounded input and* for any *initial conditions the output of the system is bounded.* This is often referred to as BIBO—bounded input bounded output—stability. We claim that a necessary and sufficient condition for a system described by a linear, constant coefficient, time invariant, integro-differential equation to be stable is that *all of the poles of $P(s)/Q(s)$—all of the zeros of $Q(s)$—must lie in the left half-plane.* When we speak of the left half-plane we will always mean the region LHP $\equiv \{s|\Re(s) < 0\}$. The right half-plane will always mean the region RHP $\equiv \{s|\Re(s) \geq 0\}$ unless we specify the open right half-plane—the region $\{s|\Re(s) > 0\}$.

We show that our condition is necessary. First, suppose that $Q(s)$ has a real zero of order n, z, in the open right half-plane and assume that

$f(t) \equiv 0$. We find that $Y(s) = P(s)/Q(s)$, where the coefficients of $P(s)$ are related to the initial conditions on $y(t)$. The partial fraction expansion of $Y(s)$ is:

$$Y(s) = \frac{a}{(s-z)^n} + \cdots$$

where $a \neq 0$. Under this condition, $y(t) = a(t^{n-1}/(n-1)!)e^{zt}$ which is a function whose magnitude is unbounded.

Now suppose that $Q(s)$ has two complex conjugate zeros in the right half-plane. Then the partial fraction expansion of $Y(s) = P(s)/Q(s)$ is:

$$\frac{\alpha s + \beta}{as^2 + bs + c} + \cdots$$

where $b^2 - 4ac < 0$. As we found on p. 18, the inverse Laplace transform of such a term grows like $e^{\mathcal{R}(\text{pole})t}$. As the real part of the pole is positive in our case, we find that the magnitude of inverse Laplace transform grows without bound. (The case of multiple roots can be handled similarly.)

Suppose that the $Q(s)$ has a single zero, z, on the imaginary axis. Suppose, for example, that $Q(s)$ has a zero at 0—that $Q(s) = s\tilde{Q}(s)$. Let the initial conditions on the ODE be identically zero. Let the function $f(t) \equiv 1$. Then:

$$Y(s) = \frac{F(s)}{Q(s)} = \frac{1/s}{s\tilde{Q}(s)} \overset{\text{partial fractions}}{=} \frac{a}{s^2} + \cdots$$

Clearly then:

$$y(t) = at + \cdots$$

This function is not bounded as $t \to \infty$ even though the input is. Similarly if $Q(s)$ has an imaginary zero at $z = j\omega$, then using $f(t) = \sin(\omega t)$ will lead one to a solution whose maximum approaches infinity linearly. (See Problem 8.) Because the instability exhibited by systems with poles of multiplicity one on the imaginary axis does not generally lead the system's output to increase without bound, such systems are said to be *marginally stable*. Note that marginally stable systems are actually unstable.

The proof that our condition is sufficient to guarantee stability is more complicated, and we do not present a proof of the sufficiency of the condition.

1.6.3 *Why We Obsess about Stability*

We have seen that if a system is not stable, then for some bounded input the output of the system becomes unbounded. One might think, however, that as long as one avoids the particular inputs that cause the system to behave badly, then one can make use of the unstable system. Let us consider a simple example to see why this is not a practical solution.

An Unstable System—An Example

Consider a system that satisfies the equation:

$$y''(t) = y(t).$$

The general solution of this equation (found using Laplace transforms, for example) is:

$$y(t) = \frac{y(0) + y'(0)}{2}e^t + \frac{y(0) - y'(0)}{2}e^{-t}.$$

Suppose that one believes that one has set the initial conditions to $y(0) = 1, y'(0) = -1$. For these initial conditions, the solution is $y(t) = e^{-t}$—a perfectly nice solution. The difficulty with this system is that *practically* it is well nigh impossible to be sure that $y(0) = 1$ and $y'(0) = -1$. Suppose that $y(0) = 1 + \epsilon$ and $y'(0) = -1 + \epsilon'$ where both ϵ and ϵ' are extremely small. The solution of the equation with these initial conditions is

$$y(t) = \frac{\epsilon + \epsilon'}{2}e^t + \left(1 + \frac{\epsilon - \epsilon'}{2}\right)e^{-t}.$$

The effect of this infinitesimal change in the initial conditions hardly affects the coefficient of the e^{-t}. However, now we find that the growing exponential e^t has a nonzero coefficient. As t gets large, we find that the part of the solution that we wanted, e^{-t}, is swamped by the part of the solution that came from a tiny imprecision in our initial conditions.

Because of this tendency of even very small mistakes to have very large consequences in unstable systems, such systems *cannot generally be used.* Another problem with such systems is that in many cases the possibility of unbounded output is itself a very bad sign. (One does not want one's car's speed to increase without bound!) For these reasons we do not generally use or want unstable systems—though we will make use of marginally stable systems.

1.6.4 *The Tacoma Narrows Bridge—a Brief Case History*

On November 7, 1940 the four month old Tacoma Narrows bridge collapsed. In the aftermath many explanations were proposed for the collapse [Kou96]. As a bridge is a very complicated structure, nobody has ever been able to say with certainty what caused the collapse of the Tacoma Narrow Bridge. It seems that what caused the bridge to tear itself apart was the wind exciting a *nonlinearly* resonant mode in the bridge [BS91]. (This is somewhat akin to the *linearly* resonant mode in the spring-mass system of Page 22.)

The Tacoma Narrows bridge serves to remind engineers of the importance of treating unstable systems—and even marginally stable systems—with the respect and caution that they deserve.

1.7 MATLAB®

Throughout this book we will use MATLAB as a very fancy calculator. We now give a very brief overview of how MATLAB is used. (We make use of the commands found in the fourth edition of the student edition of MATLAB.)

1.7.1 *Assignments*

In MATLAB, one makes assignments by writing `variable = object` where object may be a number, an array, or various other objects about which we will hear more later. Note that when using MATLAB neither variables nor arrays need to be declared. Here are a number of examples of legal assignments.

(1) `A = 3`. This assigns the value 3 to the variable `A`. It also causes MATLAB to print:

```
A =

     3
```

(2) Generally, if one wishes to suppress printing one ends the assignment with a semi-colon. The command `A = 3;` also assigns three to the variable `A`, but it does not cause anything to be printed.

(3) MATLAB prints the value of a variable if one types the variable's name. If one types `A`, MATLAB responds with:

```
A =

     3
```

(4) `B = [3, 4 ,5];` assigns the array `[3, 4, 5]` to `B`. To refer to the individual elements of `B` one refers to `B(1)` to `B(2)`, and to `B(3)`. Arrays in MATLAB always start from element number `1`. It is worth noting that the commas in the assignment statement are optional. If one leaves a space between two numbers, MATLAB assume that the two numbers are distinct elements of the array.

(5) `C = [3, 4; 5, 6]` assigns the two dimensional matrix to the variable `C`. Additionally, because there was no semi-colon after the assignment, MATLAB prints:

```
C =

     3     4
     5     6
```

One can refer to the elements of this array as one would refer to elements of a matrix. To refer to the second element in the first row one refers to `B(1,2)`. Typing `B(1,2)` causes MATLAB to respond with:

```
ans =

     4
```

MATLAB prints `ans` because it is not responding with the value of an entire named object—4 is not the value of `B`; it is the value of a component of `B`.

1.7.2 *Commands*

As we have already seen, we are often interested in finding the roots of a polynomial—for example, the roots of the denominator of the Laplace transform of some function. MATLAB has a command for this purpose—`roots`[11]. Suppose one's polynomial is:

$$c_1 x^n + c_2 x^{n-1} + \cdots + c_n x + c_{n+1}.$$

To find the roots of this polynomial, one defines the array

[11]This command does not work very well for high order polynomials because the roots of an equation are inherently difficult to calculate.

$$\texttt{C = } [c_1, c_2, \ldots, c_n, c_{n+1}],$$

and then one makes the assignment/function call `R = roots(C)`. The result is that the roots of the polynomial are now stored in the array `R`.

Another command that is useful is the **`residue`** command. The **`residue`** command gives the partial fraction expansion of a rational function. One enters two array `B` and `A` which are the coefficients of the numerator polynomial and the denominator polynomial respectively. One uses this command by writing `[R P K] = residue(B,A)`. The vector `P` contains the poles of the various fractions that appear in the partial fraction expansion. Note that if a pole appears more than once then the fraction connected to the nth occurrence of the pole is $1/(s-p)^n$. The vector `R` is the vector of coefficients—the nth coefficient is the numerator of the fraction associated with the nth item in the vector `P`. The vector `K` is the vector of coefficients of the polynomial than one gets if one starts with a fraction whose numerator has degree[12] greater than or equal to the degree of the denominator. We note that **`residue`** gives all the poles separately—it separates fractions of the form $(As+B)/((s-a)(s-\overline{a})$ into fractions of the form $C/(s-a)+D/(s-\overline{a})$.

How to Use **`residue`**—An Example

Suppose that one would like to find the inverse transform of the function $\mathcal{L}(f(t))(s) = 1/(s^2-1)$. One makes the assignments `B = 1` and `A = [1, 0, -1]`. Then one performs the operation `[R P K] = residue(B,A)`. MATLAB responds with:

```
R =
    -0.5000
     0.5000


P =
    -1.0000
     1.0000


K =
      []
```

[12]The degree of a polynomial is the highest power that occurs in the polynomial. Thus, the degree of $s+1$ is one, while the degree of s^3+1 is three. Polynomials of degree n are also sometimes said to be of *order* n.

That means that the array K is empty and that the fraction is equal to:

$$\frac{1}{s^2-1} = \frac{-0.5000}{s-(-1.0000)} + \frac{0.5000}{s-1.0000}.$$

Clearly the inverse Laplace transform of the function is $(e^t - e^{-t})/2$. There are many useful MATLAB commands, and we will see some of them in the course of this book.

1.8 Exercises

(1) Find the Laplace transform of:

(a)

$$\cos(\omega t)u(t), \omega > 0$$

Note that $u(\omega t) = u(t)$ if $\omega > 0$.

(b)

$$f(t) = te^{-t}\cos(t)u(t)$$

(c)

$$f(t) = t^2 e^{-t} u(t)$$

(d)

$$f(t) = e^{-t}\sin(2t)u(t)$$

(2) Find the function whose Laplace transform is:

(a)

$$F(s) = \frac{s+1}{s^2+1}$$

(b)

$$F(s) = \frac{1}{s+1}$$

(c)

$$F(s) = \frac{s}{s^2+1}$$

(d)

$$F(s) = \frac{1}{(s+1)^2}$$

(e)

$$F(s) = \frac{s+1}{s^2+2s+2}$$

Do each problem both "by hand" and by using the MATLAB `residue` command.

(3) Use the Laplace transform to solve the differential equation:

$$y''(t) + 5y'(t) + 4y(t) = 1$$

subject to the initial conditions:

$$y(0) = y'(0) = 0$$

(4) Use the Laplace transform to solve the integral equation:

$$\int_0^t y(r)\,dr = -y(t) + u(t)$$

where $u(t)$ is the unit step function.

(5) Use the Laplace transform to solve the integral equation:

$$\int_0^t y(r)\,dr = -y(t) + \sin(\omega t)u(t)$$

where $u(t)$ is the unit step function.

(6) Use the Laplace transform to solve the integro-differential equation:

$$3\int_0^t y(r)\,dr + 4y(t) + y'(t) = 0$$

subject to the initial condition $y(0) = 1$.

(7) Show the Laplace transform of the function $f(t) = u(t)/\sqrt{t}$ is $\sqrt{\pi/s}$ for all $s > 0$. Use the definition of the Laplace transform, the substitution $y = \sqrt{st}$, and the fact that $\int_0^\infty e^{-y^2}\,dy = \sqrt{\pi}/2$.

(8) Suppose that the relation between the input to a system, $f(t)$, and the output of the system, $y(t)$, satisfies:

$$Y(s) = \frac{1}{s^2+1}F(s)$$

By considering $f(t) = \sin(t)$, show that the system is not BIBO stable.

(9) (A Generalization of the final value theorem [Glu03].) Suppose that $g(t) = h(t)u(t)$ and that for all $t \geq T$ we know that $g(t) = g(t-T)$.

(a) Show that

$$\mathcal{L}(g(t) - g(t-T))(s) = \int_0^T e^{-st} g(t)\, dt.$$

(b) Using the properties of the Laplace transform, show that:

$$\mathcal{L}(g(t) - g(t-T))(s) = (1 - e^{-Ts})G(s).$$

(c) Show that:

$$G(s) = \frac{\int_0^T e^{-st} g(t)\, dt}{1 - e^{-Ts}}.$$

(d) Show that:

$$\lim_{s \to 0} sG(s) = \frac{1}{T} \int_0^T g(t)\, dt.$$

Thus, when a function has a final value, $\lim_{s \to 0^+} sG(s) = \lim_{t \to \infty} g(t)$. When a function is periodic for non-negative time, $\lim_{s \to 0} sG(s) = \text{average}(g(t))$. As the average of a periodic function is, in some sense, its representative value, one can say that in both cases the limit tends to a value that can reasonably be said to be a representative value for the function.

(10) This exercise makes use of the results of Exercise 9.

(a) Calculate

$$\lim_{s \to 0} s \left\{ \frac{2}{s} + \frac{s}{s^2 + 9} \right\}.$$

(b) Explain the significance of this limit in the time domain.

(11) Let $y(t) = (1 - e^{-2t} \sin(t))u(t)$.

(a) Calculate $Y(s)$.

(b) Use the final value theorem to calculate $y(\infty)$.

(c) Use the initial value theorem to calculate $y(0^+)$.

(d) Use the properties of the Laplace transform and the answer to the previous section to calculate $\mathcal{L}(y'(t))(s)$.

(e) Use the initial value theorem to calculate $y'(0^+)$.

(f) Check your answer to the previous section by making use of the definition of $y(t)$.

(12) (A particularly interesting Laplace transform.) In this exercise, you verify that for $s > 0$ the Laplace transform of:

$$x(t) = \frac{1}{t^2+1}u(t)$$

is:

$$X(s) = \left(\int_s^\infty \frac{\sin(y)}{y}\,dy\right)\cos(s) - \left(\int_s^\infty \frac{\cos(y)}{y}\,dy\right)\sin(s). \quad (1.6)$$

(a) Start by making use of the fact that $(t^2+1)x(t) = u(t)$ to show that:

$$X''(s) + X(s) = 1/s.$$

(b) Show that the general solution of the associated homogeneous differential equation is $X_\text{h}(s) = A\sin(s) + B\cos(s)$.

(c) Check that:

$$X_\text{p}(s) = \left(\int_a^s \frac{\cos(y)}{y}\,dy\right)\sin(s) - \left(\int_a^s \frac{\sin(y)}{y}\,dy\right)\cos(s)$$

is a particular solution of the inhomogeneous differential equation.

(d) Make use of the definition of the Laplace transform to show that for the given $x(t)$:

$$\lim_{s\to\infty} X(s) = 0.$$

(e) Note that the general solution of the inhomogeneous differential equation is $X_\text{h}(s) + X_\text{p}(s)$, and make use of the fact that $\lim_{s\to\infty} X(s) = 0$ to find the values of the constants A and B.

(f) Make use of these values of A and B to show that $X(s)$ as given in (1.6) is the Laplace transform of $x(t)$.

Chapter 2

Transfer Functions

2.1 Transfer Functions

As discussed in the first chapter, systems of interest to us will generally be described by integro-differential equations. Consider the Laplace transforms of the input and output of such a system, $X(s)$ and $Y(s)$ respectively. Assume that we want to set some combination of derivatives and integrals of the output to some combination of derivatives and integrals of the input. We find that the relation between $X(s)$ and $Y(s)$ is:

$$Y(s) = P(s)/Q(s) + X(s)R(s)/Q(s)$$

where the polynomial $P(s)$ is related to the initial conditions to which the system is subject, and $R(s)$ and $Q(s)$ are related to the form of the integro-differential equation.

In §1.6.2 we saw that for a system to be stable it is necessary that all of the zeros of $Q(s)$ be in the left half-plane. It is crucial that any system that one uses be stable. The output of an unstable system—as we have seen—will almost invariably tend to "run away."

Assuming that we are dealing with a stable system, we find that the portion of the response that is related to the initial conditions to which the system is subject, $P(s)/Q(s)$, has all of its poles in the left half-plane. By considering the partial fraction expansion of this function, it is clear that the function to which it corresponds decays exponentially quickly. Thus, after a "little while" the initial conditions will no longer affect the output of the system. When we consider systems, we neglect the influence of the initial conditions; we set $P(s) \equiv 0$.

Assuming that $P(s) \equiv 0$, we find that the ratio of the Laplace transform of the output to the Laplace transform of the input—is:

$$T(s) = \frac{Y(s)}{X(s)} = \frac{R(s)}{Q(s)}.$$

We call $T(s)$ the *transfer function* of the system. It is remarkable that a transfer function *exists*.

Clearly, a system is BIBO stable if all of its transfer function's poles are located in the left half-plane. It is unstable if any of its transfer function's poles are located in the right half-plane (including the imaginary axis). Finally, an unstable system is said to be marginally stable if its transfer function has one or more first order poles on the imaginary axis and the rest of its poles are in the left half-plane.

A High-Pass Filter—An Example

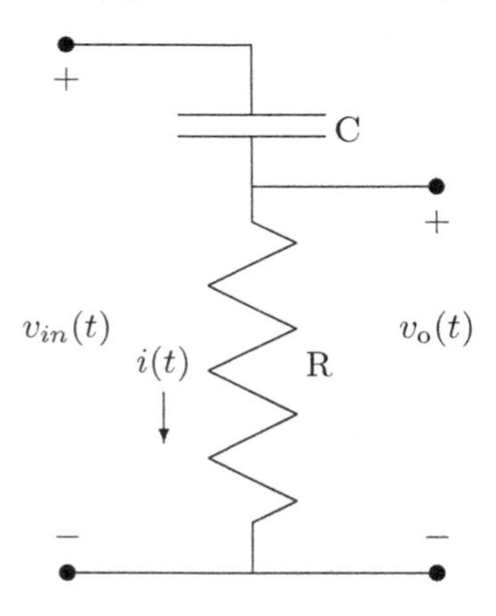

Fig. 2.1 A Simple High-Pass Filter

Consider the high-pass filter of Figure 2.1. The simplest way of describing how it functions is to consider the current flowing through the circuit at time t, $i(t)$. If we assume that the initial charge on the capacitor is zero, then using Ohm's law and Kirchoff's voltage law we find that:

$$\overbrace{\frac{1}{C}\int_0^t i(y)\,dy}^{\text{voltage across the capacitor}} + \overbrace{Ri(t)}^{\text{voltage across the resistor}} = \overbrace{v_{in}(t)}^{\text{input voltage}}.$$

We would like to find the system transfer function, so we must rewrite this in terms of the voltage at the output—$v_o(t)$. As $v_o(t) = Ri(t)$, we find that:

$$v_o(t) + \frac{1}{RC}\int_0^t v_o(y)\,dy = v_{in}(t).$$

Taking Laplace transforms, we find that:

$$V_o(s) + \frac{1}{sRC}V_o(s) = V_{in}(s),$$

or that:

$$T(s) = \frac{V_o(s)}{V_{in}(s)} = \frac{sRC}{1+sRC}.$$

After discussing what we mean by the frequency response of a system, we explain (on p. 41) why this filter is called a high-pass filter.

2.2 The Frequency Response of a System

Suppose that one has a system described by the transfer function:

$$T(s) = \frac{V_o(s)}{V_{in}(s)} = \frac{R(s)}{Q(s)},$$

and assume that the system is stable—that $Q(s)$ has no zeros in the right half-plane (including the imaginary axis). Let the input to the system be $\sin(\omega t)u(t)$. Then the output of the system is:

$$V_o(s) = \frac{R(s)}{Q(s)}\frac{\omega}{s^2+\omega^2}.$$

The partial fraction expansion of this expression is:

$$\frac{R(s)}{Q(s)}\frac{\omega}{s^2+\omega^2} = \frac{As+B}{s^2+\omega^2} + \cdots \tag{2.1}$$

where the ellipsis represents terms corresponding to the poles in the left half-plane. The terms that we have ignored all correspond to terms that contain damped exponentials. If we are interested in the long term behavior of the system, we do not need to understand any more about the ignored terms; $\frac{As+B}{s^2+\omega^2}$ is the only term of importance to us.

Let us find A and B. Rewriting (2.1) we find that for all $s \neq \pm j\omega$:

$$\omega \frac{R(s)}{Q(s)} = As + B + (s^2 + \omega^2)(\cdots)$$

where $\pm j\omega$ is not a pole of the terms contained in the ellipsis. Considering the limit of the above expression as $s \to \pm j\omega$, we find that:

$$\omega \frac{R(\pm j\omega)}{Q(\pm j\omega)} = A(\pm j\omega) + B + 0(\cdots).$$

As the polynomials $Q(s)$ and $R(s)$ have real coefficients, we find that $Q(-j\omega) = \overline{Q(j\omega)}$ and $R(-j\omega) = \overline{R(j\omega)}$. We see that:

$$\begin{aligned} A(j\omega) + B &= \omega T(j\omega) \\ A(-j\omega) + B &= \omega \overline{T(j\omega)}. \end{aligned}$$

Solving for A and B we find that:

$$A = \Im(T(j\omega)), B = \omega \Re(T(j\omega)).$$

Combining the above, we find that the transform of the behavior of the system after all of the transients[1] have died down is:

$$\mathcal{L}(\text{Steady State}(t))(s) = \frac{\Im(T(j\omega))s}{s^2 + \omega^2} + \frac{\Re(T(j\omega))\omega}{s^2 + \omega^2}.$$

The inverse Laplace transform of this expression is:

$$\begin{aligned} \text{Steady State}(t) &= \Im(T(j\omega))\cos(\omega t) + \Re(T(j\omega))\sin(\omega t) \\ &= |T(j\omega)| \left(\frac{\Im(T(j\omega))}{|T(j\omega)|} \cos(\omega t) + \frac{\Im(T(j\omega))}{|T(j\omega)|} \sin(\omega t) \right). \end{aligned}$$

This "steady state" output is sometimes called the "sinusoidal steady state" to remind us of the fact that our output is not constant—it is the steady state in the sense that it is the output of the system after all the transients have "died down."

If $\alpha^2 + \beta^2 = 1$, then the expression $\alpha \sin(\phi) + \beta \cos(\phi)$ can be written as $\sin(\phi + \theta)$ where θ satisfies:

$$\begin{aligned} \cos(\theta) &= \alpha \\ \sin(\theta) &= \beta. \end{aligned}$$

[1] A transient signal is a signal that tends to zero as $t \to \infty$.

(Thus, $\theta = \arctan(\beta/\alpha)$ or $\theta = \arctan(\beta/\alpha) + 180^o$ depending on the quadrant in which (α, β) lies.) Clearly:

$$\frac{\Im(T(j\omega))}{|T(j\omega)|}\cos(\omega t) + \frac{\Im(T(j\omega))}{|T(j\omega)|}\sin(\omega t)$$

is of this form where:

$$\theta = \arg(T(j\omega)).$$

Thus, we find that:

$$\text{Steady State}(t) = |T(j\omega)|\sin(\omega t + \arg(T(j\omega))).$$

We see that $T(j\omega)$'s magnitude gives the steady state amplification that a sine wave with angular frequency ω sees, and its argument gives the steady state phase-shift that a sine wave input sees. The function $T(j\omega)$ is called the *frequency response* of the system.

The Frequency Response of the High-pass Filter—An Example

Let us consider the high-pass filter of the previous example. The transfer function of the system was:

$$T(s) = \frac{sRC}{1 + sRC}.$$

We find that the frequency response is:

$$T(j\omega) = \frac{j\omega RC}{1 + j\omega RC}.$$

We find that for low frequency inputs, for $\omega \approx 0$, the amplification is near zero—the signals are "removed" by the filter. For high frequency signals, for:

$$\omega >> \frac{1}{RC},$$

the frequency response is almost exactly one—the signals are passed without change. This is, of course, why the filter is called a high-pass filter. The filter allows high frequency signals to pass while it "kills" low frequency signals.

Suppose that rather than using a sine wave as input, one uses a cosine. Then proceeding just as we did above, one finds that:

$$\frac{R(s)}{Q(s)}\frac{s}{s^2+\omega^2} = \frac{As+B}{s^2+\omega^2} + \cdots$$

Proceeding as we did in the case of the sine, we find that:

$$A = \Re(T(j\omega)), B = -\omega\Im(T(j\omega)).$$

Finally, we find that the steady state output of the system is:

$$\text{Steady State}(t) = |T(j\omega)|\cos(\omega t + \arg(T(j\omega))).$$

We find that the system's response to a cosine is essentially the same as its response to a sine. Since we are dealing with a time invariant system, this is as it should be. We also note that $\cos(0t) = 1$. Thus, $|T(j0)| = |T(0)|$ gives the steady state amplification that a constant input "sees". As $\sin(\omega t + \phi) = \sin(\omega t)\cos(\phi) + \cos(\omega t)\sin(\phi)$, it is clear that the output of our system to a sine-wave with an arbitrary phase shift ϕ is:

$$|T(j\omega)|\sin(\omega t + \phi + \arg(T(j\omega))).$$

2.3 Bode Plots

It is common practice in engineering to plot the frequency response of a system as two separate plots—one of the magnitude of the frequency response and one of the phase of the frequency response. The x-axis of both plots is the frequency axis (with frequency measured in radians per second) with the frequency marked on a logarithmic scale. The magnitude plot is generally given in decibels[2]—dB. By definition the number of decibels is twenty times the common logarithm[3] of the magnitude. The phase plot gives the phase in degrees. The two plots collectively are referred to as the *Bode plots* of the system[4].

Why use logarithmic scales? Because they make our life easier. How do they make our life easier? We find that we often need to multiply transfer

[2]After Alexander Graham Bell (1847–1922) who named the Bel after himself. The decibel is one tenth of a Bel [Wik].

[3]I.e. $\log_{10}(x)$—as opposed to the natural logarithm, $\ln(x) = \log_e(x)$.

[4]Named after their inventor, Hendrik Bode (pronounced bōdē—with a hard o and a hard e). Bode invented the Bode plots in 1938 while employed by the Bell Telephone Laboratories [Lew92].

functions. If we start with the transfer functions $T_1(s)$ and $T_2(s)$ we find that the product of their frequency responses in dB is:

$$\begin{aligned}(T_1(j\omega)T_2(j\omega))_{\text{dB}} &= 20\log_{10}|T_1(j\omega)T_2(j\omega)| \\ &= 20\log_{10}|T_1(j\omega)| + 20\log_{10}|T_2(j\omega)| \\ &= (T_1(j\omega))_{\text{dB}} + (T_2(j\omega))_{\text{dB}}.\end{aligned}$$

Measuring in decibels allows us to add graphs when we multiply transfer functions. Phases also add when functions are multiplied, so to find the phase plot of the combined system we also need only add the phase plots of the separate systems. We scale the frequency logarithmically because of the fact that our transfer functions generally behave like powers of s. Consider $(j\omega)^n$ measured in dB. We find that $((j\omega)^n)_{\text{dB}} = 20\log_{10}|\omega|^n = 20n\log_{10}|\omega|$ dB. If one of our axes is *logarithmically* scaled frequency and the other is dB, then this is a straight line with slope $20n$. The distance from a frequency ω to the frequency 10ω is one decade (by definition). If a polynomial "looks like" s^n in some regions, then its magnitude plot has the slope $20n$ dB/dec. Thus, each time the frequency increases by a factor of 10—what we engineers call a "decade"—the amplitude in dB increases by $20n$. From the slope of of the magnitude of the frequency response one can *read off* some of the characteristics of the system.

When using MATLAB one defines objects that are transfer functions by giving the numerator polynomial and the denominator polynomial to the MATLAB function `tf`. For example, one defines the transfer function:

$$T(s) = \frac{0.001s}{0.001s + 1}$$

by typing:

```
T = tf([0.001 0],[0.001 1])
```

MATLAB responds with:

```
Transfer function:
  0.001 s
----------
.001 s + 1
```

In order to find the Bode plots of the system, one need only type `bode(T)`. MATLAB will open a second window and will present both plots.

The Bode Plots of Our High-Pass Filter—An Example

Consider the high-pass filter of the previous example with $RC = .001$. Define T as above, and type `bode(T)`. MATLAB produces Figure 2.2. We see that at high frequencies the magni-

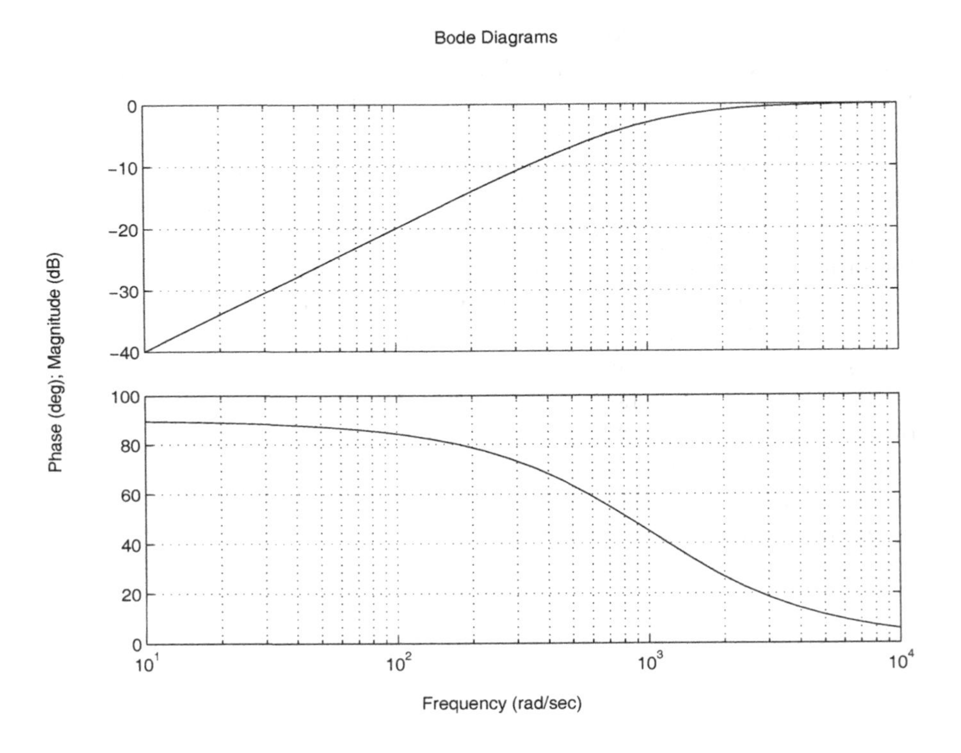

Fig. 2.2 The Bode Plots of the High-Pass Filter

tude of the response is approximately $0\,\text{dB} = 1$. We note that in the decade between 1 rad/sec and 10 rad/sec the magnitude rose from -40dB to -20dB—a slope of 20dB/dec. This slope corresponds to the fact that at such low frequencies the frequency response looks like $j\omega$. At low frequencies the phase is near 90° which is also what one would expect.

2.4 The Time Response of Certain "Typical" Systems

Many of the systems and subsystems that we will encounter will be either first or second order systems—systems whose denominator is of degree one or two respectively—or will be systems that can be approximated by first or second order systems. We now consider the responses one expects from such systems.

We consider the relation between the various parameters that describe the system and the response of the system to a unit step, $u(t)$. We would like to determine the relationship between the system's *rise time*, the system's *overshoot*, and the parameters that describe the system. We define the rise time as the time it takes the system to go from 10% of its final value to the first point at which it always stays within 90% of its final value. Let:

$$v_{\text{o}\ max} = \max_{t \geq 0} v_\text{o}(t), \text{ and } v_{\text{o}\ final} = \lim_{t \to \infty} v_\text{o}(t).$$

We define the overshoot to be:

$$\text{Overshoot} = \frac{v_{\text{o}\ max} - v_{\text{o}\ final}}{v_{\text{o}\ final}} \times 100\%.$$

(If the voltage tends to its final value without exceeding its final value, then let $v_{\text{o}\ max} = v_{\text{o}\ final}$.)

2.4.1 *First Order Systems*

Consider a component whose transfer function is first order—of the form:

$$T(s) = \frac{C}{\tau s + 1}.$$

What will the response of this system be when the input is a unit step function, $u(t)$?

As the Laplace transform of $u(t)$ is $1/s$, from the final value theorem we know that:

$$\lim_{t \to \infty} v_\text{o}(t) = \lim_{s \to 0} sT(s)\frac{1}{s} = \lim_{s \to 0} T(s) = C.$$

Intuitively we can see this another way. Looking at what is happening from the "component's point of view," we find that for large time the component looks back on an uninterrupted input of 1—a signal at 0 rad/sec (also often referred to as DC). We have already seen that the amplification at 0 rad/sec—at DC—is $T(0)$. Thus the output must tend to C.

The component's behavior as it goes from its initial state to its steady state is called the *transient response* of the component. Consider the transient response of a first order system to a unit step function. The Laplace transform of the response of the system is:

$$V_o(s) \equiv \mathcal{L}(v_o(t))(s) = \frac{C}{s(\tau s + 1)} = \frac{C}{s} - \frac{C\tau}{\tau s + 1}.$$

Calculating the inverse Laplace transform we find:

$$v_o(t) = C(1 - e^{-t/\tau})u(t).$$

Let us find the rise time of this system. The steady state output of the system is C. Because the output rises monotonically, we know that the times at which the output of the system reaches 10% of its final value (for the first time) and the time at which it reaches 90% of its output (for the last time)—$t_{10\%}$ and $t_{90\%}$ respectively—are well defined. They are equal to:

$$\begin{aligned} t_{10\%} &= -\tau \ln(1 - .1) \\ t_{90\%} &= -\tau \ln(1 - .9). \end{aligned}$$

The rise time is $t_{90\%} - t_{10\%} = \ln(9)\tau \approx 2.2\tau \approx 3\tau$. We usually say that the rise time of a first order system is roughly 3τ. Because the output rises monotonically to its final value, there is no overshoot in this system.

In many ways a first order system has a very nice response. The output attains its final value by rising up to the final value without any overshoot or any oscillations. As we shall see, second order systems do not always behave so nicely.

2.4.2 *Second Order Systems*

Consider a component whose transfer functions has the form:

$$T(s) = \frac{\omega_n^2}{s^2 + 2\zeta\omega_n s + \omega_n^2}, \qquad \omega_n, \zeta > 0.$$

This is a reasonably typical transfer function for a second order system. The constant ω_n is called the natural frequency of the system, and ζ is called the damping factor of the system. As $T(0) = 1$ the system provides neither amplification nor attenuation at DC; the steady state response of this system to a unit step function will be 1.

If the input to the system is $u(t)$, then the Laplace transform of the output of the system is:

$$\mathcal{L}(v_o(t))(s) = \frac{\omega_n^2}{(s^2 + 2\zeta\omega_n s + \omega_n^2)}\frac{1}{s} = \frac{-(s + \zeta\omega_n) - \zeta\omega_n}{(s + \zeta\omega_n)^2 + (\omega_n^2 - \zeta^2\omega_n^2)} + \frac{1}{s}.$$

To find the inverse Laplace transform of this function, we must know whether the discriminant $\Delta \equiv \omega_n^2 - \zeta^2\omega_n^2$ is positive, zero, or negative. We consider each case.

If $\Delta > 0$, then $\zeta < 1$ and the inverse Laplace transform is:

$$v_o(t) = u(t)\left(1 - e^{-\zeta\omega_n t}\left(\cos(\omega_n\sqrt{1-\zeta^2}t) + \frac{\zeta\sin(\omega_n\sqrt{1-\zeta^2}t)}{\sqrt{1-\zeta^2}}\right)\right) \quad (2.2)$$

$$= u(t)\left(1 - \frac{e^{-\zeta\omega_n t}}{\sqrt{1-\zeta^2}}\sin(\omega_n\sqrt{1-\zeta^2}t + \phi)\right), \quad (2.3)$$

$$\phi = \arctan\left(\frac{\sqrt{1-\zeta^2}}{\zeta}\right).$$

Such systems are called *under-damped* because they have too little damping to prevent oscillations.

If $\Delta = 0$, then $\zeta = 1$ and the inverse Laplace transform is:

$$v_o(t) = \left(1 - e^{-\omega_n t} - \omega_n t e^{-\omega_n t}\right)u(t).$$

Such systems are called *critically damped* because they have just enough damping to stop oscillations from occurring.

If $\Delta < 0$, then $\zeta > 1$ and the inverse Laplace transform is:

$$u(t)\left(1 - e^{-\zeta\omega_n t}\cosh(\omega_n\sqrt{\zeta^2-1}t) - \frac{\zeta}{\sqrt{1-\zeta^2}}e^{-\zeta\omega_n t}\sinh(\omega_n\sqrt{\zeta^2-1}t)\right).$$

Note that both $\sinh(\alpha t)$ and $\cosh(\alpha t)$ behave like $e^{\alpha t}/2$ as $t \to \infty$. Thus the inverse Laplace transform looks like $u(t)$ + terms of order $e^{-\omega_n(\zeta - \sqrt{\zeta^2-1})}$. Clearly this tends to 1 as $t \to \infty$. Also the solution here is clearly non-oscillatory. Such systems are called *over-damped.*

Let us consider the relation between the parameters which define the second order system, ω_n and ζ, and the time response of the system for the under-damped system—the system for which we will have the most use. In order to find a useful formula for the rise time, it is reasonable to try to find the rise time by thinking of the system output as $1 - e^{-\omega_n\zeta t}$ and ignoring the effects of the sinusoidal term. Using this formula it is clear that the rise time is $3/(\omega_n\zeta)$. This formula is only approximately correct, but it is simple and widely used. Next, we would like to find the overshoot. Differentiating (2.2) we find that for $t > 0$:

$$v_o'(t) = \frac{\omega_n}{\sqrt{1-\zeta^2}} e^{-\zeta\omega_n t} \sin(\sqrt{1-\zeta^2}\omega_n t).$$

The first zero of this function occurs when $\sqrt{1-\zeta^2}\omega_n t = \pi$. This gives the maximum of the output. (If one takes a later zero of the derivative, one finds that the exponentially decaying term causes the net result to be smaller.) We find that the maximum value of the output is:

$$\begin{aligned}\max_{t\geq 0} v_o(t) &= 1 - \frac{1}{\sqrt{1-\zeta^2}} e^{-\frac{\pi\zeta}{\sqrt{1-\zeta^2}}} \sin\left(\pi + \arctan\left(\frac{\sqrt{1-\zeta^2}}{\zeta}\right)\right)\\ &= 1 + e^{-\frac{\pi\zeta}{\sqrt{1-\zeta^2}}}.\end{aligned}$$

Thus the percent overshoot is:

$$\text{percent overshoot} = e^{-\frac{\pi\zeta}{\sqrt{1-\zeta^2}}} \times 100\%. \tag{2.4}$$

2.5 Three Important Devices and Their Transfer Functions

Though we are not trying to present a nuts and bolts view of control engineering in this book, we describe three systems of interest to the control engineer—the operational amplifier (op amp), the DC motor, and the simple satellite. We make no attempt to derive their properties from first principles; we describe them, and we relate their transfer functions to their *a priori* known properties.

2.5.1 *The Operational Amplifier (op amp)*

A diagram of an op amp is given in Figure 2.3. The relation between the inputs and the output of the op amp is:

$$V_o(s) = \frac{A}{s\tau + 1} (V_+(s) - V_-(s)).$$

The op amp is a difference amplifier—it amplifies the difference between its two input. The two inputs, $v_-(t)$ and $v_+(t)$, are (assumed to be) measured by "ideal sensors." The "sensors" measure the voltages at the inputs

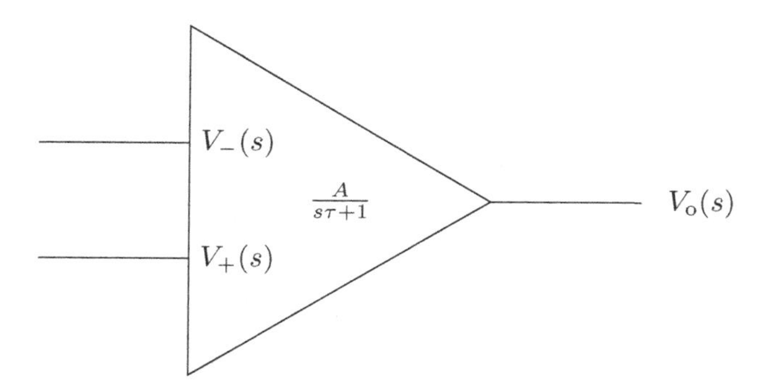

Fig. 2.3 An Operational Amplifier

without in any way affecting the inputs—in particular, no current flows into the inputs of the op amp. The frequency response of the op amp is (approximately) $A/(j\omega\tau + 1)$. If $\omega \approx 0$, then the frequency response is almost a pure amplification A. Generally speaking A is quite large ($> 50,000$) and τ is moderately small (τ is often several hundredths). An op amp is usually part of a larger system like the amplifier in the following example.

A Simple Op Amp Amplifier—An Example

Consider the simple amplifier of Figure 2.4. We find that:

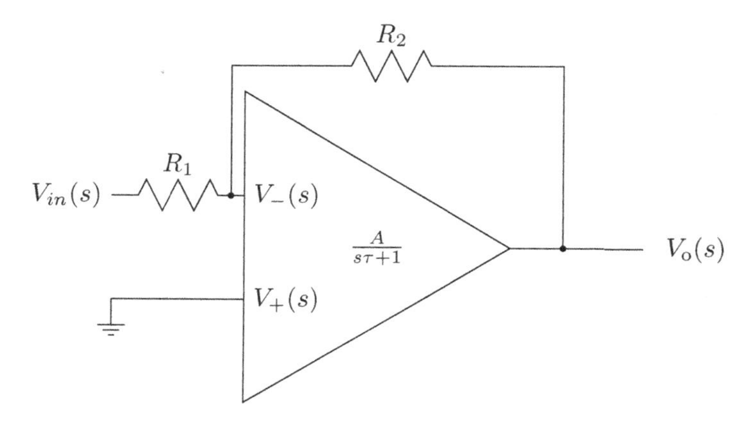

Fig. 2.4 A Simple Amplifier

$$\begin{aligned} V_o(s) &= \frac{A}{s\tau+1}(V_+ - V_-) \\ &= \frac{A}{s\tau+1}\left(0 - \left((V_o - V_{in})\frac{R_1}{R_1+R_2} + V_{in}\right)\right) \\ &= -\frac{A}{s\tau+1}\left(V_o\frac{R_1}{R_1+R_2} + V_{in}\frac{R_2}{R_1+R2}\right). \end{aligned}$$

At this point, we can continue in two different ways. If all that we want to know is how our amplifier reacts at reasonably low frequencies, it is sufficient to consider $s = j\omega$ and to approximate. At reasonably low frequencies, the right-hand side is multiplied by the very large number $A/(j\omega\tau + 1)$. The left-hand side is a (presumably) reasonably small number $V_o(j\omega)$. Assuming that the left hand is "small enough," we find that to a good approximation:

$$V_o\frac{R_1}{R_1+R_2} + V_{in}\frac{R_2}{R_1+R2} \approx 0 \Rightarrow V_o \approx -\frac{R_2}{R_1}V_{in}.$$

If we would like to be more precise, we find that:

$$V_o(s)/V_{in}(s) = -\frac{R_2}{R_1}\frac{1}{\left(1 + \frac{R_1+R_2}{R_1}\frac{\tau s+1}{A}\right)}$$

Note that when the frequency response is considered, when we consider $s = j\omega$, we find that as long as $\tau j\omega$ is not too large, the frequency response is very near $-R_2/R_1$ as it should be.

Let us define the nominal gain of the amplifier by $G = R_2/R_1$. Let us define the bandwidth of our amplifier, W, as the frequency at which the magnitude of the frequency response falls to $G/\sqrt{2}$. We find that:

$$\left|\frac{V_o(jW)}{V_{in}(jW)}\right| = \frac{R_2}{R_1}\frac{1}{\sqrt{2}} \Rightarrow \left|1 + \frac{R_1+R_2}{R_1}\frac{\tau jW + 1}{A}\right| = \sqrt{2}.$$

Assuming that W is large enough that $\tau Wj + 1 \approx \tau Wj$, we find that the bandwidth is fixed by the equation:

$$\frac{R_1+R_2}{R_1}\frac{\tau W}{A} = 1.$$

If G is reasonably large, then:

$$1 = \frac{R_1+R_2}{R_1}\frac{\tau W}{A} = (1+G)\frac{\tau W}{A} \approx GW\frac{\tau}{A}.$$

This implies that:

$$GW \approx A/\tau = \text{constant}.$$

That is, the product of the (nominal) gain and the (approximate) bandwidth is a constant that depends on the characteristics of the op amp. This constant, called the gain-bandwidth product, is one of the op amp characteristics that manufacturers generally specify.

2.5.2 *The DC Motor*

A DC motor has as its input a voltage $v_{in}(t)$ and as its output the angle of the motors shaft, $\phi(t)$. We take two feature of the motor as central:

(1) If $v_{in}(t) = Cu(t)$, then the steady state speed of the motor, $\phi'(t)$, is proportional to C .
(2) The motor takes a finite amount of time to settle down to its final speed.

A simple transfer function which gives this behavior is:

$$\frac{\Phi(s)}{V_{in}(s)} = \frac{K}{s(s\tau + 1)}$$

where $\Phi(s) = \mathcal{L}(\phi(t))(s)$. This is the simplest transfer function that is generally used to describe a DC motor.

Consider the response of the motor to a step input. Let $v_{in}(t) = Cu(t)$. Then $V_{in}(s) = C/s$. We find that

$$\mathcal{L}(\phi(t))(s) = \frac{CK}{s^2(\tau + 1)} \overset{\text{partial fraction expansion}}{=} CK\left(\frac{1}{s^2} - \frac{\tau}{s} + \frac{\tau^2}{s\tau + 1}\right).$$

With a step input, the output of the motor for $t > 0$ is:

$$\phi(t) = CK\left(t - \tau + \tau^2 e^{-t/\tau}\right) \tag{2.5}$$

$$\phi'(t) = CK\left(1 - \tau e^{-t/\tau}\right).$$

Considering the second equation, we find that after the transient term has died down the motor speed is CK. Thus, the constant K in the motor's transfer function gives the ratio of the speed to the input voltage. From the form of $\phi'(t)$, we see that the rise time of the motor speed is approximately 3τ. The constant τ is called the motor's time constant.

Evaluating the Step Response Using MATLAB®—An Example

Suppose we have a motor whose transfer function is:

$$T(s) = \frac{5}{s(s/2+1)}$$

and we want to know how the motor's speed reacts to a unit step input, $u(t)$. To find the relation between the derivative of the motors output to its input we need only multiply the transfer function of the motor by s—which is the same thing as differentiating the output of the motor. (Note that from (2.5) we see that $\phi(0) = 0$.)

We would like to evaluate the step response of the system characterized by the transfer function $5/(s/2 + 1)$. MATLAB has a simple command that performs this evaluation. First we ask MATLAB to assign the transfer function to a variable, and then we ask MATLAB to find the step response. We proceed as follows:

```
V = tf([5],[1/2 1])
```

to which MATLAB replies:

```
Transfer Function:
    5
---------
0.5 s + 1
```

Then we type:

```
step(V)
```

To this MATLAB replies with Figure 2.5. Of course the command `step` can be used with any transfer function.

2.5.3 *The "Simple Satellite"*

Suppose that one is interested in controlling the angular position of a simple satellite—considered here as a disk with moment of inertia I, whose angle relative to some fixed ray emanating from its center is θ, whose angular

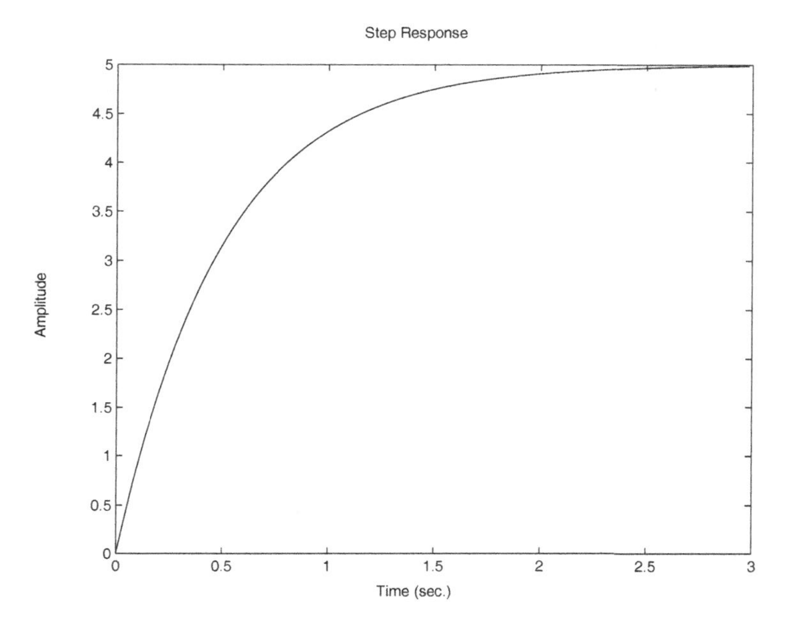

Fig. 2.5 The Step Response of a Motor

velocity is $\omega = \dot{\theta}$, whose angular momentum is $L = I\omega$, and whose "input" is a torque τ. We know that:

$$\frac{d}{dt}L = I\dot{\omega} = I\ddot{\theta} = \tau.$$

Assuming that the initial condition of the satellite is $\theta(0) = \dot{\theta}(0) = 0$, we find that:

$$Is^2\mathcal{L}(\omega(t))(s) = \mathcal{L}(\tau(t))(s)$$

or that the transfer function of the satellite is:

$$T(s) = \frac{1}{Is^2}.$$

To keep our calculations simple we often take $I = 1$.

2.6 Zeros in the Right Half-Plane

We have seen that it is *crucial* that a transfer function's poles be in the left half-plane. We now consider the interesting, if somewhat less important, question: "What happens when a transfer function has positive zeros?" (To explore this issue a bit further, see [HB07].) The answer to this question does not directly affect the stability of the system. What does it affect?

Consider a system whose transfer function is $T(s)$ and that satisfies:

(1) The system is stable,
(2) $T(0) = 1$,
(3) and $T(a) = 0, \quad a > 0$.

We prove that

Theorem 11 *The response of a system satisfying 1 – 3 to a unit step input tends to* 1 *as* $t \to \infty$, *but it suffers from undershoot—it descends below zero at some point.*

PROOF: Let the output of the system to a unit step be denoted by $y(t)$. Then:

$$Y(s) = T(s)(1/s) = A/s + \text{terms whose poles are in the LHP.}$$

The inverse Laplace transform of this function is $y(t) = A +$ terms that decay in time. Thus, the limit of $y(t)$ as $t \to \infty$ exists and is finite and can be calculated by using the final value theorem. We find that:

$$\lim_{t\to\infty} y(t) = \lim_{s\to 0+} sY(s) = \lim_{s\to 0+} T(s) = 1.$$

As $T(a) = 0$, we know that $Y(a) = T(a)/a = 0$. Let's consider the definition of $Y(a)$. It is:

$$Y(a) = \int_0^\infty e^{-at} y(t)\, dt,$$

and this integral equals zero. As $y(t)$ is positive for large t, and as the exponential term is always positive, the only way this integral can be zero is if $y(t) < 0$ for some values of $t > 0$.

If we think of a system for which $T(0) = 1$ as a system that wants to output a "one" when its input is a step function, then we see that when the transfer function has a right half-plane zero the output behaves oddly.

Despite "wanting" to tend to one, the output becomes negative at some point. It is almost as though we gave the system the command to go up one unit and it answered, "Fine—but first I am going to go down a bit!"

A System with One Zero in the RHP—An Example

Consider the system whose transfer function is:

$$T(s) = \frac{1-s}{(s+1)^2}.$$

As all the poles are in the LHP, the system is stable. Also, $T(0) = 1$ and $T(1) = 0$. Thus, the system must "suffer" from undershoot. Let $y(t)$ be the system's step response. Using the initial value theorem, it is easy to see that:

$$y(0^+) = \lim_{s\to\infty} sY(s) = \lim_{s\to\infty} T(s) = 0.$$

Now that we know $y(0^+)$, it is easy to calculate $y'(0^+)$. Making use of the initial value theorem once again, we find that:

$$\begin{aligned} y'(0^+) &= \lim_{s\to\infty} s(sY(s) - y(0^+)) \\ &= \lim_{s\to\infty} sT(s) \\ &= \lim_{s\to\infty} \frac{s-s^2}{s^2+2s+1} \\ &= -1. \end{aligned}$$

We find that immediately after the step function is input the output must be negative. The step response of the system (as calculated by MATLAB) is given in Figure 2.6.

2.7 Block Diagrams and How to Manipulate Them

A central problem of control theory is to determine how the *properties of the (often reasonably simple) subsystems* of a system and the *interconnections between these subsystems* determine the properties of the system as a whole. We are often interested in systems that employ *feedback*—systems where some fraction of the output is fed back to the system's input to help control the output itself. Throughout this book, we use block diagrams to describe systems in terms of their subsystems. In this section, we describe how block diagrams are to be used, understood, and manipulated.

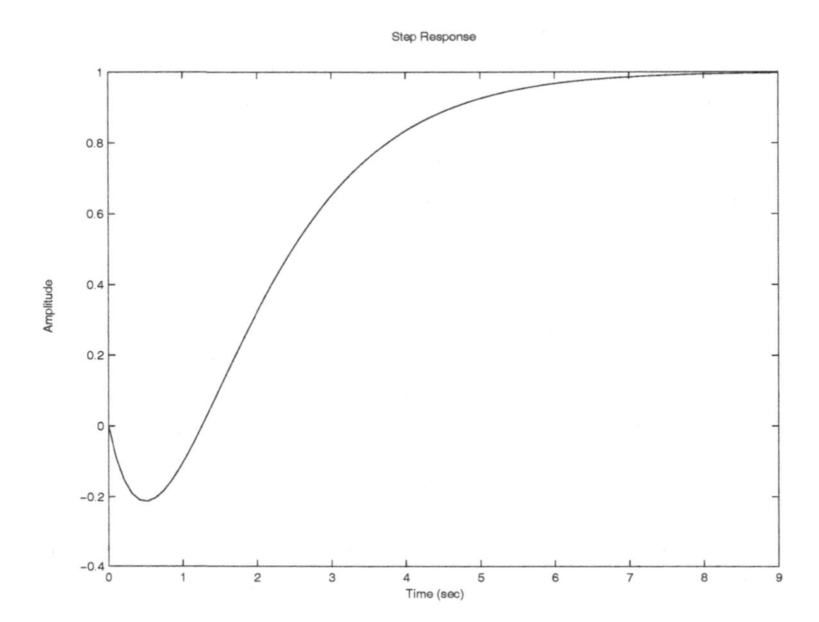

Fig. 2.6 The step response of a system with a single positive zero. Despite the fact that for all positive time the input to the system is one, the output dips below zero before rising up towards one.

The fundamental components of the block diagram are the differencer, the arrow, and the "block." A differencer is generally represented by a circle with pluses and minuses that tell us which signals are to be added and which are to be subtracted. Arrows tell us which way signals flow, and blocks generally perform an operation on an input signal and output an output signal.

The Block Diagram of a Motor Controller—An Example

Consider a simple motor controller that consists of a motor, a tachometer (to measure the motor's shaft speed and output an electrical signal that can be used by the rest of the system), a differencer (to calculate the difference between the motor's actual speed and the desired speed), and an amplifier (to amplify the difference between the actual speed and the desired speed). This can be shown in a block diagram. (See Figure 2.7.) This pictorial description of the motor controller gives one a feeling for the interconnections between the subsystems of the motor controller. From

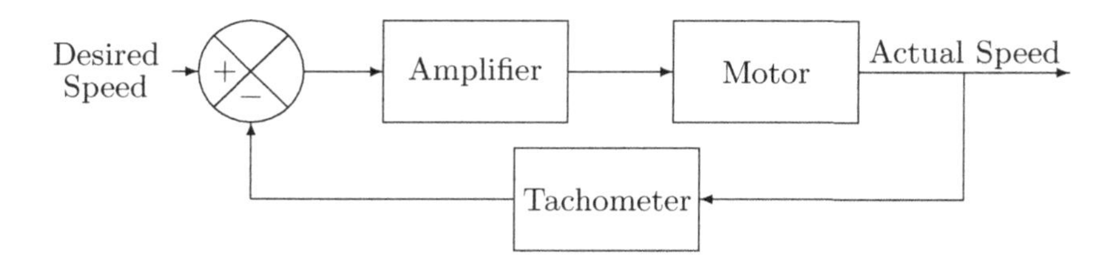

Fig. 2.7 A Simple Motor Controller

the diagram, we see that the input to the amplifier is the difference between the desired speed and the measured speed. The amplified difference is the input to the motor.

Generally speaking, our diagrams give more information than Figure 2.7 does. Rather than labeling each block with its purpose, we usually label each block with the transfer function of the unit represented by the block (and perhaps the unit's name as well). Let us consider a simple DC motor receiving an input $v_{in}(t)$.

A DC Motor—An Example

We consider a simple system that consists of the input voltage to a DC motor, the motor, and the output shaft position of the motor. Assume that the transfer function of the motor is $T(s) = 5/(s(s/2+1))$. The block diagram of the system is given in Figure 2.8. Because the item in the block is a transfer function we interpret

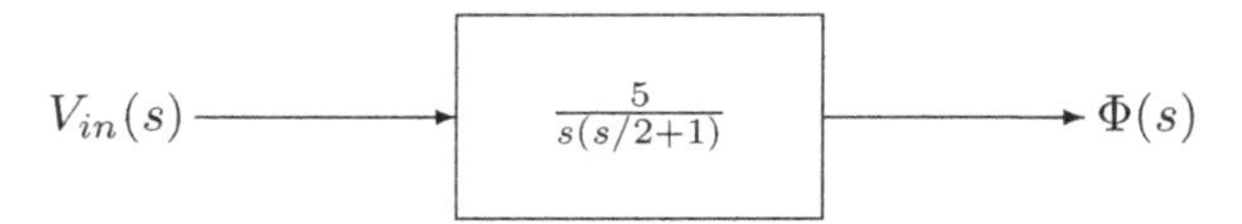

Fig. 2.8 A DC Motor

the picture to mean that

$$\Phi(s) = \frac{5}{s(s/2+1)} V_{in}(s).$$

Let us consider what happens when blocks are interconnected in various ways. Consider the upper figure in Figure 2.9 in which two blocks are connected in series. It is easy to see that the two block diagrams in Figure

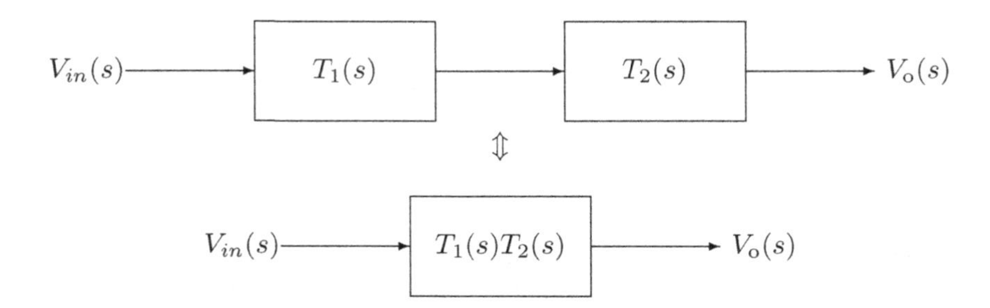

Fig. 2.9 The Series Connection of Two Blocks

2.9 are equivalent. The Laplace transform of the signal between the two blocks of the first diagram is clearly just the input times the first transfer function—$V_{in}(s)T_1(s)$. The Laplace transform of the output of the second block is just $T_2(s)$ times the input of the second block. That gives us $V_o(s) = T_1(s)T_2(s)V_{in}(s)$. This is exactly what the second block diagram in the figure shows. Considering only the relation between their input and their output, the two block diagrams in Figure 2.9 are equivalent.

Next consider the most important connection in control theory—the feedback connection. Consider Figure 2.10. How can we find the output of the lower diagram in terms of its input? We follow the output "around the loop backwards." Clearly, $V_o(s) = G_p(s)\alpha(s)$. Also, $\alpha(s) = V_{in}(s) - \beta(s)$. Since $\beta(s) = H(s)V_o(s)$, we find that:

$$V_o(s) = G_p(s)\left(V_{in}(s) - H(s)V_o(s)\right) \Leftrightarrow \frac{V_o(s)}{V_{in}(s)} = \frac{G_p(s)}{1 + G_p(s)H(s)}.$$

This shows that the upper block diagram is equivalent to the lower one. The importance of this equivalence cannot be overemphasized.

2.8 A Final Example

We would like to design a system to control the angle of the motor shaft of a DC motor. A simple way of doing this is to use a feedback system like that of Figure 2.11. One takes the input signal—which gives the *desired*

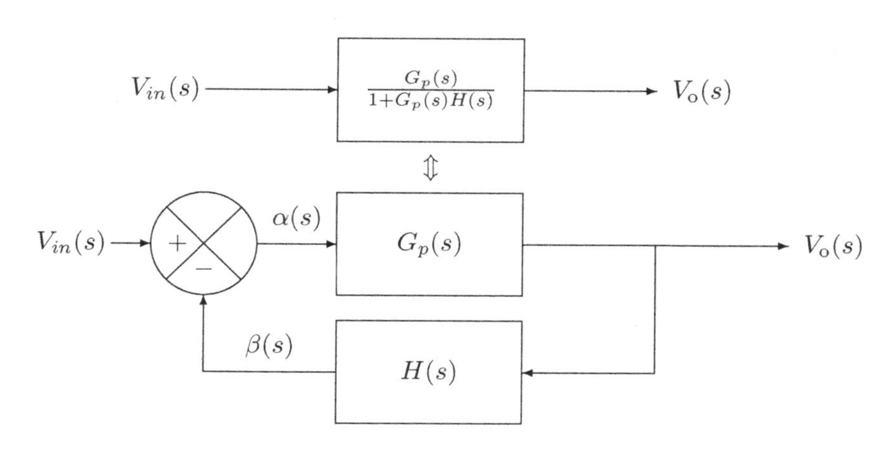

Fig. 2.10 A Simple Feedback Circuit

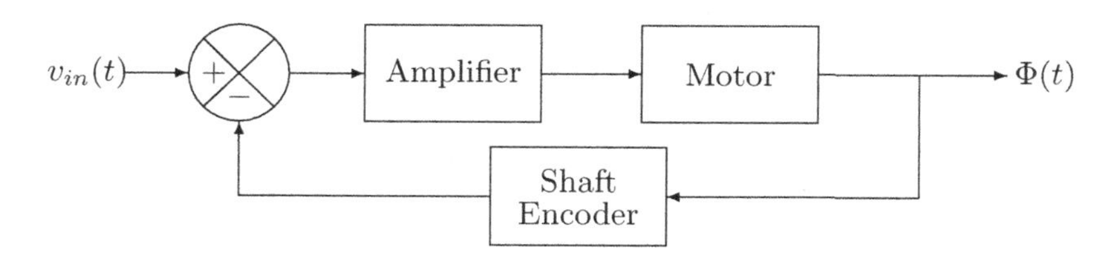

Fig. 2.11 Another Simple Motor Controller

position of the motor shaft—and considers the difference between it and the output of the shaft encoder—the *actual* position of the motor shaft. Then one amplifies this difference, and one uses this value to drive the motor. This should give negative feedback—if the motor shaft is turned too much in one direction, the feedback should act to restore the motor to its correct position.

Let us move to a more quantitative description of the system. Suppose that the motor's transfer function is:

$$T_M(s) = \frac{5}{s(0.5s+1)},$$

the shaft encoder's transfer function is $T_S(s) = 1$, and the amplifier's transfer function is $T_{\mathrm{A}}(s) = K$ where we choose K later. The block diagram of the motor controller is given in Figure 2.12.

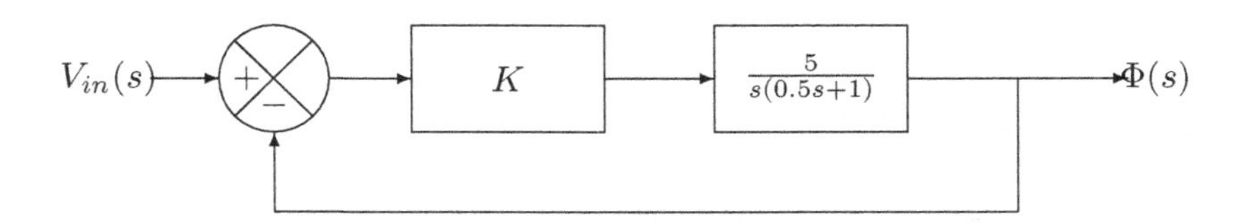

Fig. 2.12 A Quantitative Block Diagram of a Simple Motor Controller

As the two blocks on top are connected in series, they can be combined in to one block whose transfer function is:

$$\frac{5K}{(s(0.5s+1)}.$$

Now we have a simple feedback system and we find that the transfer function of the whole system is:

$$T(s) = \frac{\frac{5K}{s(0.5s+1)}}{1 + \frac{5K}{s(0.5s+1)}} = \frac{5K}{s(0.5s+1) + 5K} = \frac{10K}{s^2 + 2s + 10K}.$$

This is precisely the form of the second order system discussed on p. 46. We find that $\omega_n = \sqrt{10K}$ and $\zeta = 1/\sqrt{10K}$. From what we have already seen, this means that as long as $K \leq 0.1$ the system's step response will not oscillate at all, while for $K > 0.1$ the system's step response will oscillate about its steady state value for a little while.

For values of $K \leq 0.1$, the decaying part of the solution decays like $e^{\alpha t}$ where $\alpha = \omega_n \sqrt{\zeta^2 - 1} - \zeta\omega_n$. For small K we find that[5]:

$$\alpha = \sqrt{10K}\sqrt{\frac{1}{10K} - 1} - 1 = \sqrt{1 - 10K} - 1 \overset{\sqrt{1+x} \approx 1 + x/2}{\approx} -5K.$$

We see that (at least for small values of K) the system gets faster as K gets larger. We also see that past a certain point increasing K causes the system to start oscillating about its steady state value. This is rather typical of feedback systems—up to a point increasing the gain helps the system, and past that point it starts hurting the system. In Figures 2.13 – 2.15 we show

[5]As the first two terms in the Taylor series of $\sqrt{1+x}$ about zero are 1 and $x/2$, for small $|x|$ we find that $\sqrt{1+x} \approx 1 + x/2$.

the step response of the system for $K = 0.05, 0.1$, and $K = 0.2$ respectively.

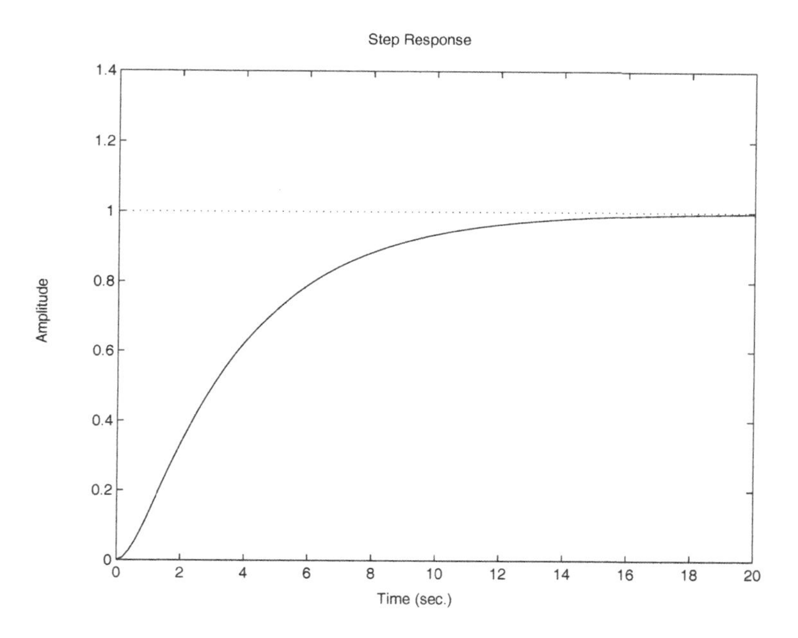

Fig. 2.13 The Step Response of a Controlled Motor—$K = 0.05$

2.9 Exercises

(1) (a) If the relation between a system's input, $x(t)$, and its output,$y(t)$, is described by the differential equation:

$$y''(t) + y'(t) + y(t) = x(t)$$

what is the system's transfer function?

(b) Using the (MATLAB) `bode` command, plot the frequency response of the system.

(2) Using the information about the time response of low-order systems, approximate the response of the following systems to a unit step input, and sketch the estimated response:

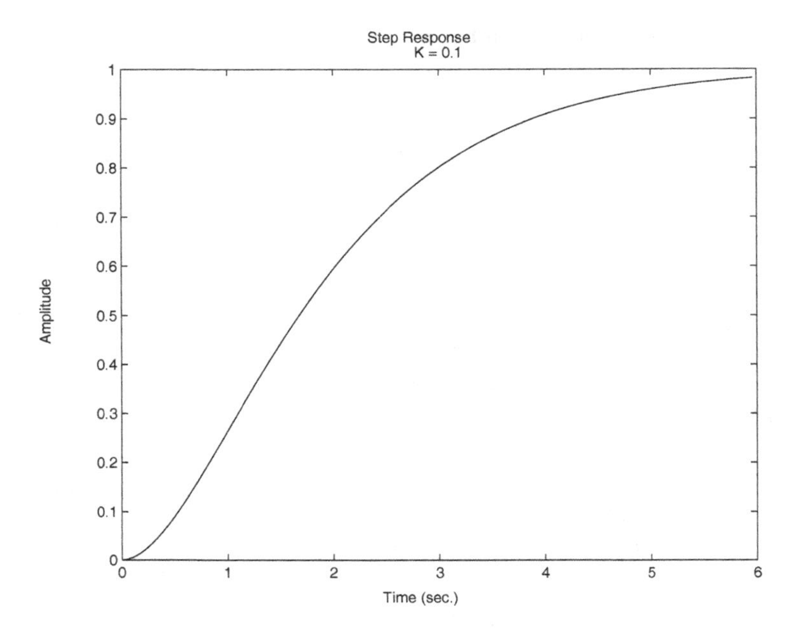

Fig. 2.14 The Step Response of a Controlled Motor—$K = 0.1$

(a)

$$T(s) = \frac{5}{s+5}$$

(b)

$$T(s) = \frac{10}{s+2}$$

(c)

$$T(s) = \frac{5}{s^2 + 10s + 50}$$

(d) Check the results of 2a-2c using MATLAB.

(3) Use the result of 2c and your knowledge of the properties of the Laplace transform to estimate the output of the system:

$$T(s) = \frac{5s}{s^2 + 10s + 50}$$

to a unit step input.

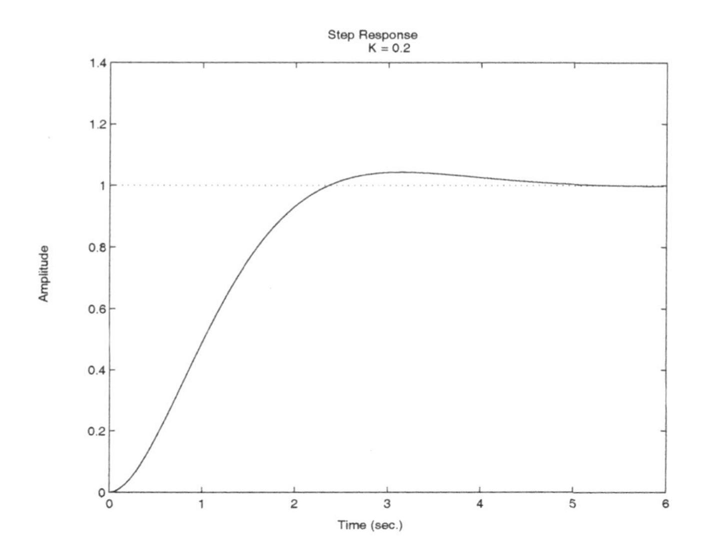

Fig. 2.15 The Step Response of a Controlled Motor—$K = 0.2$

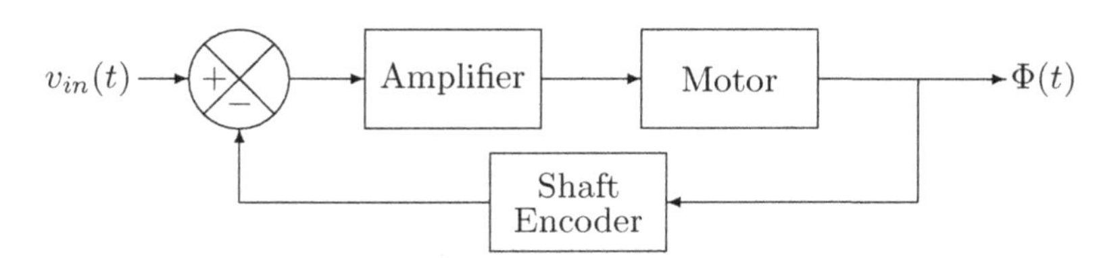

Fig. 2.16 A Simple Motor Controller

(4) Suppose that one is given the system of Figure 2.16 and that the transfer function of the motor is:

$$T_M(s) = \frac{1}{s(s/2+1)},$$

the transfer function of the shaft encoder is 2, and the gain of the amplifier is 1. Find the transfer function of the entire system, $T(s)$. Also, please estimate the response of the system to a step input.

(5) Suppose that one has an under-damped system whose transfer

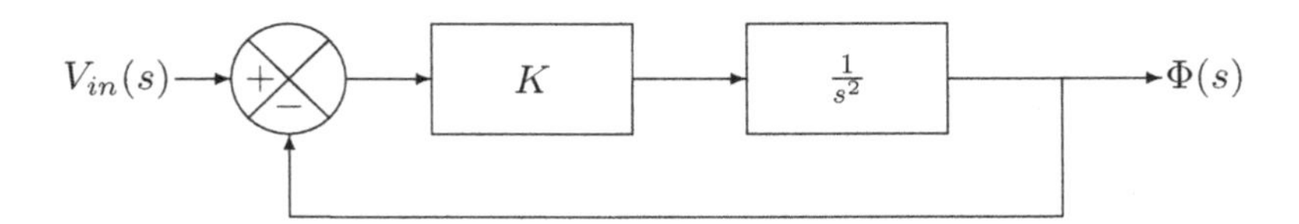

Fig. 2.17 A "Mildly" Unstable Controller for a Simple Satellite

function is:

$$T(s) = \frac{\omega_n^2}{s^2 + 2\zeta\omega_n s + \omega_n^2}.$$

In what region must the poles of the system be located in order for the system to satisfy the constraints:

(a) The rise time must be less than or equal to T seconds.
(b) The percent overshoot must be less than $O \times 100$ percent.

What does this tell us about the significance of the ratio between the imaginary part of the poles to the real part of the poles?

(6) (a) Find the transfer function the system of Figure 2.17. This figure depicts a first attempt at controlling the angular position of a simple satellite.
(b) Show that this system is unstable for all $K > 0$. Find a bounded function which when input to the system causes the output of the system–the angular position of the satellite–to be unbounded. That is, show that the system is not BIBO stable. (This is another example of a resonant system.)
(c) Find the unit step response of the system. Note that though the response is bounded, it is not what one would like to see from a properly designed system.

(7) Consider a system whose transfer function is:

$$T(s) = \frac{\omega_n^2}{s^2 + 2\zeta\omega_n s + \omega_n^2}.$$

Show that if $\zeta = 0$ the system is unstable, while if $\zeta > 0$ the system is stable. This shows that an undamped second order system is not stable.

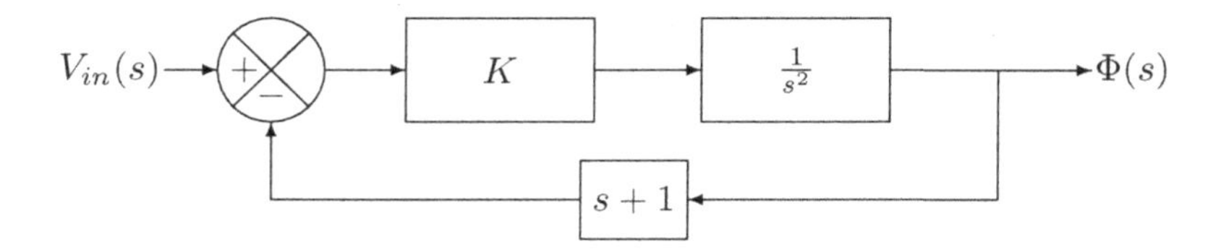

Fig. 2.18 A Controller for a Simple Satellite

(8) (a) Find the transfer function of the system of Figure 2.18.
(b) Is this system stable?
(c) What is the inverse Laplace transform of $(1+s)\Phi(s)$ (neglecting any initial conditions) in terms of $\phi(t)$?

(9) (a) What is the step response of the system whose transfer function is

$$H(s) = \frac{2-s}{s^2+3s+2}?$$

(b) What is odd about this step response?
(c) What property of the transfer function causes this behavior?

(10) (a) What is the step response of the system whose transfer function is

$$H(s) = \frac{(1-s)^2}{(1+s)^3}?$$

(b) What is odd about this step response?
(c) What property of the transfer function causes this behavior?

(11) Let

$$T(s) = \frac{2-s}{(s+3)(s+4)}$$

be the transfer function of a system, and let $y(t)$ be the step response of the system—be the response of the system to a unit step function, $u(t)$.

(a) Calculate $\lim_{t\to\infty} y(t)$ without calculating $y(t)$.
(b) Calculate $y(0^+)$ without calculating $y(t)$.
(c) Calculate $y'(0^+)$ without calculating $y(t)$.

(d) Calculate $y(t)$, and use the results of the preceding sections to check that you have calculated $y(t)$ correctly.

(12) Which of the following transfer functions correspond to stable systems, to marginally stable systems, and to unstable systems (that are not marginally stable)?

(a)

$$T_1(s) = \frac{s-1}{s^2+1}$$

(b)

$$T_2(s) = \frac{s-1}{s^2+3s+2}$$

(c)

$$T_3(s) = \frac{1}{(s^2+1)(s^2+4)}$$

(d)

$$T_4(s) = \frac{1}{(s^2+3)^2}$$

Chapter 3

Feedback—An Introduction

3.1 Why Feedback—A First View

We now have the tools that are necessary to consider the pros and cons of feedback systems. Consider the system of Figure 3.1. What are the

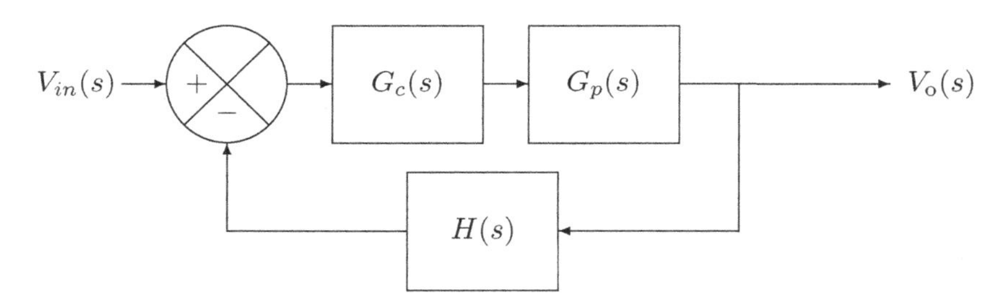

Fig. 3.1 A Generic Feedback Circuit

"physical implications" of the diagram? Generally speaking $G_c(s)$ is the transfer function of an object that we as system designers have added to the system in order to improve the system's performance—it is added to *compensate* the system's shortcomings. The transfer function of the object whose output we are supposed to be controlling is $G_p(s)$. It is the *given* physical plant, and generally speaking we cannot change $G_p(s)$. The feedback network's transfer function is $H(s)$. We also design the feedback network.

Though the labeling in the block diagram makes it seem as though all of the inputs and output are of the same sort (say voltages), this need not be the case. When it is the case, it often means that some blocks consist of more than one element. For example, a block that is supposed to be a motor but accepts a voltage input and returns a voltage output clearly consists of a motor and some sort of position (or velocity) to voltage transducer.

Frequently our goal is to design a system whose output tracks its input as closely as possible. That is, we would like to minimize $|v_{in}(t) - v_o(t)|$. In this case it is often advantageous to set $H(s)$ to 1. In this way, the signal that controls our "plant" is directly related to the signal that we would like to minimize. When $H(s) = 1$, we say that we have a *unity feedback* system.

As we have already seen, one way of trying to improve a system's performance is to set $G_c(s) = K$—to make our compensator a pure gain. For the rest of this section, we assume that our compensator is a pure gain.

Consider the transfer function of our system, $T(s)$, assuming that $G_c(s) = K$ and $H(s) = 1$. We find that the frequency response of our system is:

$$T(j\omega) = \frac{KG_p(j\omega)}{1 + KG_p(j\omega)} = \frac{1}{1 + 1/(KG_p(j\omega))}.$$

Supposing that $G_p(jw)$ is nonzero, we see that for large enough K the transfer function will be close to one—and this is *exactly* what we want. If the gain is large enough the system tracks its input very closely in the (sinusoidal) steady state. As we saw in the previous chapter, generally speaking at sufficiently high gains the system's performance start to degrade. This is one of the many tradeoffs that a system designer must deal with—frequently what helps cure one problem creates a different problem.

3.2 Sensitivity

We have seen one of feedback's main uses. We can "trade" the addition of a large gain in the loop for a system whose transfer function is very nearly one at many frequencies. It is not (or at least is should not be) sufficient to explain why feedback itself is good—we must explain why it is better than other methods of achieving our goal. Our goal is to get the output of our system to track its input in a reasonable fashion; we would like an overall transfer function that is as near as possible to 1. Why should we employ feedback techniques—why don't we consider a system like that of Figure 3.2? If we assume that all of the zeros of $G_p(s)$ are in the left-half plane (something which—for reasons discussed in §2.6 and that will be discussed in §6.3.1—is often desirable), then we can let $G_c(s) = 1/G_p(s)$. This is the transfer function of a stable system, and it gives us exactly the result that

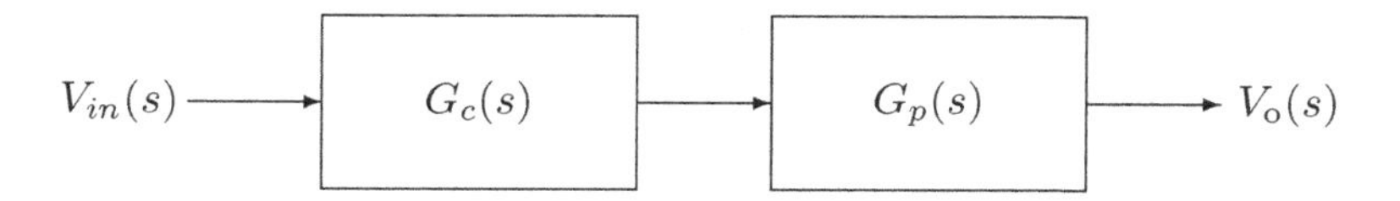

Fig. 3.2 Open-Loop Compensation

we wanted—the overall transfer function of the system is now precisely 1!

To understand why this more direct method of compensation is often not practical, it is helpful to understand the concept of sensitivity. Suppose we have a value a that is affected by a value b. Suppose that we make a small change in b—we change it to $b + \Delta b$. As a depends on b, we will find that the current value of a is $a + \Delta a$. If we measure all changes in relation to the initial values of the measured quantities, then we we find that the relative change in a is $\Delta a/a$ and the relative change in b is $\Delta b/b$. The ratio of the relative change in a to the relative change in b is just:

$$\frac{\Delta a/a}{\Delta b/b}.$$

We define the sensitivity of a to b to be the ratio of the relative change in a to the relative change in b for an infinitesimal change in b. That is the sensitivity of a to b, denoted by S_b^a, is defined as:

$$S_b^a \equiv \lim_{\Delta b \to 0} \frac{\Delta a/a}{\Delta b/b} = \lim_{\Delta b \to 0} \frac{\Delta a}{\Delta b}\frac{b}{a} = \frac{\partial a}{\partial b}\frac{b}{a}.$$

If the sensitivity of a to b is S_b^a, then if b increases by a small amount, say $x\%$, a will increase by approximately $S_b^a x\%$.

Consider the sensitivity of our open-loop system to changes in $G_p(s)$. Suppose that we assumed that the transfer function of the given system $G_p(s)$ was $G_0(s)$. Our compensator is $G_c(s) = 1/G_0(s)$. The system's transfer function is $T(s) = G_p(s)/G_0(s)$. The sensitivity of $T(s)$ to changes is $G_p(s)$ is:

$$S_{G_p(s)}^{T(s)} = \frac{\partial T(s)}{\partial G_p(s)}\frac{G_p(s)}{T(s)} = \frac{1}{G_0(s)} G_p(s) \frac{G_0(s)}{G_p(s)} = 1.$$

A small change in the actual value of the $G_p(s)$ makes an approximately proportional change in $T(s)$.

Now consider the sensitivity of the transfer function of a system that uses a large gain and feedback to implement a system whose overall transfer function is very nearly one. Consider the system of the previous section whose transfer function is $T(s) = KG_p(s)/(1 + KG_p(s))$. We find that the sensitivity of $T(s)$ to $G_p(s)$ is given by:

$$\begin{aligned} S_{G_p(s)}^{T(s)} &= \frac{\partial T(s)}{\partial G_p(s)} \frac{G_p(s)}{T(s)} \\ &= \frac{K(1 + KG_p(s)) - KKG_p(s)}{(1 + KG_p(s))^2} G_p(s) \frac{1 + KG_p(s)}{KG_p(s)} \\ &= \frac{1}{1 + KG_p(s)}. \end{aligned}$$

So long as $KG_p(s)$ is large, the sensitivity of $T(s)$ to changes is $G_p(s)$ is quite small. Closed-loop feedback can be made much less sensitive to changes in the given function $G_p(s)$ than open-loop feedback. Since we do not generally know exactly what the characteristics of the system to be controlled are, it is often beneficial to utilize closed-loop feedback.

3.3 More about Sensitivity

Once again, we consider the feedback system of Figure 3.1. We would like to determine the sensitivity of the system to its various components. As we have seen previously:

$$T(s) = \frac{G_c(s)G_p(s)}{1 + G_c(s)G_p(s)H(s)}.$$

Making use of the definition of the sensitivity, we find that:

$$\begin{aligned} S_{G_c(s)}^{T(s)} &= \frac{G_p(s)}{(1 + G_c(s)G_p(s)H(s))^2} \frac{G_c(s)}{T(s)} = \frac{1}{1 + G_c(s)G_p(s)H(s)} \\ S_{G_p(s)}^{T(s)} &= \frac{G_c(s)}{(1 + G_c(s)G_p(s)H(s))^2} \frac{G_p(s)}{T(s)} = \frac{1}{1 + G_c(s)G_p(s)H(s)} \\ S_{H(s)}^{T(s)} &= -\frac{(G_c(s)G_p(s))^2}{(1 + G_c(s)G_p(s)H(s))^2} \frac{H(s)}{T(s)} = -\frac{G_c(s)G_p(s)H(s)}{1 + G_c(s)G_p(s)H(s)} \end{aligned}$$

Assume that $G_p(s)$ and $H(s)$ are function's whose values remain "reasonable" and that $|G_c(s)|$ is relatively large. Then we find that $S_{G_c(s)}^{T(s)} \approx 0$, $S_{G_p(s)}^{T(s)} \approx 0$, and $S_{H(s)}^{T(s)} \approx -1$.

What are the implications of these sensitivities? We have found that if the compensator gain is reasonably large, then the system transfer function is relatively insensitive to the compensator gain and to the transfer function of the item being controlled. On the other hand, we have found that the system transfer function is relatively sensitive to the feedback. How does all this help us?

In many cases, we are given an item to control without being given a precise description of the item. Often a precise description is not possible—as for example when one must design a general purpose controller to be used with any item from a large class of items. In such cases, the insensitivity to the item to be controlled is a very important feature of our technique. The sensitivity to the feedback element is not hard to deal with. Generally speaking, the feedback is electronic in nature. It is reasonably easy to purchase high precision electrical and electronic components. By using feedback, we exchange the need for an accurate model of the system to be controlled—a model that will often be impossible to find—for an accurate model of a feedback network that we build—something that is achievable.

3.4 A Simple Example

Suppose that we would like to control the speed of a motor. Assume that motor's transfer function (the transfer function from voltage to *speed* in revolutions per second) is:

$$G_p(s) = \frac{K}{\tau s + 1}.$$

Suppose that the nominal values of K and τ are 5 and 0.1 respectively. Consider two controllers for the motor. The first controller is shown in Figure 3.3 and the second in Figure 3.4.

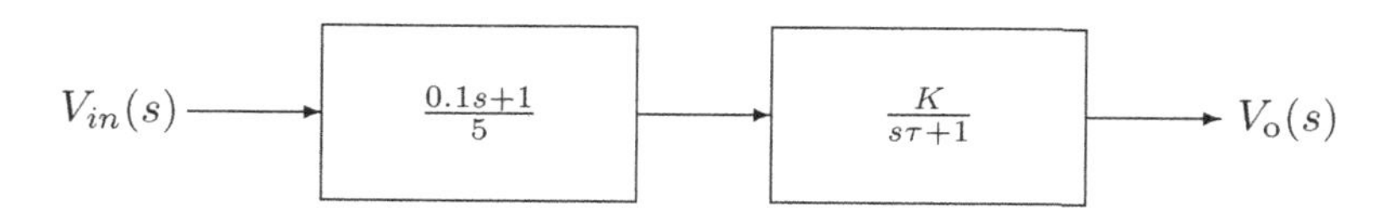

Fig. 3.3 An Open-Loop Motor Speed Controller

The transfer function of the first system—the open-loop system—is:

$$T_{ol}(s) = \frac{K}{5}\frac{0.1s+1}{\tau s+1}.$$

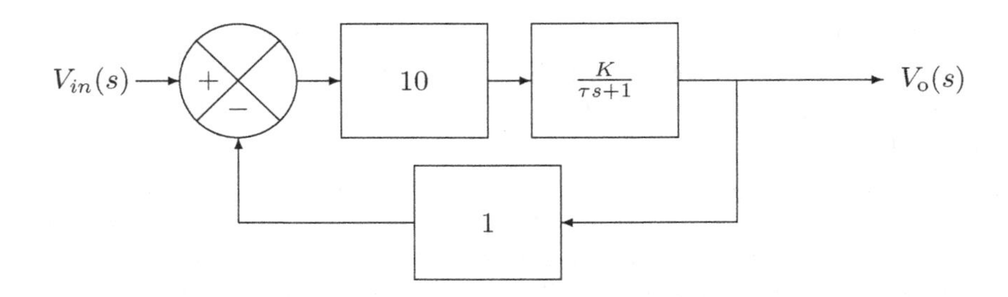

Fig. 3.4 A Closed-Loop Motor Speed Controller

The transfer function of the second system—the unity feedback closed loop system—is:

$$T_{cl}(s) = \frac{10K}{\tau s + 1 + 10K}.$$

Consider the step response of each system for $K = 7, \tau = 0.15$. These results are plotted in Figures 3.5 and 3.6. We see that the system with feedback performs *much* better than the system without feedback. The system with feedback is faster than the system without feedback, and the system with feedback comes very close to stabilizing on the precise value we input.

In order to understand the response of the system to a step input, let us consider the sensitivity of the gain of the system at DC, $T(0)$, to changes in the parameters K and τ. Using the definition of sensitivity, we find that:

$$\begin{aligned}
S_K^{T_{ol}(0)} &= 0.2 \\
S_\tau^{T_{ol}(0)} &= 0 \\
S_K^{T_{cl}(0)} &= \frac{10}{(1+10K)^2} \overset{K=5}{\approx} 0.004 \\
S_\tau^{T_{cl}(0)} &= 0.
\end{aligned}$$

We see that the DC gain is much more sensitive to changes in K in the open-loop system.

3.5 System Behavior at DC

Frequently we must design a system to track its input. If the input is a unit step, then as $t \to \infty$ the system should settle down to some constant value. In the best case scenario, the system settles on the input value. Sometimes,

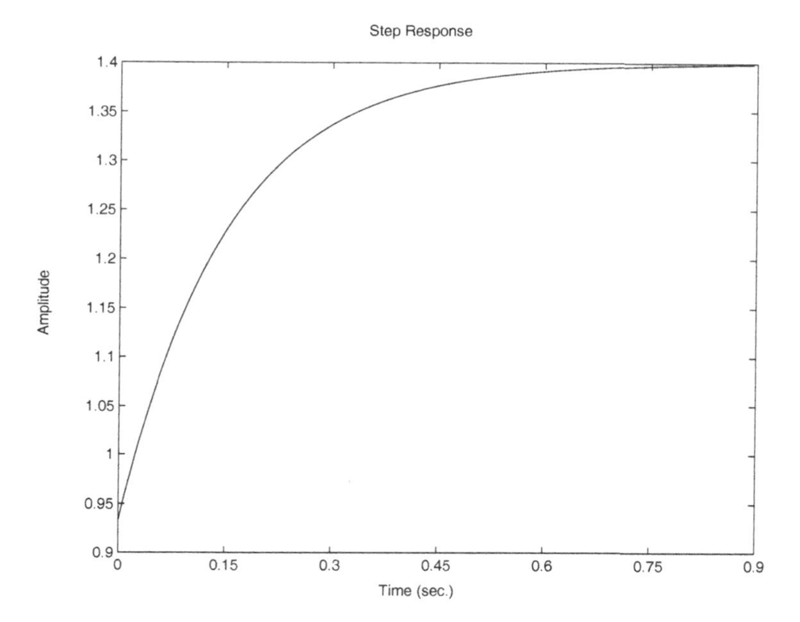

Fig. 3.5 The Step Response of the Open-Loop System

the system will not settle at all, and sometimes the system will settle on a value other than the input value.

Once again we consider a system like that of Figure 3.1. The transfer function of the system is:

$$T(s) = \frac{G_c(s)G_p(s)}{1 + G_c(s)G_p(s)H(s)}.$$

We see that $T(0)$—which controls the behavior at DC—is:

$$T(0) = \frac{G_c(0)G_p(0)}{1 + G_c(0)G_p(0)H(0)}$$

as long as all of these numbers exist.

What is the implication of this formula? Suppose that we are dealing with a unity feedback system. Then we find that:

$$T(0) = \frac{G_c(0)G_p(0)}{1 + G_c(0)G_p(0)}.$$

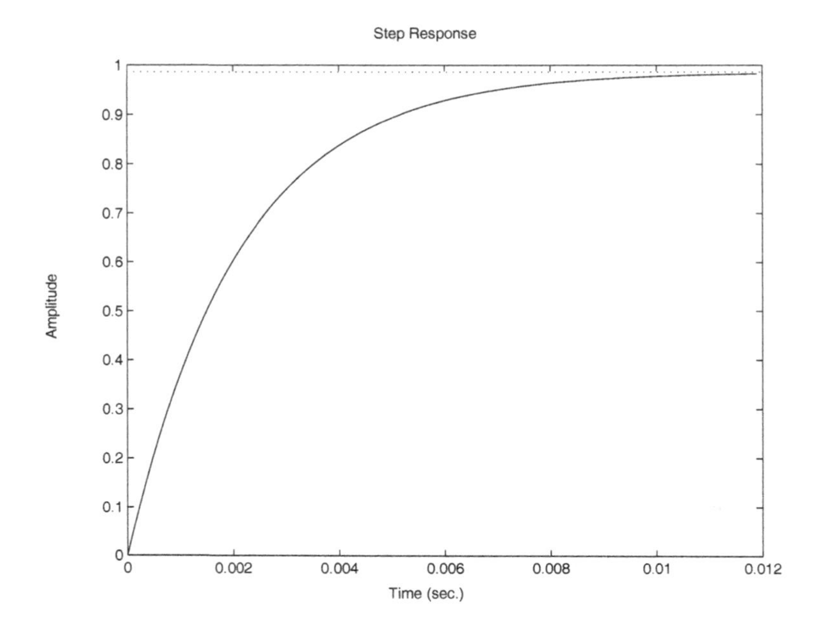

Fig. 3.6 The Step Response of the Closed-Loop System

If $G_c(0)G_p(0)$ is large, then $T(0) \approx 1$. Once again, we find that making $G_c(s)$ large generally improves the system's performance at DC.

We see something else too. Suppose that $G_c(s)G_p(s)$ is of the form:

$$G_c(s)G_p(s) = \frac{1}{s^n}\frac{P(s)}{Q(s)} \qquad n \geq 1$$

where $P(s)$ and $Q(s)$ are polynomials, all of the roots of $Q(s)$ are in the left half-plane, and both $P(0) \neq 0$ and $Q(0) \neq 0$. We define the system's order to be the value of n—the number of poles at the origin. We find that:

$$\begin{aligned}
\lim_{s\to 0} T(s) &= \lim_{s\to 0} \frac{\frac{1}{s^n}\frac{P(s)}{Q(s)}}{1 + \frac{1}{s^n}\frac{P(s)}{Q(s)}} \\
&= \lim_{s\to 0} \frac{\frac{P(s)}{Q(s)}}{s^n + \frac{P(s)}{Q(s)}} \\
&= \frac{P(0)/Q(0)}{P(0)/Q(0)} = 1.
\end{aligned}$$

That is, if the "gain" of the elements $G_c(s)G_p(s)$ is "infinite" at DC—if the system's order is one or greater—then the steady state output of the system is the same as the constant value input to the system.

Suppose that the input to the system is a ramp, $tu(t)$. Then the system's output is:

$$\frac{P(s)}{s^n Q(s) + P(s)} \frac{1}{s^2}.$$

As by assumption $s^n Q(s) + P(s)$ has no roots at $s = 0$, the partial fraction expansion of this expression is:

$$\frac{P(s)}{s^n Q(s) + P(s)} \frac{1}{s^2} = \frac{A}{s} + \frac{B}{s^2} + \cdots .$$

where the terms that have been ignored are all exponentially decaying (as the system is stable). Multiplying both sides by s^2 one finds that:

$$\frac{P(s)}{s^n Q(s) + P(s)} = As + B + s^2(\cdots).$$

Evaluating this expression at $s = 0$ we find that $B = 1$. Differentiating once and evaluating at $s = 0$ we find that:

$$A = \begin{cases} -Q(0)/P(0) & n = 1 \\ 0 & n > 1 \end{cases}.$$

We see that if $n = 1$, the system will not track a ramp perfectly. The slope of the output will be correct, but the output will be displaced by $-Q(0)/P(0)$ units in the steady state.

Suppose that one is given an element to control, and one knows that $G_p(0)$ is neither zero nor infinity. How can one compensate the system in such a way that the resulting system tracks its input perfectly in the steady state? We have seen many times that adding a gain block before the given system should improve the performance of the system. Our current result gives us a way of getting excellent performance from the system. We must add a block whose gain at DC is infinite. What sort of block has this property? An *integrator*. Let us choose an integrator as our compensator; let $G_c(s) = 1/s$. After adding the integrator, we find that $T(0) = 1$—the system tracks its input perfectly in the long run.

Motor Control Revisited—An Example

Let us consider a system for regulating a motor's speed. We have already seen that taking $G_c(s)$ to be a pure gain improves the system performance at DC. Now we would like to see what happens if we replace the finite gain by an integrator—an element with infinite gain at DC. Consider the system of Figure 3.7. As there is one pole at zero in $G_c(s)G_p(s)$, we know that the steady state output of the system when the input is a unit step is one.

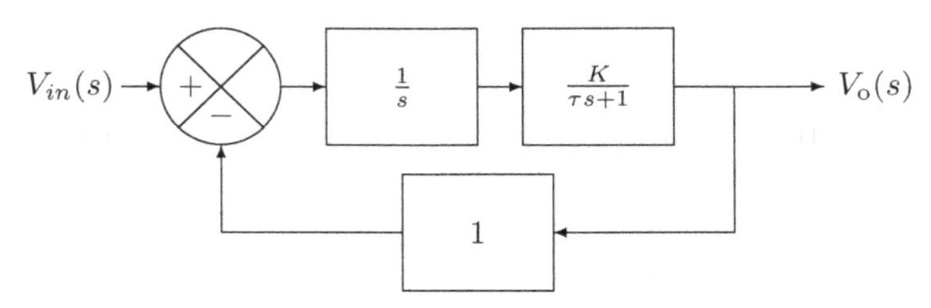

Fig. 3.7 A Motor Speed Controller with Integral Compensation

In this example it is also easy to get a feel for how integral compensation works. Without integral compensation—if $G_c(s) = C$—the output of the differencer cannot go to zero in the steady state. If the output of the differencer went to zero, the input to the motor would go to zero, and the motor would eventually stop rotating. Thus, C times the output of the differencer drives the motor. This also explains why the larger C is the better the steady state response of the motor. If C is large, then the output of the differencer must be smaller in order to drive the motor at a particular speed. If the the compensator is an integrator, then after a while the voltage on the integrator builds up to the correct value to drive the motor, and after reaching the correct voltage, the output of the differencer *must* drop to zero for proper operation. Here we have a nice example of theory and intuition leading to the same result.

3.6 Tracking Inputs of the Form $t^n u(t)$

Once again we consider a system like that of Figure 3.1, and to keep things simple, we assume that $H(s) = 1$—that we are dealing with a unity-feedback system. It will be very useful to consider the transfer function from the input to the system, $V_{in}(s)$, to the input to $G_c(s)$—which is just $V_{in}(s) - V_o(s)$, the difference between the input and the output. We refer to this difference as the error, we denote it by $e(t) \equiv v_{in}(t) - v_o(t)$, and we denote its Laplace transform by $E(s)$.

Clearly, $E(s) = V_{in}(s) - G_p(s)G_c(s)E(s)$. Thus, the transfer function from the system's input to its error is:

$$T_{V_{in}(s)\to E(s)}(s) \equiv \frac{E(s)}{V_{in}(s)} = \frac{1}{1 + G_c(s)G_p(s)}.$$

If the system is stable, all the poles of this new transfer function are in the left-half plane.

Now assume that

$$G_c(s)G_p(s) = \frac{1}{s^n}\frac{P(s)}{Q(s)}$$

where $P(s)$ and $Q(s)$ are polynomials and neither $P(s)$ nor $Q(s)$ equals zero when $s = 0$. We find that:

$$T_{V_{in}(s)\to E(s)}(s) = \frac{s^n Q(s)}{s^n Q(s) + P(s)}$$

and, assuming that the closed loop system is stable, all of this function's poles are located in the left half-plane.

Let us consider the signals that are of interest to us, $t^m u(t)$. It is easy to show that the Laplace transform of $t^m u(t)$ is $m!/s^{m+1}$. (See exercise 13.) Consider $V_{in}(s) = m!/s^{m+1}$. We find that $E(s)$ is just:

$$E(s) = T_{V_{in}(s)\to E(s)}(s)\frac{m!}{s^{m+1}} = \frac{s^n Q(s)}{s^n Q(s) + P(s)}\frac{m!}{s^{m+1}} = m!\frac{s^{n-m-1}Q(s)}{s^n Q(s) + P(s)}.$$

As the system is stable, the roots of $s^n Q(s) + P(s)$ are all in the left half-plane. If $n > m$, then the poles of $E(s)$ are precisely the poles of that polynomial, and we know that $e(t)$ decays to zero—we know that the error tends to zero. That is, as $t \to \infty$, $e(t) \to 0$, $v_o(t) \to v_{in}(t)$, and we say that as t tends to infinity the output of the system tracks its input.

In particular, we have found that if $n = 1$, a stable system will track a unit-step—$v_{in}(t) = t^0 u(t)$. If $n = 2$, a stable system will track both $v_{in}(t) = tu(t)$, a ramp input, and $v_{in}(t) = u(t)$ (and, by linearity, any linear combination of such inputs). Additionally, if $n = 3$, a stable system will track $v_{in}(t) = t^2 u(t)$, a quadratic input, $v_{in}(t) = tu(t)$, and $v_{in}(t) = u(t)$ (and, by linearity, any linear combination of such inputs).

3.7 Internal Stability

It is important to determine whether on not a system is bounded-input bounded-output stable. When dealing with systems with feedback, it is important to know more. It is important to know that even when noise and other bounded disturbances are present both the *output* and *the system's internal signals* all remain bounded—it is important to know that the system is *internally stable.* (See, for example, [DFT92].)

Figure 3.8 shows a system with feedback with two disturbing signals, $d(t)$ and $n(t)$, and with all of its relevant internal signals, $x_1(t), x_2(t)$, and $x_3(t)$. Equations 3.1 – 3.3 give the dependence of the internal signals on the input and the two disturbing signals.

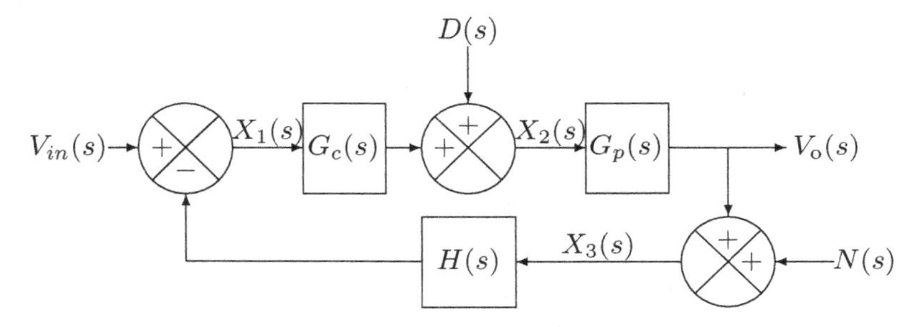

Fig. 3.8 The System with All Relevant Signals Shown

$$X_1(s) = V_{in}(s) - H(s)X_3(s). \tag{3.1}$$

$$X_2(s) = G_c(s)X_1(s) + D(s). \tag{3.2}$$

$$X_3(s) = G_p(s)X_2(s) + N(s). \tag{3.3}$$

These equations can be rewritten as:

$$\begin{bmatrix} 1 & 0 & H(s) \\ -G_c(s) & 1 & 0 \\ 0 & -G_p(s) & 1 \end{bmatrix} \begin{bmatrix} X_1(s) \\ X_2(s) \\ X_3(s) \end{bmatrix} = \begin{bmatrix} V_{in}(s) \\ D(s) \\ N(s) \end{bmatrix}.$$

Inverting the matrix is not difficult. We find that:

$$\begin{bmatrix} X_1(s) \\ X_2(s) \\ X_3(s) \end{bmatrix} = \frac{1}{\text{Den}(s)} \begin{bmatrix} 1 & -H(s)G_p(s) & -H(s) \\ G_c(s) & 1 & -H(s)G_c(s) \\ G_c(s)G_p(s) & G_p(s) & 1 \end{bmatrix} \begin{bmatrix} V_{in}(s) \\ D(s) \\ N(s) \end{bmatrix} \tag{3.4}$$

where $\text{Den}(s) = 1 + G_c(s)G_p(s)H(s)$.

After considering Figure 3.8 and (3.4), several things become clear. First of all, when $D(s) = N(s) = 0$, we find that $X_3(s) = V_o(s)$. Under this condition, we find that $X_3(s) = G_c(s)G_p(s)V_{in}(s)/\text{Den}(s)$—which leads to the transfer function of the system as it should.

In general, (3.4) relates the important internal signals to the external "inputs." If any of the functions that appear in the inverse matrix has poles in the right half-plane, then there exist bounded $v_{in}(t)$, $d(t)$, and $n(t)$ for which one or more of the internal signals is not bounded, and the system is not internally stable.

Considering the form of equations (3.4), we see that the zeros of $\text{Den}(s) = 1 + G_c(s)G_p(s)H(s)$ are among the poles of the relevant internal signals. However, it is conceivable that there could be additional poles. How could this happen? If an element of the matrix has right half-plane poles that are not identical to those in $\text{Den}(s)$, then for certain bounded inputs some internal signal will not be bounded.

Looking at (3.4) and at $\text{Den}(s)$, we find that this is a concern when a right half-plane pole of $G_c(s)$, $G_p(s)$, or $H(s)$ is canceled by a right half-plane zero of one of the two remaining transfer functions. In this case, this pole will be absent from $\text{Den}(s)$ but will be present in at least one of the other elements of the matrix. Thus, for certain bounded inputs at least one of the internal signals will be unbounded.

We find that in order for a system to be internally stable,

(1) $\text{Den}(s) = 1 + G_c(s)G_p(s)H(s)$ must not have any zeros in the right half-plane. This is the same condition we have already seen for BIBO stability.
(2) Additionally, there must not be any cancellation of right half-plane poles of $G_c(s)$, $G_p(s)$, or $H(s)$ by right half-plane zeros of the remaining two functions.

An Internally Stable System—An Example

Consider a unity-feedback system with $G_c(s) = K$ and $G_p(s) = 1/(s-1)$. The system's transfer function, $T(s)$ is:

$$T(s) = \frac{K}{s - 1 + K}.$$

For all $K > 1$, the system is BIBO stable. Additionally, as the system was stabilized without canceling right half-plane poles, the closed-loop system is internally stable.

A System that Is not Internally Stable—An Example

Now consider the system of Figure 3.9. Here a pole in the right half-plane is being canceled by a zero in the right half-plane. Calculating $V_o(s)$ as a function of $V_{in}(s)$ and $D(s)$, we find that:

$$V_o(s) = \frac{1}{s-1}\left(D(s) + \frac{s-1}{s+1}(V_{in}(s) - V_o(s))\right).$$

Solving for $V_o(s)$ we find that:

$$V_o(s) = \frac{s+1}{(s+2)(s-1)}D(s) + \frac{1}{s+1}V_{in(s)}.$$

Clearly almost any disturbance, $d(t)$, will cause $v_o(t)$ to grow without bound, and we find that though the system is BIBO stable it is *not* internally stable.

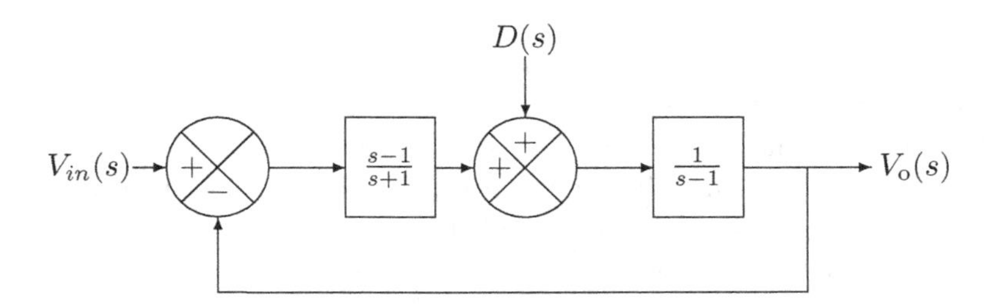

Fig. 3.9 A System that Is BIBO Stable but not Internally Stable

3.8 Disturbance Rejection

Suppose that there is reason to think that spurious signals—disturbances—are entering our system at the input to the system which we are controlling—at the input to the block containing $G_p(s)$. Then we must consider a model like that of Figure 3.10.

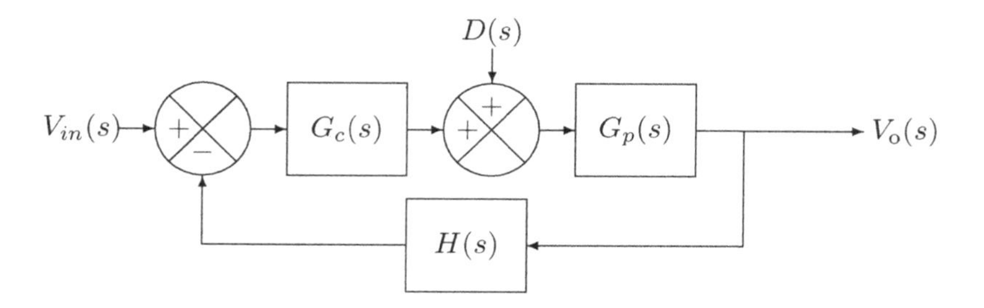

Fig. 3.10 A Feedback Circuit with a Disturbance

Making use of (3.4) of §3.7, we find that:

$$V_o(s) = \frac{G_p(s)}{1 + G_c(s)G_p(s)H(s)} D(s) + \frac{G_c(s)G_p(s)}{1 + G_c(s)G_p(s)H(s)} V_{in}(s).$$

If $G_c(s)$ is sufficiently large, then the coefficient of $D(s)$ is approximately $1/(G_c(s)H(s))$ while the coefficient of $V_{in}(s)$ is approximately $1/H(s)$. In the simple and important unity feedback case, we see that the coefficient of $D(s)$ is approximately equal to $1/G_c(s) \approx 0$ while the coefficient of $V_{in}(s)$ is approximately equal to one. That is the disturbance signal is attenuated by systems of this type. Of course, the disturbance that is attenuated is the signal that enters the system after the amplifier. Not all undesired signal are attenuated by such systems—there is no cure-all for noise problems.

3.9 Exercises

(1) Consider the simple buffer of Figure 3.11.

(a) Find the transfer function, $T(s)$, of this circuit where:

$$T(s) = \frac{V_o(s)}{V_{in}(s)}.$$

(b) Find the gain of the buffer at DC.

(c) Calculate the sensitivity of $T(s)$ to the DC gain of the amplifier, A.

(d) Evaluate the sensitivity at DC—when $s = 0$.

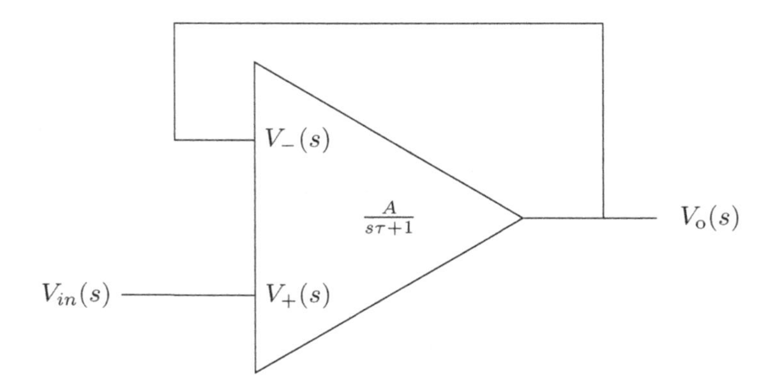

Fig. 3.11 A Simple Buffer

In this circuit we are once again trading the possibility of a large but unknown gain, A, for a lower but very well known gain, 1.

(2) Consider the system shown in Figure 3.12. Using the techniques of this chapter, determine the steady state output of the system to $u(t)$ when:

  (a) $n = 0$
  (b) $n = 1$
  (c) $n = 2$

(3) Consider the system shown in Figure 3.12.

  (a) Calculate the circuit's transfer function when:

    i. $n = 0$
    ii. $n = 1$
    iii. $n = 2$

  For which values of n is the system stable?

  (b) Considering that for some values of n the system is not stable, how do you explain the results of the previous question? (Hint: what do we know about the final value theorem?)

(4) In the system of Figure 3.12 let $n = 1$. What is the response of the system to a unit ramp, $V_{in}(s) = 1/s^2$.

(5) Consider the system of Figure 3.13. Let $G_p(s) = K$.

  (a) Find the output as a function of the input and the disturbance—

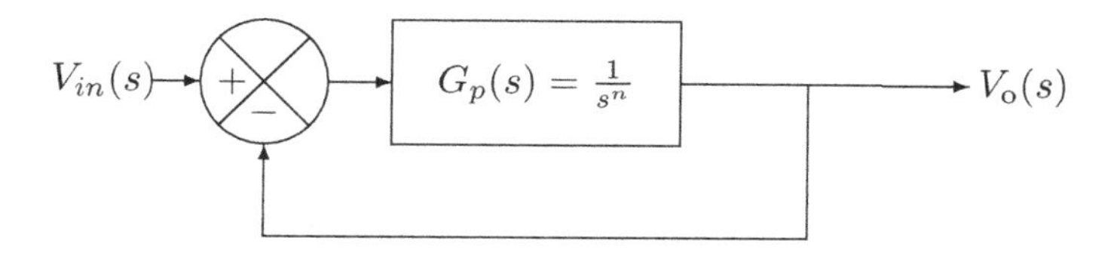

Fig. 3.12 A Simple Feedback Circuit

that is find:

$$V_o(s) = T_I(s)V_{in}(s) + T_D(s)D(s).$$

(b) If K is very large, how does the disturbance affect the output?
(c) If K is small how does the disturbance affect the output?

(6) Consider the system of Figure 3.13. Let $G_p(s) = 1/s$.

(a) Find the output as a function of the input and the disturbance—that is find:

$$V_o(s) = T_I(s)V_{in}(s) + T_D(s)D(s).$$

(b) If the disturbance is a constant value, $d(t) = cu(t)$, how does it affect the output in the steady-state?
(c) If the disturbance is very high-frequency ($\omega >> 1$), how does it affect the output in the steady-state?

(7) Please explain how Problem 5 can be used to help explain the results seen in Problem 6.

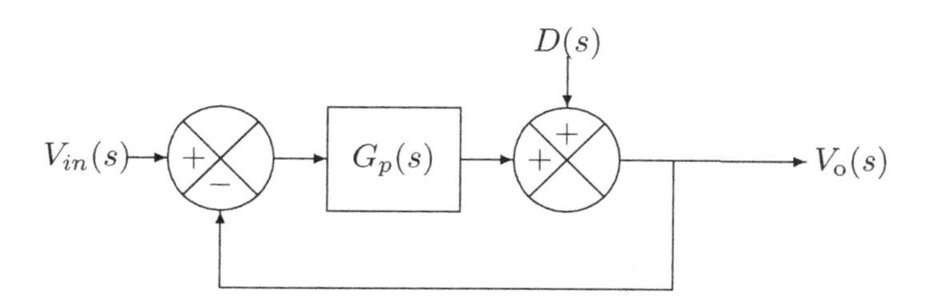

Fig. 3.13 A Simple Feedback Circuit with a Disturbance

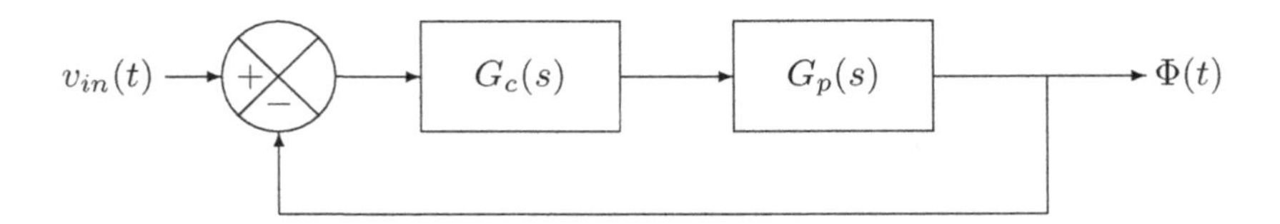

Fig. 3.14 A Simple Motor Controller

(8) Consider the system of Figure 3.14. Assume that the transfer function of the motor (from voltage to speed in radians per second) is:

$$G_p(s) = \frac{1}{s/2+1}.$$

(a) If $G_c(s) = K > 0$, what is the rise time of the system, and what is the system's response to a unit step input?

(b) If $G_c(s) = K/s, K > 1/2$, what is the system's rise time, and what is the system's response to a unit step input?

(c) Why might one prefer to use a system of the first type rather than a system of the second type?

The moral of this story is that simpler is sometimes better too.

(9) Consider a unity-feedback system for which $G_c(s) = 3/(s+2)$ and $G_p(s) = 1/(s-1)$.

(a) Is the system stable? Explain.

(b) Is the system internally stable? Explain. If the system is not internally stable, please show how a bounded input signal, either an "official" input or an undesired signal, can lead to the output of (at least) one of the blocks becoming unbounded.

(10) Consider a unity-feedback system for which $G_c(s) = s/(s+2)$ and $G_p(s) = 1/s$.

(a) Is the system stable? Explain.

(b) Is the system internally stable? Explain. If the system is not internally stable, please show how a bounded input signal, either an "official" input or an undesired signal, can lead to the output of (at least) one of the blocks becoming unbounded.

(11) In a unity-feedback system (a system for which $H(s) = 1$),

$$G_c(s) = C, \text{ and } G_p(s) = \frac{1}{s(s+a)}, \quad a > 0.$$

(a) What is the transfer function of the system, $T(s)$?
(b) What is the sensitivity of $T(s)$ to a, $\mathrm{S}_a^{T(s)}$?
(c) Calculate

$$\lim_{C \to \infty} \mathrm{S}_a^{T(s)}.$$

What does this tell you about the sensitivity of the system when the gain is large?
(d) What is $\mathrm{S}_a^{T(0)}$?
(e) Why is the answer to the previous section "intuitively obvious?" (Consider the value of $T(0)$.)

(12) Consider a unity-feedback system for which:

$$G_c(s) = \frac{1}{s^2+1}, \qquad G_p(s) = s + a.$$

(a) Show that for all $a > -1$ the system is closed-loop stable.
(b) Calculate $S_a^{T(s)}$.
(c) Calculate $S_a^{T(j)}$. Explain why (for $a > -1$) this value is precisely what one should have expected.

(13) Prove that:

$$\mathcal{L}(t^m u(t))(s) = \frac{m!}{s^{m+1}}.$$

Chapter 4

The Routh-Hurwitz Criterion

4.1 Roots of Polynomials—A Little History

As we have seen, many of the properties of a system can be determined by examining the roots of certain polynomials. If, for example,

$$G_p(s) = \frac{P(s)}{Q(s)}$$

is the transfer function of a system and $P(s)$ and $Q(s)$ are polynomials, then the system is stable if and only if the roots of $Q(s)$ are all located in the left half-plane. If the system is stable, $G_p(0) = 1$, and $P(s)$ has a positive zero, then the system suffers from undershoot.

It is well known that that the roots of a second-order polynomial of the form $as^2 + bs + c$ are:

$$s = \frac{-b \pm \sqrt{b^2 - 4ac}}{2a}.$$

Making use of this formula, it is easy to check whether or not a quadratic's roots are located in the left half-plane.

Let us briefly consider the history of methods for calculating and characterizing the roots of polynomials [Ste03b]. By about 1,600 B.C.E., the Babylonians knew how to calculate the roots of a quadratic polynomial. By the mid-1500s, some three thousand years later, formulas for finding the roots of cubic and quartic equations that only made use of sums, differences, products, quotients, and roots had been discovered, but no general formulas of this type for the roots of higher order polynomials had been found.

In 1770 – 71, J. L. Lagrange (1736 – 1813) unified the methods used to find the roots of polynomials of order less than or equal to four and showed that those methods could not be used to treat polynomials of order five or greater.

During the early part of the nineteenth century, P. Ruffini (1765 – 1822) came close to proving that no such general formula existed. In 1824, N. H. Abel (1802 – 1829) finished the work and established that no such general formula could exist.

One might think that this result implies that there is no simple analytical way for us to examine the stability of a system whose denominator is of degree greater than or equal to five. One would be mistaken. To see why, consider *precisely* what we need to know. We *need* to know whether or not a polynomial has any roots in the right half-plane. We *do not care* precisely where the roots are located. Despite the fact that there is not always an elementary way to calculate the roots of a polynomial, in the late 1800s it was shown that there is *always* an elementary way to determine whether or not some of the zeros of a polynomial are located in the right half-plane.

In the next section, we derive a necessary and sufficient condition for a polynomial to have all of its roots in the left half plane. See Exercise 11 for a simple necessary condition (that is not generally sufficient).

4.2 Theorem, Proof, and Applications

In this chapter we consider transfer functions that are quotients of polynomials in s—rational functions of s—with real coefficients. Suppose that one has a feedback system for which $G_p(s) = P_1(s)/Q_1(s)$ and $H(s) = P_2(s)/Q_2(s)$, with $P_1(s), P_2(s), Q_1(s)$ and $Q_2(s)$ polynomials in s whose coefficients are real. Then the transfer function of the system is:

$$T(s) = \frac{G_p(s)}{1 + G_p(s)H(s)} = \frac{P_1(s)Q_2(s)}{P_1(s)P_2(s) + Q_1(s)Q_2(s)}.$$

We need to know whether the roots of the polynomial are in the right half-plane. This problem was studied in the late 1800's, and a number of criteria were developed. These tests give us information about the location of the roots without telling us exactly where the roots are. One of the most helpful of the criteria is the Routh-Hurwitz criterion[1] Before discussing and proving the Routh-Hurwitz criterion, we state and prove a number of facts about polynomials.

[1]The Routh-Hurwitz criterion is named after the people who independently discovered it. The first was E. J. Routh (1831-1907). The second was A. Hurwitz (1859-1919) [OR].

(1) If $P(s) = s^n + a_1 s^{n-1} + \cdots a_n$, then $a_1 = -\sum_{i=1}^{n} s_i$ where s_i is the ith root of $P(s)$.

PROOF: From modern algebra it is known that $P(s)$ can be written as $(s - s_1)\cdots(s - s_n)$ where the s_i are the roots of $P(s)$. Multiplying out this expression and equating coefficients with our first expression for the polynomial we prove the proposition.

(2) If $P(s) = a_0 s^n + \cdots + a_n$, then:

$$\frac{a_1}{a_0} = -\sum_{i=1}^{n} s_i.$$

PROOF: We know that $P(s)/a_0$ has the same roots as $P(s)$, and we see that $P(s)/a_0$ satisfies the conditions of 1. Thus the coefficient of s^{n-1} in $P(s)/a_0$ is equal to minus the sum of the roots of the polynomial. As the coefficient of s^{n-1} is a_1/a_0, the result is proved.

(3) The roots of the polynomial $P(s) = a_0 s^n + a_1 s^{n-1} + \cdots a_n$, $a_i \in \mathcal{R}, i = 1, ...n$ are either real numbers or occur in complex conjugate pairs.

PROOF: In modern algebra it is proved that any polynomial with real coefficients can be factored into first and second order polynomials with real coefficients. A first order polynomial with real coefficients, $as + b$, clearly has one real zero. A second order polynomial with real coefficients, $as^2 + bs + c$ has two roots which are either both real or are complex conjugates of one another. As the roots of a polynomial are just the roots of its factors, we have proved the result.

(4) If $a_1/a_0 \leq 0$, then the polynomial has at least one root in the right half-plane.

PROOF: If $a_1/a_0 = 0$, then from 2 we know that the sum of the roots of the polynomial is zero. In particular, we know that the sum of the real parts of the roots is zero. This can only happen if either all of the real parts are zero—which leaves us with all of our roots in what we have defined as the right half plane, or it means that there are some roots with positive real parts and some with negative real parts—which also means that there is at least one root in the right half-plane. If $a_1/a_0 < 0$, then there is at least one root whose real part is positive.

(5) An even polynomial is a polynomial in which s appears only in even powers. An odd polynomial is a polynomial in which s appears only in odd powers. One can write any polynomial as $P(s) = P_{even}(s) + P_{odd}(s)$ where the first polynomial is even and the second is odd. Consider a polynomial with real coefficients, $P(s)$. Let $\omega \in \mathcal{R}$. If $P(j\omega) = 0$, then $P_{even}(j\omega) = P_{odd}(j\omega) = 0$. Moreover, if $j\omega$ is an nth order zero of the $P(s)$, then it is at least an nth order zero of both $P_{odd}(s)$ and of $P_{even}(s)$.

PROOF: For any even polynomial $P_{even}(j\omega) = \alpha \in \mathcal{R}$. For any odd polynomial, $P_{odd}(j\omega) = j\beta \in j\mathcal{R}$. Thus, $P(j\omega) = 0$ implies that both α and β must equal zero.
If $j\omega \neq 0$ is a zero of a polynomial with real coefficients, then from 3 we know that $-j\omega$ is also a zero. That is, $(s - j\omega)(s + j\omega) = s^2 + \omega^2$ is a factor of the polynomial. In particular, this implies that if $j\omega$ is a factor of $P(s)$, then $s^2 + \omega^2$ is a factor of $P(s), P_{odd}(s)$ and $P_{even}(s)$. Thus, one can write:

$$\frac{P(s)}{s^2 + \omega^2} = \frac{P_{even}(s)}{s^2 + \omega^2} + \frac{P_{odd}(s)}{s^2 + \omega^2}.$$

It is clear that each of these expressions is a polynomial and that the even and odd polynomials give rise to even and odd polynomials.
If $j\omega$ is a second order zero of $P(s)$, then we know that it is a zero of $P(s)/(s^2 + \omega^2)$. From our previous results, we see that it must be a zero of the even and the odd part of this polynomial. Thus, if $j\omega$ is a second order zero of $P(s)$ then:

$$\frac{P(s)}{(s^2 + \omega^2)^2} = \frac{P_{even}(s)}{(s^2 + \omega^2)^2} + \frac{P_{odd}(s)}{(s^2 + \omega^2)^2}$$

and each expression is a polynomial. This logic can be used for a zero of any order. Thus, we find that if $j\omega \neq 0$ is a zero of $P(s)$ of the nth order, then $s^2 + \omega^2$ is an nth order factor of $P(s), P_{even}(s)$, and $P_{odd}(s)$. This is equivalent to saying that $j\omega$ and $-j\omega$ are n^{th} order zeros of all three polynomials.
Consider now zeros that occur at $s = 0$. If the zero is of mth order, that $P(s)$ is of the form $a_0 s^n + \cdots a_{n-m} s^m$. Clearly

both the odd and the even parts of this polynomial have zeros of order at least m at 0.

We are now able to state and prove the Routh-Hurwitz criterion. We use an expanded version of Meinsma's proof [Mei95].

Theorem 12 *A non-constant n^{th} order polynomial $P(s) = a_0 s^n + a_1 s^{n-1} + \cdots + a_n$ has no zeros in the right half-plane if and only if a_1 is nonzero, a_0 and a_1 have the same sign, and the polynomial:*

$$Q(s) = P(s) - \frac{a_0}{a_1}\left(a_1 s^n + a_3 s^{n-2} + a_5 s^{n-4} + \cdots\right)$$

is of degree $n-1$ and has no zeros in the right half-plane.

Furthermore, the number of zeros of $Q(s)$ in the right half-plane is equal to the number of zeros of $P(s)$ in the right half-plane if a_0 and a_1 are of the same sign, and is one less if a_0 and a_1 are of different signs.

PROOF: As long as $a_1 \neq 0$, $Q(s)$ will be of degree $n-1$. If $a_1 = 0$, then from 4 we know that $P(s)$ has at least one zero in the right half-plane. We have shown that if $Q(s)$ is not of degree $n-1$, then $P(s)$ has at least one zero in the right half-plane. In the rest of the proof, we assume that $a_1 \neq 0$ and we show that the conclusions of our theorem are correct for this case too.
If the degree of $P(s)$—n—is even, then $Q(s) = P(s) - (a_0/a_1)sP_{odd}(s)$, while if n is odd, then $Q(s) = P(s) - (a_0/a_1)sP_{even}(s)$. Define:

$$Q_\eta(s) = P(s) - \eta\left(a_1 s^n + a_3 s^{n-2} + \cdots\right).$$

For $\eta = 0$, we have $Q_0(s) = P(s)$. For $\eta = a_0/a_1$, we find that $Q_{a_0/a_1}(s) = Q(s)$. Suppose that n is even. Then, $Q_\eta(s) = (P_{even}(s) - \eta s P_{odd}(s)) + P_{odd}(s)$. Clearly the term in parentheses is even and $P_{odd}(s)$ is odd. Also, any joint zeros of $P_{even}(s)$ and $P_{odd}(s)$ are zeros of $Q_\eta(s)$. From property 5, we know that any m^{th} order imaginary zero of Q_η must be at least an mth order zero of $P_{odd}(s)$ and an mth order zero of the term in parentheses. But one of the terms in parentheses is $sP_{odd}(s)$—and we know that our zero is at least an mth order zero of this term. Hence, it must also be at least an mth order zero of $P_{even}(s)$. We have shown that the imaginary zeros of $Q_\eta(s)$ are precisely the shared imaginary zeros of $P_{even}(s)$ and $P_{odd}(s)$. We see that $P(s)$ and

$Q_\eta(s)$ have the same imaginary zeros. We find that $Q_\eta(s)$ can be written:

$$Q_\eta(s) = R(s)T_\eta(s)$$

where the zeros of $R(s)$ are the imaginary zeros of $Q_\eta(s)$—which are independent of η. Clearly, T_η cannot have any imaginary zeros. From this it follows that as one goes from $P(s)$ to $Q(s)$, no zeros of $Q_\eta(s)$ cross the imaginary axis.
If n is odd, the same result holds. The proof is basically the same, but we start from the identity $Q_\eta = (P_{odd}(s) - \eta s P_{even}(s)) + P_{even}(s)$.
Because the degree of $Q(s)$ is one less than the degree of $P(s)$, we know that in going from $P(s)$ to $Q(s)$ we "lose" one zero. We must find out where that zero "goes." Consider $Q_\eta(s)$. We find that:

$$Q_\eta(s) = (a_0 - \eta a_1)s^n + a_1 s^{n-1} + \cdots .$$

As $\eta \to a_0/a_1$, it is clear that $n-1$ of the zeros of this polynomial tend to the $n-1$ zeros of $Q(s)$. We must determine what happens to the nth zero. As $\eta \to a_0/a_1$, the coefficient of s^n tends to zero. In order for this term to have any effect, the magnitude of s must be large. For large s, only the highest order terms in the polynomial make an effective contribution—the term $(a_0 - \eta a_1)s^n + a_1 s^{n-1}$ is dominant for large values of s. The equation $(a_0 - \eta a_1)s^n + a_1 s^{n-1} = 0$ has n solutions. One solution is $-a_1/(a_0 - \eta a_1)$, and the other $n-1$ solutions are zero. The zero solutions correspond to the (relatively small) solutions of the equation $Q(s) = 0$. The other solution is the one we need to study now. As $\eta \to a_0/a_1$ from zero, this solution goes off to infinity. If a_0 and a_1 have the same sign, then the solution goes off towards $-\infty$—thus the zero that was lost must have always been in the left half-plane (since no zeros can cross the imaginary axis). If the signs are different, then $-a_1/(a_0 - \eta a_1)$ tends towards $+\infty$ and the zero must always have been in the right half-plane.

Repeated use of this theorem allows us to tell how many zeros of a given polynomial are in the right half-plane and how many are in the left half-plane.

The Use of the Routh-Hurwitz Criterion—An Example

(1) Let us start with $P(s) = s^2 - 5s + 6$. We see that $Q(s) = P(s) - (1/-5)s(-5s) = -5s + 6$. From the fact that a_0 is positive and a_1 is negative, we know that the number of zeros of $P(s)$ in the right half-plane is one more than the number of zeros of $-5s + 6$ in the right half-plane. As this has one zero in the right half-plane, we find that $P(s)$ has two zeros in the right half-plane. This is correct as $P(s) = (s-2)(s-3)$.

(2) Now consider $P(s) = s^4 + 3s^3 - 3s^2 + 3s + 1$. We find that $Q_1(s) = P(s) - (1/3)s(3s^3 + 3s) = 3s^3 - 4s^2 + 3s + 1$. From the fact that a_0 and a_1 have the same sign, we know that $P(s)$ and $Q_1(s)$ have the same number of zeros in the right half-plane. Now let us consider $Q_1(s)$ as our new "$P(s)$" and continue using the theorem.
We see that our new a_0 and a_1 have opposite signs. Also our new $Q(s)$—which we call $Q_2(s)$ satisfies:

$$Q_2(s) = Q_1(s) - (3/-4)s(-4s^2 + 1) = -4s^2 + (15/4)s + 1.$$

Thus, we know that $Q_1(s)$ has one more zero in the right half-plane than $Q_2(s)$ has.
Continuing, we find that $Q_3(s) = Q_2(s) - (4/15)s(15/4s) = 15/4s + 1$. As the a_0 and a_1 of $Q_2(s)$ are of the opposite signs, we find that $Q_2(s)$ has one more zero in the right half-plane than does $Q_3(s)$. Finally it is clear that $Q_3(s)$ has no zeros in the right half-plane.
Combining all of this information, we find that $P(s)$ has two zeros in the right half-plane. We can use MATLAB to verify that this is so. We use the `roots` command. We find that:

```
roots([1 3 -3 2 1])
ans =
     -3.9387
      0.5963 + 0.8028 i
      0.5963 - 0.8028 i
     -2.539
```

There are indeed two roots in the right half-plane (and they are complex conjugates of one another).

(3) Finally, consider $P(s) = s^3 + s^2 + 1$. We find that $P_{even} = s^2 + 1$. Thus, $Q_1(s) = P(s) - 1 \cdot sP_{even}(s) = s^2 - s + 1$. We see

that $P(s)$ has the same number of zeros in the right half-plane that $Q_1(s)$ has. It is clear that $Q_2(s) = Q_1(s) - s \cdot s = -s + 1$. We see that $Q_1(s)$ has one more right half-plane zero than $Q_2(s)$. As $Q_2(s)$ has one positive real zero, we see that $P(s)$ has two zeros in the right half-plane. Using MATLAB it is simple to verify this result. We find that:

```
roots([1 1 0 1])
ans =
     -1.4656
      0.2328 + 0.7926 i
      0.2328 - 0.7926 i
```

Once again, we see that the answer checks.

When using the Routh-Hurwitz criterion to determine the number of zeros a polynomial has in the right half-plane, we find that at each step we calculate "$Q(s)$" according to the rule:

$$\begin{aligned}
Q(s) &= P(s) - (a_0/a_1)s(a_1 s^{n-1} + a_3 s^{n-3} + \cdots) \\
&= (a_0 s^n + a^2 s^{n-2} + \cdots) + (a_1 s^{n-1} + a_3 s^{n-3} + \cdots) \\
&\qquad -(a_0/a_1)s(a_1 s^{n-1} + a_3 s^{n-3} + \cdots) \\
&= (a_1 s^{n-1} + a_3 s^{n-3} + \cdots) \\
&\qquad +((a_2 - (a_0/a_1)a_3)s^{n-2} + (a_4 - (a_0/a_1)a_5)s^{n-2} + \cdots) \quad (4.1)
\end{aligned}$$

Also, we know that the next "$P(s)$" is just the current $Q(s)$. This allows us to calculate the coefficients in a recursive fashion and to list all of the coefficients in a neat tabular form known as the Routh array as follows.

The Routh Array—An Example

Let us consider $P(s) = s^4 + 3s^3 - 3s^2 + 3s + 1$ again. This time we write down the odd and even parts separately in an array as:

$$\begin{matrix} a_0 & a_2 & a_4 \\ a_1 & a_3 & 0 \end{matrix} = \begin{matrix} 1 & -3 & 1 \\ 3 & 3 & 0 \end{matrix}$$

The coefficient of the highest power of s always goes in the leftmost column of the first row. The next row starts with the next lower power of s. By looking at these two rows, one can reconstruct the polynomial. It is clear from (4.1) that the matrix for $Q_1(s)$ starts with the last row of the matrix for $P(s)$ and has as its second row:

$$a_2 - (a_0/a_1)a_3 \; a_4 - (a_0/a_1)a_5 \; 0 \; = \; -3 - (1/3)3 = -4 \; 1 \; 0 \, .$$

Of course, the same procedures work to write down each of the $Q_i(s)$ where one just uses the values that correspond to the previous $Q_i(s)$.

As the last row of $Q_i(s)$ is the first row of $Q_{i+1}(s)$, there is no need to write down separate matrices. Rather one writes down one large matrix. Proceeding this way, one finds that the matrix that corresponds to our problem is:

$$\begin{array}{rcc} 1 & -3 & 1 \\ 3 & 3 & 0 \\ -4 & 1 & 0 \\ 15/4 & 0 & 0 \\ 1 & 0 & 0 \end{array}$$

This matrix is called the *Routh array*. Each adjacent pair of rows corresponds to a polynomial. The first two rows correspond to $P(s)$. The second and third rows correspond to $Q_1(s)$, and so forth. Additionally the last row represents the last polynomial—which is just a constant. We know that the number of right half-plane zeros of $P(s)$ is the same as the number of right half plane zeros of $Q_1(s)$ because the signs of a_0 and a_1—the first elements of the first two rows—are the same. We note that our a_0 and a_1 will always be the first two elements in the two rows that correspond to our current polynomial. Thus, we find that $Q_1(s)$ has one more right half-plane zero than $Q_2(s)$ because 3 and -4 have different signs. Also, we know that $Q_2(s)$ has one more zero in the right half-plane than $Q_3(s)$ because -4 and $15/4$ have different signs. Finally, we find that $Q_3(s)$ has the same number of right half-plane zeros as $Q_4(s)$. But $Q_4(s)$ is a constant and cannot have zeros in the right half-plane. Thus, we find that $Q_3(s)$ has no right half-plane zeros, $Q_2(s)$ has one zero in the right half-plane, $Q_1(s)$ has two zeros in the right half-plane, and, finally, $P(s)$ has two zeros in the right half-plane. In fact, we see that all that one needs to do to see how many right half-plane zeros $P(s)$ has is *to count the number of sign changes in the first column of the Routh array.*

If at any stage one finds that "$Q(s)$" is not of degree one less than "$P(s)$," then one cannot proceed to the next "$Q(s)$." In terms of the Routh array, one finds that it terminates prematurely. In this case, we cannot use this technique to determine the number of right half-plane zeros. If, however, the degree of "$Q(s)$" is more than one lower than the degree of "$P(s)$," then our current $a_1 = 0$. If this is so, then we know that $P(s)$ has at least one zero in the right half-plane. We see that if one cannot build the complete Routh array, then the original polynomial had at least one zero in the right half-plane.

Premature Termination of the Routh Array—An Example

In general, the Routh array has one more row than the order of the polynomial with which it is associated. Consider $P(s) = s^5 + s^4 + 2s^3 + s^2 - 3s - 2$. As this is a fifth order polynomial, we need to find a six row Routh array. Calculating the array, we find:

$$\begin{array}{ccc} 1 & 2 & -3 \\ 1 & 1 & -2 \\ 1 & -1 & 0 \\ 2 & -2 & 0 \\ 0 & 0 & 0 \end{array}$$

We see that the array terminates early. What can we learn from the Routh array in this case? First, from our theorem we know that the polynomial must have at least one zero in the right half-plane. But we can actually say more. By making use of the array and our theorem, we find that $P(s)$ has the same number of zeros in the right half plane as does the polynomial $Q_1(s) = s^4 + s^3 + s^2 - 1s - 2$. Furthermore, we know that $Q_1(s)$ has the same number of zeros in the right half plane as has $Q_2(s) = s^3 + 2s^2 - s - 2$. Finally this has the same number of zero in the right half-plane as $Q_3(s) = 2s^2 - 2$. As we know that the zeros of $Q_3(s)$ are ± 1, we see that $P(s)$ must have one zero in the right half-plane. Using MATLAB, we find that:

```
roots([1 1 2 1 -3 -2])
ans =
    -0.1424 - 1.6662 i
    -0.1424 + 1.6662 i
    -1.0000
```

```
-0.7152
 1.0000
```

This shows us precisely one zero in the right half-plane—as it should.

As we have already explained, if the Routh array terminates prematurely, we know that there is at least one zero of $P(s)$ in the right half-plane. Sometimes even if the Routh array terminates prematurely, we are able to get all of the information that we want from it. We just have to work a little harder for the information.

4.3 A Design Example

Suppose that we would like to control the shaft angle of a DC motor. We have seen (p. 51) that the transfer function from the input voltage of a DC motor to its output shaft angle is of the form:

$$G_p(s) = \frac{K}{s(s\tau + 1)}.$$

Suppose that we have a motor whose parameters are $K = 10\frac{\text{deg}}{\text{Volt}\cdot\text{sec}}$, and $\tau = 1\,\text{sec}$. Let $v_{in}(t)$ be measured in volts and let the shaft angle, $\Phi(t)$, be measured in degrees.

Let us assume that we are using this motor to position a pen and that we are not planning to use this motor to do anything very quickly. Moreover, assume that we *do not want* the motor system to respond to fast signals—we want the motor system to treat such signals as unwanted noise.

We can express the constraints on our system by the following three requirements:

- Every five millivolts of input voltage should correspond to 1° of arc. (This condition is necessary to pin down the relationship between the input in Volts and the desired output in degrees.)
- The steady-state response to a ramp input ($v_{in} = tu(t)$) must be correct to within 10°.
- The output to a one volt input at 20 rad/s (i.e. $v_{in}(t) = \sin(20t)u(t)$) must remain between $\pm 7.5^\circ$ of arc.

We could try to construct the system using the system of Figure 4.1. The $1/200$ attenuator is included so that in the comparison between the input which is in Volts and the output which is in degrees, if the output is the desired one then the output of the differencing block will be zero. Let

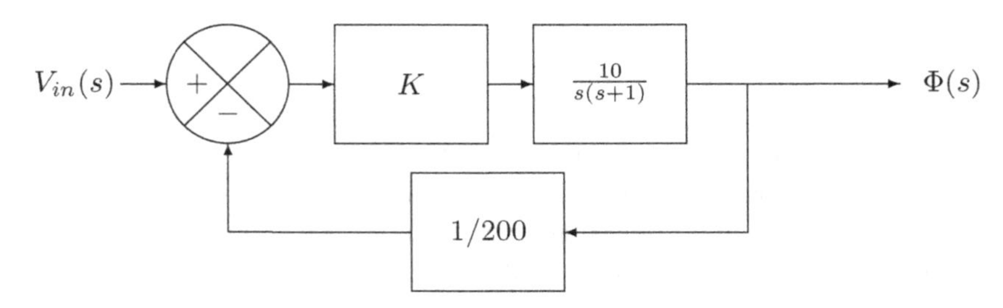

Fig. 4.1 A Simple Motor Controller

us see if it is possible for this system to meet all of the requirements. The transfer function of this system is:

$$T(s) = \frac{\frac{10K}{s(s+1)}}{1 + \frac{1}{200}\frac{10K}{s(s+1)}} = \frac{10K}{s(s+1) + \frac{K}{20}}.$$

We find that the DC gain of the system is $T(0) = 200$—as it should be to meet the first requirement. Next, we consider the second requirement. A ramp input leads to the output:

$$V_o(s) = \frac{10K}{s(s+1) + \frac{K}{20}} \frac{1}{s^2} \overset{\text{partial fraction expansion}}{=} \frac{A}{s^2} + \frac{B}{s} + \frac{Cs + D}{s(s+1) + \frac{K}{20}}.$$

It is easy enough to calculate that $A = 200$ and $B = \frac{-4,000}{K}$. Thus:

$$v_o(t) = \left(200t - \frac{4000}{K} + \text{exponentially decaying terms}\right) u(t).$$

We see that $-4000/K$ is the steady state error when the input is a ramp. Thus, we find that the response to a ramp is better the larger K is. To keep the error within ten degrees requires that K be at least 400. Because we find that as K increases the transfer function, $T(s)$, tends to one in ever larger regions, we take K to be as small as we can; we take $K = 400$. Then the transfer function of the system is:

$$T(s) = \frac{4000}{s^2 + s + 20}.$$

We find that $T(20j) = 4000/(-380 + 20j)$. Thus the error due to a 20 rad/s input of one volt will be on the order of 10°. This is almost good enough. We must find a slightly better solution.

One way to force the response of the system to relatively high frequency signals to decrease is to add an integrator to the system as we have done in Figure 4.2 rather than a simple amplifier as we have done in Figure 4.1. The transfer function of the new system is:

$$T(s) = \frac{2000K}{200s^2(s+1) + 10K}.$$

Let us use the Routh-Hurwitz test to see if this system is—or can be—stable. The polynomial in the denominator is $200s^3 + 200s^2 + 10K$. The Routh array that corresponds to this polynomial is:

$$\begin{array}{cc} 200 & 0 \\ 200 & 10K \\ -10K & 0 \\ 10K & 0 \end{array}$$

We see that there are always two roots in the right half-plane—no matter how big or small $K > 0$ is. Thus, we cannot use this scheme to deal with the problem.

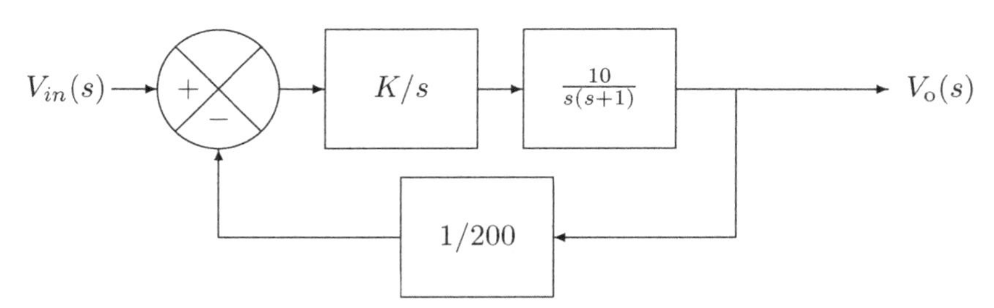

Fig. 4.2 A Simple Motor Controller with Integral Compensation

We see that using a simple amplifier is not enough to give us a system with the performance that we need. Also, we see that adding an integrator to the system—though it should help the system to respond less to "high frequency" signals—makes the system unstable. We now try to add a part that is a compromise between the two techniques—we add a low-pass filter. We replace the block whose transfer function is K/s by a block[2] whose

[2]To see that a block whose transfer function is:

$$G_p(s) = \frac{K}{\tau s + 1}$$

is a low-pass filter consider the magnitude of the frequency response of the block. We

transfer function is

$$K/(\tau s + 1)$$

(where $\tau > 0$ is a new parameter). The transfer function of the system with this block added is:

$$T(s) = \frac{2000K}{200(\tau s + 1)s(s + 1) + 10K} = \frac{2000K}{200\tau s^3 + 200(\tau + 1)s^2 + 200s + 10K}.$$

Let us examine the stability of this system. The Routh array that corresponds to the denominator is:

$$\begin{array}{ll} 200\tau & 200 \\ 200(\tau + 1) & 10K \\ 200 - \frac{10K\tau}{\tau+1} & 0 \\ 10K & 0 \end{array}$$

As long as:

$$200 - \frac{10K\tau}{\tau + 1} > 0 \Leftrightarrow 200 > (10K - 200)\tau$$

we have a stable system. This condition requires that for stability we must have:

$$\tau < \frac{200}{10K - 200}. \tag{4.2}$$

This condition says that the value of τ cannot be too large, and this is reasonable. We have already seen that the block K/s leads to a system that is unstable for any value of K. If τ is very large then $K/(\tau s + 1) \approx (K/\tau)/s$, and this is known to lead to an unstable system.

Let us see if our system meets the requirements now. If the system input is a ramp then the output is:

$$V_o(s) = \frac{2000K}{200(\tau s+1)s(s+1)+10K}\frac{1}{s^2} = \frac{A}{s^2} + \frac{B}{s} + \frac{Cs^2 + Ds + E}{200(\tau s+1)s(s+1)+10K}.$$

A simple calculation shows that $A = 200$ and $B = -4000/K$. To achieve the steady-state accuracy that we require, $K \geq 400$. Let $K = 400$. Pick τ

find that:

$$|G_p(j\omega)| = \frac{|K|}{\sqrt{\tau^2\omega^2 + 1}}.$$

Clearly this function has a maximum at $\omega = 0$ and decreases monotonically for $\omega > 0$.

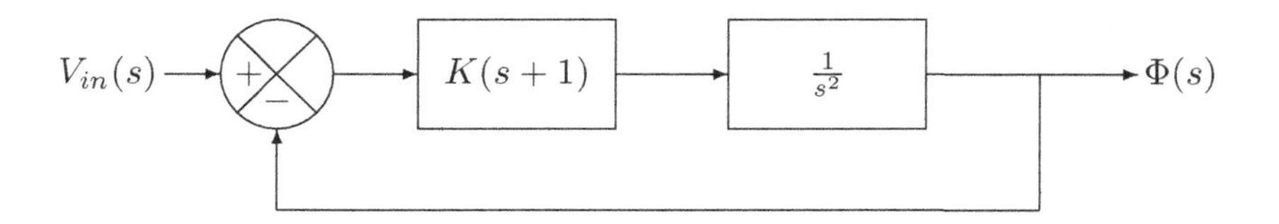

Fig. 4.3 A Controller for a Simple Satellite

according to (4.2). We find that $\tau < 1/19$. Let $\tau = 1/20$. We must now check the response to a 20 rad/s signal is not too large. We find that:

$$T(20j) = \frac{4000}{(j+1)(20j+1)20j+20} = \frac{4000}{-400-380j}.$$

We find that $|T(20j)| = 7.25$. We are just able to meet our requirements using this system.

4.4 Exercises

(1) Using the Routh array, determine which of the following polynomials have zeros in the right half-plane. Where possible, determine the number of zeros in the right half-plane as well.

  (a) $s^2 + 3s + 1$
  (b) $s^3 + 1$
  (c) $s^4 + 3s^2 + 1$
  (d) $s^4 + s^3 + s^2 + s + 1$
  (e) $s^4 + s^3 + 1$

(2) Explain Problem 1a without recourse to the Routh-Hurwitz theorem.

(3) Explain Problem 1c without recourse to the Routh-Hurwitz theorem. (Hint: how is Problem 1a related to Problem 1c?)

(4) Recall that the transfer function of a simple satellite is $G_p(s) = 1/s^2$. Use the Routh array to show that the system of Figure 4.3, a system for which the controller has the transfer function $G_c(s) = K(s+1)$, is stable for all $K > 0$.

(5) Prove that the system of Figure 4.4 is unstable for all values $a, b > 0$.

Also, prove that for all such values of a and b the system has two poles in the right half-plane.

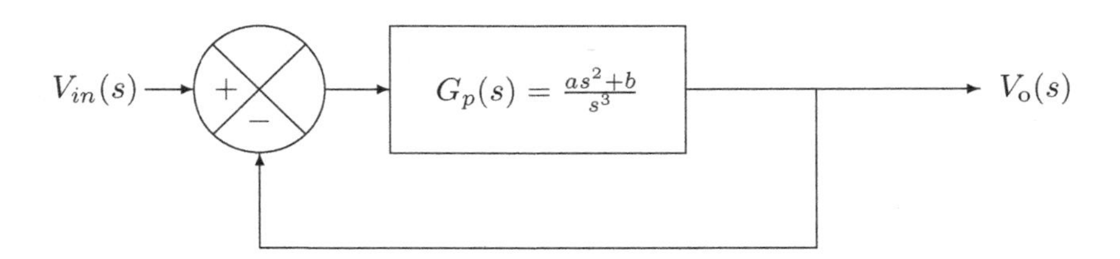

Fig. 4.4 A Simple Feedback Circuit

(6) (a) Use the Routh array to find the values of $K > 0$ for which the system of Figure 4.5 is stable when $a > 0$.
(b) Use the Routh array to show that if $a = 0$, then the system of Figure 4.5 is stable for all $K > 0$.
(c) What does this say about the effect of zeros in the right half-plane? (We will see more about this effect in 6.3.1.)

(7) Consider the system of Figure 4.5 but replace the transfer function $\frac{(s-a)}{s^2+s+1}$ by the transfer function:

$$G_p(s) = \frac{(s-a)(s+3)}{s^2+3s+2}.$$

(a) Use the Routh array to find the values of $K > 0$ for which the system of Figure 4.5 is stable when $a > 0$.
(b) Use the Routh array to show that if $a = 0$, then the system of Figure 4.5 is stable for all $K > 0$.
(c) What does this say about the effect of zeros in the right half-plane? (We will see more about this effect in 6.3.1.)

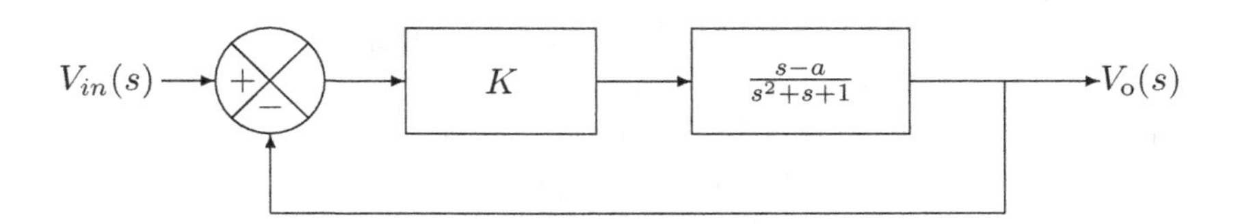

Fig. 4.5 A Controller for a System with a Zero in the RHP

(8) (a) Use the Routh array to show that the roots of $s^2 + as + b$ are all in the left half-plane if and only if a and b are positive.
(b) Prove this result without recourse to the Routh-Hurwitz theorem.

(9) Use the Routh array to find the range of gains, K, for which the system of Figure 4.5 is stable when $G_p(s)$ is replaced by:

$$G_p(s) = \frac{s^2 + 201s + 10100}{s(s^2 + s + 1)}.$$

(10) Consider the system of Figure 4.5, but let:

$$G_p(s) = \frac{s + 1}{(s + 2)(s - 1)}.$$

Use the Routh array to determine the range of K for which the system is stable.

(11) Prove that if a polynomial of the form $P(s) = a_0 s^n + a_1 s^{n-1} + \cdots a_n$ with real coefficients has no zeros in the right half-plane (including the imaginary axis), then all the coefficients, $a_k, k = 0, \ldots, n$ have the same sign and no coefficient is zero. You may wish to proceed as follows.

(a) First consider polynomials with real coefficients of the form $Q(s) = s^n + a_1 s^{n-1} + \cdots + a_n$, and note that if all the zeros are real and in the left half-plane, then $Q(s) = (s - r_1) \cdots (s - r_n)$ and all the r_k are negative. Multiply out the factors. What can be said about the coefficients of the polynomial?

(b) Next, note that if the roots of a polynomial with real coefficients are complex, then the roots occur in complex conjugate pairs. If a root is in the left half-plane, then it is of the form $r = \alpha + j\beta, \quad \alpha < 0, \beta \neq 0$. Thus its complex conjugate is of the form $\overline{r} = \alpha - j\beta$ and is also in the left half-plane. Consider $(s - r)(s - \overline{r})$. Show that all of this quadratic's coefficients are positive.

(c) Use the results of the preceding sections to show that if all of the zeros of $P(s)$ are in the left half-plan, then all of its coefficients are positive.

(d) Now consider a polynomial of the form $P(s) = a_0 s^n + a_1 s^{n-1} + \cdots + a_n$. By dividing the polynomial by a_0, show that if all of the roots of such a polynomial are in the left half-plane, then all of the coefficients have the same sign and none are zero.

(12) Use the results of exercise 11 to show that none of the following polynomials has all of its zeros in the left half-plane.

(a) $P(s) = s^2 - s + 1$.
(b) $P(s) = s^3 + 3s^2 + 1$.
(c) $P(s) = s^4 + 6s^3 + 2s + 4$.

(13) Use the results of exercise 11 to show that a unity feedback system for which:

$$G_c(s) = K, \qquad G_p(s) = \frac{1}{s^2(s+5)}$$

is not stable.

(14) It is well known that the Maclaurin series for e^s is:

$$e^s = 1 + s + s^2/2! + \cdots + s^n/n! + \cdots.$$

It is also well known that e^s has no zeros. Use the Routh array to determine which of the following partial sums have zeros in the right half-plane.

(a) $p_2(s) = 1 + s + s^2/2$.
(b) $p_3(s) = 1 + s + s^2/2 + s^3/6$.
(c) $p_4(s) = 1 + s + s^2/2 + s^3/6 + s^4/24$.
(d) $p_5(s) = 1 + s + s^2/2 + s^3/6 + s^4/24 + s^5/120$.

Chapter 5

The Principle of the Argument and Its Consequences

5.1 More about Poles in the Right Half Plane

If a transfer function that is a rational function of s has any poles in the right half plane, the transfer function represents an unstable system. We state that this is more generally true—a pole or poles in the right half plane is the sign of an unstable system. No poles in the right half plane is the sign of a stable system.

Much of control theory is devoted to answering the questions "*does the transfer function of the system of interest have any poles in the right half plane? Might small changes to the system cause poles to migrate to the right half-plane?*" Many of the techniques used to determine the stability of a system assume that the system's transfer function is a rational function of s and that the polynomials in the numerator and the denominator have real coefficients. Any system whose component parts are real gains, integrators and differentiator is of this type. The examples of the previous chapters make it clear that many systems are (at least approximately) of this type.

Some techniques allow other classes of analytic functions as well. *All the techniques we consider assume that the transfer function is analytic except for isolated poles.* Though this is a strong restriction on the type of system one may use, many practical systems can be described by transfer functions of this type. The most commonly used block whose transfer function is not a rational function of s is a delay block. As we saw in the first chapter (Theorem 7, p. 8) the transfer function that corresponds to a delay of T seconds is e^{-Ts}.

In the preceding chapter, we examined the Routh-Hurwitz criterion. We saw how one can determine the stability of a system whose transfer function is a rational function of s. In this chapter, we consider a technique that works as long as the transfer function is analytic in s except for isolated poles—the Nyquist plot[1]. We will see that this new techniques leads us to Bode plots[2], and we will see how easy-to-use Bode plots are.

5.2 The Principle of the Argument

The major theoretical tool used in this chapter is the principle of the argument. Let N be the number of zeros in a (reasonable) region Ω and let P be the number of poles in Ω. We say that a curve encircles the origin once if the curve goes around the origin once in the counterclockwise direction. If it goes around the origin in the clockwise direction then we say that it encircles the origin -1 times. The principal of the argument states:

Theorem 13 *Let $f(s)$ be a function that is analytic (except for a finite number of poles) in a neighborhood Ω and that has neither poles nor zeros on the boundary of Ω, $\partial\Omega$. Then the number of times that the function $f(s)$ encircles the origin, E, as s traverses $\partial\Omega$ in the counterclockwise direction is equal to the number of zeros in Ω less the number of poles in Ω. That is:*

$$N - P = E.$$

We defer the proof of this theorem to the next section. We consider a simple example to help fix the ideas presented in the theorem.

The Function $1/s$—An Example

We apply our theorem to the function $f(s) = 1/s$ and the region $\Omega = \left\{ s \middle| |s| \leq 1 \right\}$. The question we consider—as it will always be in this chapter—is how many poles and zeros are located *inside* the region Ω. To answer this question we use the principle of the argument. We map the boundary of the region of interest to us, Ω, using the function about which we want information, $f(s)$.

We want to map the boundary of the unit disk—the unit circle—using the function $1/s$. Let us describe the unit circle as the points $s = e^{j\theta}$ as θ goes from 0 to 2π. This causes us to traverse the unit circle counterclockwise (as seen in the left half of Figure 5.1). Then $1/s = e^{j(-\theta)}$. Thus the second circle

[1]Named after its inventor, Harry Nyquist (1889-1976). Nyquist invented the plot in 1932 while employed by the Bell Telephone Laboratories [Lew92].

[2]Named after their inventor, Hendrik Bode. Bode invented the Bode plots in 1938 while employed by the Bell Telephone Laboratories [Lew92].

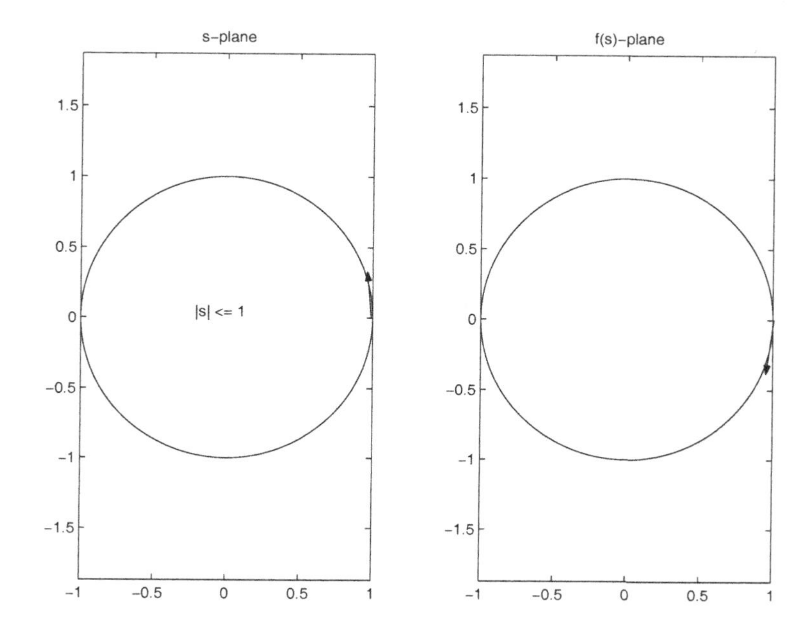

Fig. 5.1 The Domain, Ω, Its Boundary, $\partial\Omega$, and the Image of Its Boundary, $f(\partial\Omega)$

is traversed in the clockwise direction (as seen in the right half of Figure 5.1). We see that in this example $E = -1$—there is minus one encirclement. This means that the number of zeros of $1/s$ inside Ω less the number of poles of $1/s$ inside Ω is equal to minus 1. This is true; $1/s$ has no zeros and one pole in Ω.

5.3 The Proof of the Principle of the Argument

The proof of the principle of the argument relies on equalities A and B:

$$E \stackrel{A}{=} \frac{1}{2\pi j} \oint_{\partial\Omega} \frac{f'(s)}{f(s)} ds \stackrel{B}{=} N - P.$$

We prove each of the equalities separately.

Let us start with equality A. We must show that the number of encirclements of the origin is equal to the given integral. We know that $(f'(s)/f(s))\, ds = d(\ln(f(s)))$. Also, $f(s) = |f(s)|e^{j\Phi(f(s))}$ where $\Phi(f(s))$ is $f(s)$'s phase given in radians. We see that $\ln(f(s)) = \ln(|f(s)|) +$

$\ln(e^{j\Phi(f(s))})$ and this in turn is equal to $\ln(|f(s)|) + j\Phi(f(s))$. The integral of $d(\ln(f(s)))$ around a closed curve should be the change in $\ln(f(s)) = \ln(|f(s)|) + j\Phi(f(s))$ over that curve. The change in the logarithm of the magnitude is zero—the magnitude of the function will be the same at both the "beginning" and the "end" of the *closed* curve. The only change will be in j times the phase of the function—which is $2\pi j$ times the number of encirclements of zero.

Let us formalize this proof. Let $s(t)$ be a function from the interval $[a, b]$ to $\partial\Omega$ which parameterizes the curve $\partial\Omega$. Then:

$$\begin{aligned}\frac{1}{2\pi j}\oint_{\partial\Omega}\frac{f'(s)}{f(s)}\,ds &= \frac{1}{2\pi j}\int_a^b \frac{f'(s(t))}{f(s(t))}s'(t)\,dt\\ &= \frac{1}{2\pi j}\int_a^b (\ln(f(s(t))))'\,dt\\ &= \frac{1}{2\pi j}\int_a^b \left(\ln(|f(s(t))|) + \ln(e^{j\Phi(f(s(t)))})\right)'\,dt\\ &= \frac{1}{2\pi}\left(\Phi(f(s(b))) - \Phi(f(s(a)))\right)\\ &= \left.\frac{\Phi(f(s(t)))}{2\pi}\right|_{t=a}^{t=b} = E.\end{aligned}$$

In order to prove equality B, we note that $h(s) = f'(s)/f(s)$ is analytic at every point at which $f(s) \neq 0$ and $f(s) \neq \infty$. That is, away from the poles and zeros of $f(s)$ the function $h(s)$ is analytic. Thus, to calculate the value of $\oint_{\partial\Omega} h(s)\,ds$ we need only consider the residues of $h(s)$ at its poles. Let us find these residues.

We note that if an analytic function, $f(s)$, has N_z zeros at a point z, then the function can be written $f(s) = (s-z)^{N_z}g(s)$ where $g(z) \neq 0$. Thus near a zero we find that:

$$h(s) = \frac{f'(s)}{f(s)} = \frac{N_z(s-z)^{N_z-1}g(s) + (s-z)^{N_z}g'(s)}{(s-z)^{N_z}g(s)} = \frac{N_z}{s-z} + \frac{g'(s)}{g(s)}.$$

As $g'(s)/g(s)$ is analytic in a neighborhood of the zero, we find that the residue at z is N_z.

Similarly, if $f(s)$ has P_z poles at the point z, then $f(s) = (s-z)^{-P_z}g(s)$ where $g(z) \neq \infty, g(s) \neq 0$. Thus, near a pole we find that:

$$h(s) = \frac{f'(s)}{f(s)} = \frac{-P_z(s-z)^{-P-1}g(s) + (s-z)^{-P_z}g'(s)}{(s-z)^{-P_z}g(s)} = \frac{-P_z}{s-z} + \frac{g'(s)}{g(s)}.$$

As $g'(s)/g(s)$ is analytic in a neighborhood of the pole, we find that the residue at z is $-P_z$.

Adding up all the residues, we find that:

$$\oint_{\partial\Omega} \frac{f'(s)}{f(s)}\, ds = 2\pi j \sum \text{residues} = 2\pi j \left(\sum_i N_{z_i} - \sum_k P_{z_k} \right) = 2\pi j (N - P)$$

which was to be shown.

5.4 How are Encirclements Measured?

From the proof of the principle of the argument, we see how to define an encirclement properly. When tracing out the image of the boundary of the region of interest start, one starts from some point and then one traces out the rest of the image. Let the point from which one starts be defined as 0°—that is measure all angles relative to the ray which connects the origin to the initial point and continues out to infinity. Then keep track of the angle of the current point. When one finishes tracing out the boundary, one will have returned to one's initial point. Upon one's return to the initial point, one will find that one has swept out $n360^o$ where n may be any integer. Let us consider two simple examples.

A Function with No Poles or Zeros in the Unit Disk—An Example

Consider the function:

$$f(s) = \frac{1}{s^2 - 5s + 6}.$$

As the numerator is one, this function cannot have any zeros. Thus, we know that $N = 0$ for any region whatsoever. Let us map the boundary of our region—which is just the unit circle—using our function, $f(s)$. The region and the image of the boundary (under the mapping $f(s)$) are given in Figure 5.2. In the plot of the region we started at 0° (with respect to the positive x-axis as is customary) and proceeded counterclockwise in a circle. We plotted the first quarter circle using periods, the next using pluses, the next using `o`'s and the last quarter circle using `x`'s. Similarly, when plotting the image we start with dots and proceed to plot the image. Let us consider how many times we encircle the origin in this case.

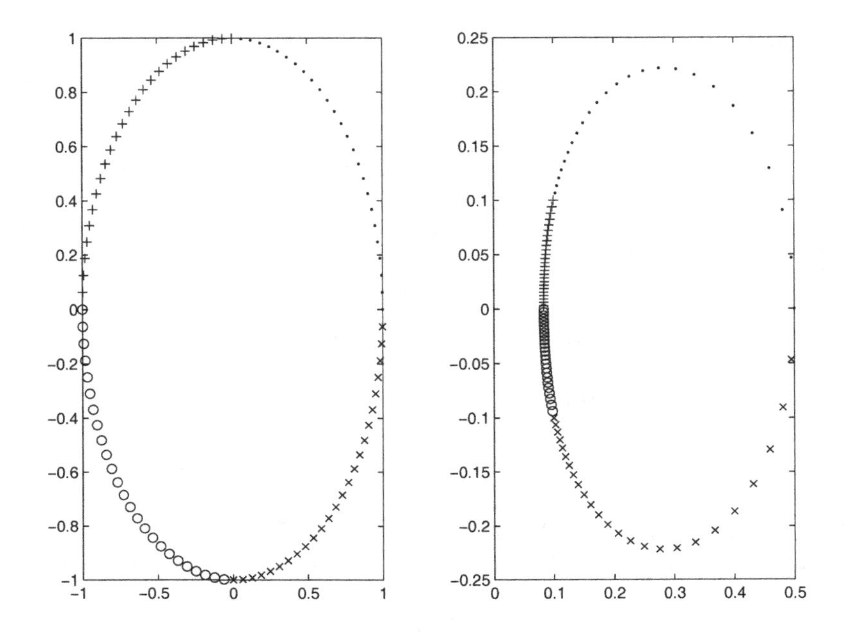

Fig. 5.2 The Boundary, $|s| = 1$, and Its Image under $f(s)$

We find that the first point in the image is at 0°—which is convenient. As we proceed with the dots we find that the angle is increasing. At the top of the image—which is approximately $f(s) = 0.275 + j0.222$, we find that the angle is 39°. After that point, however, we find that the angle decreases. At the point at which the curve switches from pluses to o's, we find that the angle is back down to 0°. At the image's lowest point—at approximately $0.275 - j0.222$ the angle is -39°. However, after this point the angle increases. When we finish tracing out the image, we find that the angle is 0°. As we know that $N = 0$ and we have now seen that $E = 0$, we find that $P = 0$—$f(s)$ has no poles in the unit disk.

A Function with One Pole in the Region—An Example

Let us consider the function:

$$f(s) = \frac{1}{s^2 - 5s + 6}$$

once again. This time, however, we choose as our region:

$$|s| \leq 2.5.$$

We have plotted both the boundary of the region and the image of the boundary under $f(s)$ is Figure 5.3. This time the first dot

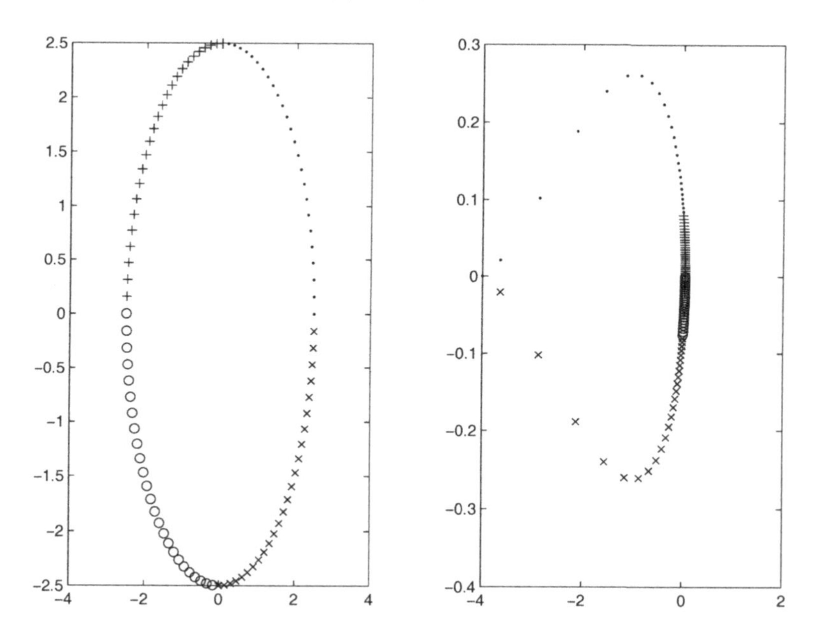

Fig. 5.3 The Boundary, $|s| = 2.5$, and Its Image under $f(s)$

in the image occurs at the point $-4 + j0$. This is our starting point and we say that the (relative) angle of this point is 0°. As we proceed we find that the highest point in the image occurs at $-0.86 + j0.26$. This has an (relative) angle of -16.8°; the minus sign means that we have moved in the clockwise direction. Proceeding from the dots to the pluses and from the pluses to the o's, we find that at the point at which we next cross the real axis—at 0.04 as a quick calculation shows—the angle is now -180°. We are now on the opposite side of the origin. Continuing, we find that as we finish the plot of the image of the boundary—and return to the point from which we started—the angle has decreased to -360°. That is, we have encircled the origin -1 times. As the numerator of $f(s)$ is one, we know that

$N = 0$. Thus, $P = 1$. Indeed, the function:

$$f(s) = \frac{1}{s^2 - 5s + 6} = \frac{1}{(s-3)(s-2)}$$

has one pole in the region $|s| \leq 2.5$.

5.5 First Applications to Control Theory

The transfer function of our generic linear system is:

$$T(s) = \frac{G_p(s)}{1 + G_p(s)H(s)}.$$

Assuming that $G_p(s)H(s)$ is an analytic function (except for a finite number of poles)—something that will always be true for the linear systems in this book—we can use the principle of the argument to determine the presence or absence of zeros of $1 + G_p(s)H(s)$ in the right half-plane.

$G_p(s)H(s) = 1/s$—An Example

We would like to determine the number of zeros that $1 + G_p(s)H(s)$ has in the right half-plane. We note that $1 + G_p(s)H(s)$ is a rational function of s. Thus, it has a finite number of poles. This means that all of its poles that lie in the right half-plane are located in some region of the form:

$$\Omega_R = \{s|\mathcal{R}(s) \geq 0, |s| < R\}.$$

Thus, if one can show that for all sufficiently large R the number of zeros of $1+G_p(s)H(s)$ inside Ω_R remains fixed, then one knows that this is the number of zeros of $1 + G_p(s)H(s)$ in the entire right half-plane.

In our case, there is a problem with the application of the principle of the argument—0, which is a pole of $1 + G_p(s)H(s)$, is on the border of Ω_R. To avoid this pole we take a somewhat different region:

$$\Omega_{\rho,R} = \{s|\mathcal{R}(s) \geq 0, |s| \geq \rho, |s| \leq R\}.$$

This region is shown in Figure 5.4. This region excludes the origin. As $1+G_p(s)H(s)$ is an analytic function of s, if the origin is a pole, then there are no poles or zeros "near" zero. Thus, if for all sufficiently small values of ρ and for all sufficiently large

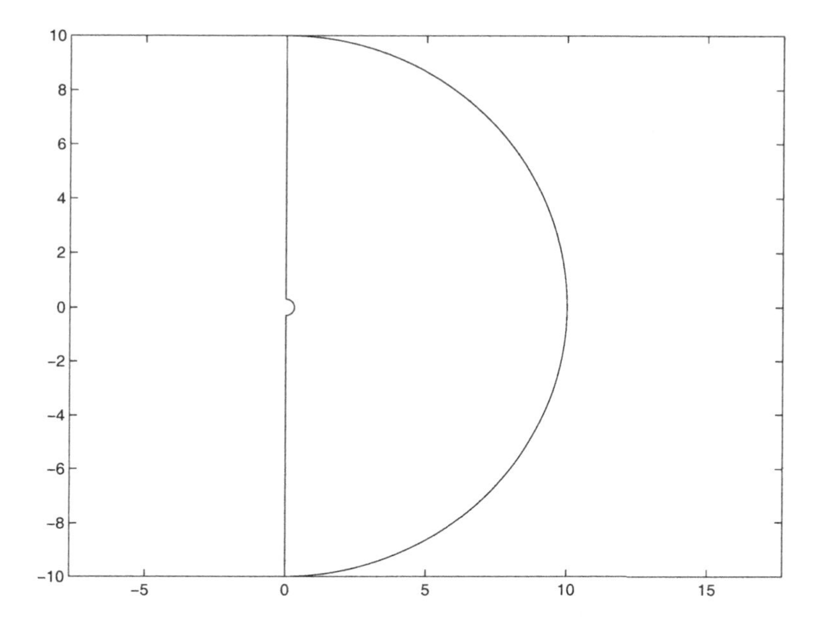

Fig. 5.4 The Region $\Omega_{\rho,R}$ Where $\rho = .3$ and $R = 10$

values of R one finds that there are N zeros in $\Omega_{\rho,R}$, then there are N zeros in the right half-plane.

In our example, it is clear that:

$$\lim_{s\to\infty} 1 + G_p(s)H(s) = 1.$$

For all sufficiently large R, the large semi-circle is mapped into a neighborhood of 1. When mapping the imaginary axis, we find that $1 + G_p(j\omega)H(j\omega) = 1 + 1/(j\omega)$. That is, the points on the imaginary axis are all mapped to points whose real value is one. Finally the points on the small semi-circle $\rho e^{j\theta}$ where θ varies from $\pi/2$ to $-\pi/2$ are mapped into the points $1 + (1/\rho)e^{-j\theta}$–which is a large semi circle in the right half-plane.

Putting all of these facts together we find that the left-hand figure in Figure 5.5 is mapped into the right-hand figure of Figure 5.5 by $1 + G_p(s)H(s)$. As it is clear that no matter how small ρ is and no matter how big R is the second figure never encircles the origin, it is clear that the number of zeros in $\Omega_{\rho,R}$ is equal to the number of poles in the region. As there are no poles in the region, there are no zeros either. Hence, the system is stable.

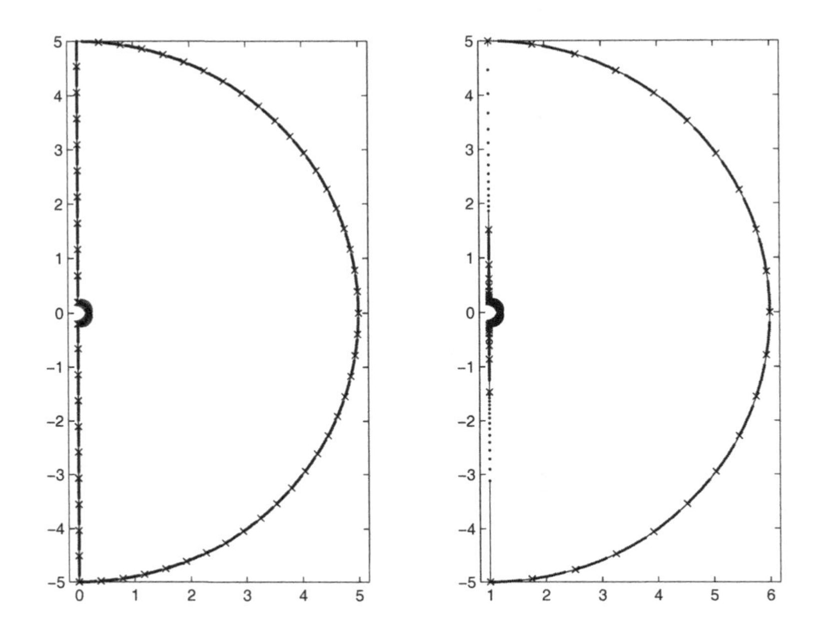

Fig. 5.5 The Region $\Omega_{\rho,R}$ and Its Image under $1 + 1/s$

Though in our previous example we considered the function $1 + G_p(s)H(s)$, generally we consider the function $G_p(s)H(s)$. The number of times that $1 + G_p(s)H(s)$ encircles 0 is clearly equal to the number of time $G_p(s)H(s)$ encircle -1. *We generally check encirclements of* -1 *to determine the stability of our systems.* The plot of $G_p(s)H(s)$ on the "border" of the left-half plane (with possible detours around poles on the imaginary axis) is called the *Nyquist plot* of the system.

5.6 Systems with Low-Pass Open-Loop Transfer Functions

We have seen that the Nyquist plot can be used to determine whether a closed loop system will be stable given the transfer function of the system when the "loop" is opened (as shown in Figure 5.6)—$G_p(s)H(s)$. What do we mean when we say $G_p(s)H(s)$ is low-pass? We mean that:

$$\lim_{s\to\infty} G_p(s)H(s) = 0.$$

If $G_p(s)H(s) = P(s)/Q(s)$ is a rational function of s, then the statement "$G_p(s)H(s)$ is low-pass" is equivalent to the statement $\deg(P(s)) < \deg(Q(s))$. In terms of the Nyquist diagram this says that the semi-circle "at infinity"—which satisfies $|s| = R >> 1$—is (essentially) mapped into the origin. Also for large values of ω, we find that $G_p(j\omega)H(j\omega) \to 0$. Thus, *there is no need to explicitly consider the semi-circle at infinity.* In low-pass systems the semi-circle at infinity cannot present a problem. In order to determine the stability of a low-pass system it is sufficient to consider $s = j\omega$ (including, of course, small semi-circles around poles (or zeros) at the origin and any imaginary poles (or zeros) of $1 + G_p(s)H(s)$). In fact, because the Taylor series associated with $G_p(s)H(s)$ always has real coefficients, we find that the region $\partial\Omega$—which is symmetric about the real axis—is mapped into a region symmetric about the real axis by $G_p(s)H(s)$. Thus, it is sufficient to calculate the image of the upper half of $\partial\Omega$ under the mapping $G_p(s)H(s)$ (as from it we can infer the image of the lower half of $f(\partial\Omega)$).

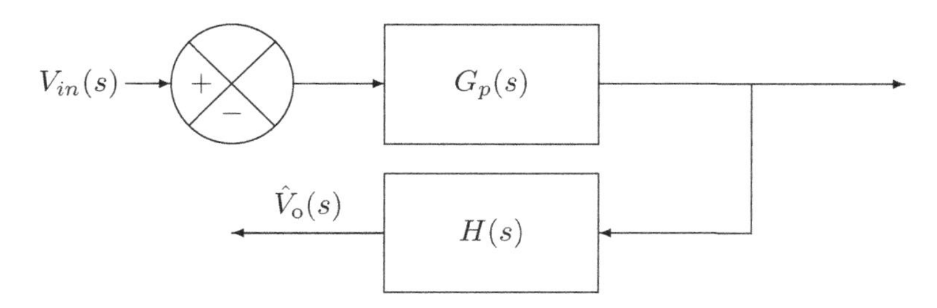

Fig. 5.6 A Simple Feedback Circuit with the "Loop" Opened and the Output Moved to $\hat{V}_o(s)$

$G_p(s)H(s) = 1/s^5$—An Example

When $G_p(s)H(s) = 1/s^5$, we are dealing with a low-pass function. We need not concern ourselves with the behavior for $|s| >> 1$. It is clear that such numbers are mapped (very nearly) into the origin. We see that for $s = j\omega$, $G_p(j\omega)H(j\omega) = 1/(j\omega^5)$. Also, for $s = \rho e^{j\theta}, \theta : \pi/2 \to -\pi/2$, we find that $G_p(s)H(s) = \rho^{-5}e^{-5j\theta}$. Thus the small semi-circle is mapped into five large semi-circles—into two and one half circles—as shown in Figure 5.7. In Figure 5.7 the image of the small semi-circle overlaps itself a number of times. On the left-hand side it overlaps itself once, and on the right hand side it overlaps twice. We see that -1 is encircled twice—which means that there are two more zeros of $1 + 1/s^5$ in the right half-plane than there are poles in the right half-plane. This is clearly correct as:

$$1 + \frac{1}{s^5} = \frac{s^5 + 1}{s^5} = 0 \Leftrightarrow s^5 = -1.$$

Two of the fifth roots of -1 are located in the right half-plane, and $1+1/s^5$ clearly has no poles in the right half-plane (excluding the origin which is explicitly not part of Ω).

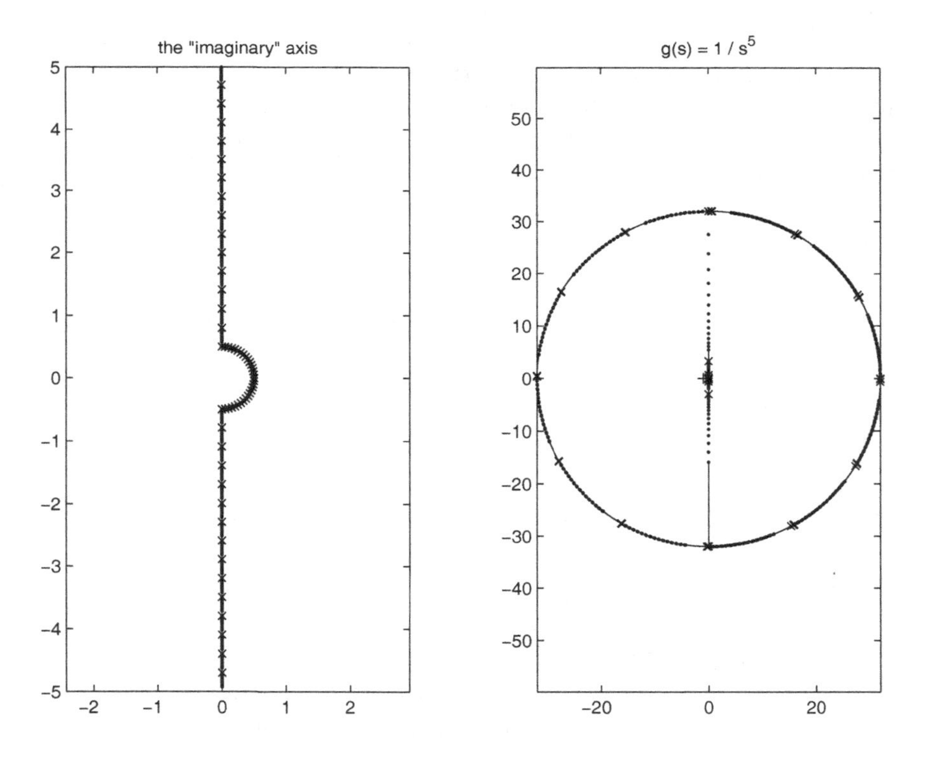

Fig. 5.7 The "Modified" Imaginary Axis and Its Image under $1/s^5$

Let us consider the general effect of poles at the origin on the Nyquist plot. If one has a transfer function of the form:

$$G_p(s)H(s) = c\frac{P(s)}{s^n Q(s)}, \ cP(0)/Q(0) = d > 0,$$

one finds that near zero $G_p(s)H(s)$ looks like d/s^n. As there is a pole at the origin, one must put a small semi-circle near the origin. The semi-circle is described by $\rho e^{j\theta}, \theta \in [\pi/2, -\pi/2]$. The image of this semi-circle under $G_p(s)H(s)$ is (approximately) $(d/\rho^n)e^{-jn\theta}$. Thus, the image of the small semi-circle is (approximately) n very large semi-circles that are traversed in the counter-clockwise direction starting at the point $(d/\rho^n)e^{jn\pi/2}$. Thus, if $n = 1$, the semi-circle starts from (approximately) the point $(d/\rho)e^{-j\pi/2}$–from the negative imaginary axis. When $n = 1$, the semi-circle (mostly) circles through the right half-plane. For $n = 2$, the small semi-circle generates an almost (or possibly slightly more than) complete circle. This circle can cause encirclements of -1 and drive a system into instability. For $n \geq 3$, the image of the small semi-circle will encircle -1. Unless the image of the rest of the boundary encircles -1 in such a way as to "unwind" the encirclements, the system *will* be unstable. Though increasing the gain at low frequencies should improve the performance of our system at DC, once again we find that increasing the gain too much can cause the system to be unstable. We consider some examples.

$G_p(s)H(s) = \frac{s+1}{s(s+2)}$—An Example

Consider the transfer function—$G_p(s)H(s) = \frac{s+1}{s(s+2)}$. This function is low-pass, has one pole at zero, and:

$$\left.\frac{s+1}{s+2}\right|_{s=0} = \frac{1}{2} > 0.$$

We see that the the small semi-circle around the origin is mapped into a large semi-circle that is largely in the right half-plane. To get more information about the behavior of the function when s is near the origin, let us approximate $G_p(s)H(s)$ by making use of the fact[3] that:

$$\frac{1}{1+x} \approx 1 - x, \ |x| << 1.$$

We find that:

$$\frac{s+1}{s(s+2)} = \frac{1}{2s}\frac{s+1}{1+s/2} \approx \frac{1}{2s}(s+1)(1-s/2) \approx \frac{1}{2s}(1+s/2) = \frac{1}{2s}+1/4.$$

We see that as $s \to 0$, $G_p(s)H(s)$ behaves like $1/(2s) + 1/4$. In particular, if $s = j\omega$, then the function looks like $-j\omega/2 + 1/4$. This is a straight line with real part equal to $1/4$ that tends to $-j\infty$ as $\omega \to 0^+$.

Additionally, when $s = j\omega$, we find that:

$$G_p(j\omega)H(j\omega) = \frac{j\omega + 1}{j\omega(j\omega + 2)} = \frac{\omega - j(\omega^2 + 2)}{\omega(\omega^2 + 4)}.$$

Thus, for $\omega > 0$ we find that $G_p(j\omega)H(j\omega)$ always has positive real part and negative imaginary part. Without drawing anything it is clear that there can be no encirclements of -1—our plot is always in the right half-plane. In Figure 5.8 we see both the curve we are mapping and its image under $G_p(s)H(s)$. There are no encirclements of -1. Thus, $N = P$. As $G_p(s)H(s)$ has no poles in the right half-plane, we find that $1 + G_p(s)H(s)$ has no zeros in the right half-plane either; the closed-loop system is stable.

$G_p(s)H(s) = \frac{s+3}{s(s+2)}$—An Example

Here we consider a slightly different transfer function—instead of having $s + 1$ in the numerator, we have $s + 3$. The small semi-circle around the origin is still mapped into a large

[3]The series:

$$1 - x + x^2 - \cdots + (-x)^n + \cdots$$

is a geometric series with ratio $-x$. If $|x| < 1$, the series' sum is:

$$\frac{1}{1+x}.$$

If $|x| << 1$, then the series is well approximated by its first two terms—that is, for small value of $|x|$, we find that:

$$1 - x \approx \frac{1}{1+x}.$$

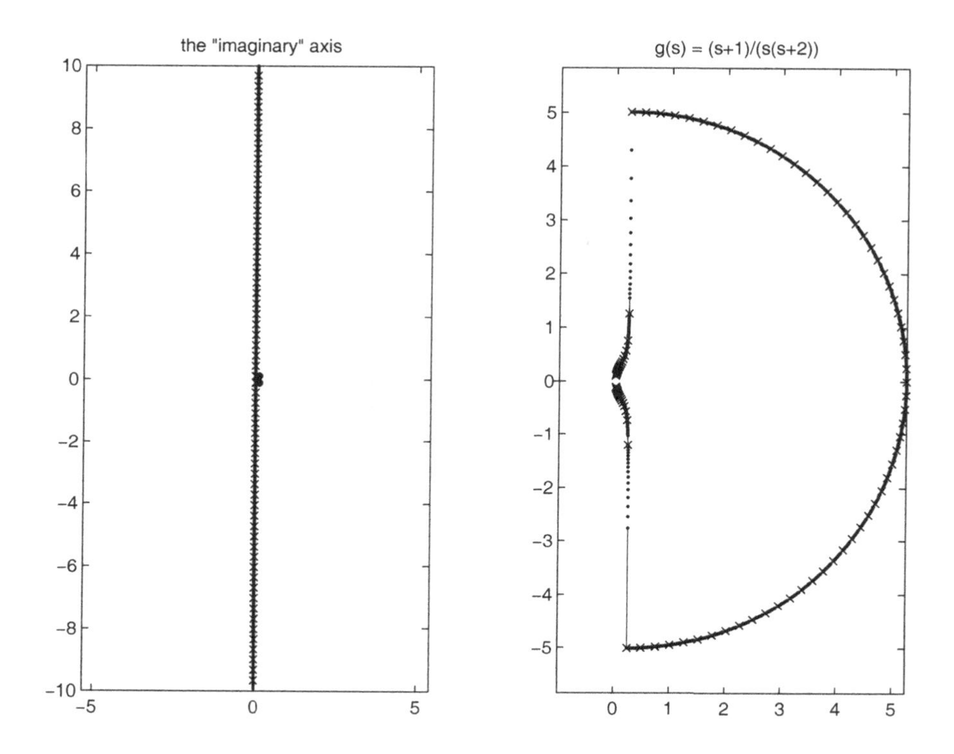

Fig. 5.8 The Imaginary Axis (with Detour) and Its Image under $\frac{s+1}{s(s+2)}$

semi-circle as it was previously. Also:

$$\begin{aligned}\angle G_p(j\omega)H(j\omega) &= \angle(j\omega+3) - \angle j\omega - \angle(j\omega+2)\\ &= \tan^{-1}(\omega/3) - 90^{\text{o}} - \tan^{-1}(\omega/2).\end{aligned}$$

This function is equal to 90° when $\omega = 0$. For $\omega > 0$ the function is less that -90°, but it cannot ever decrease to -180°. Thus, for $\omega > 0$ we find that $\mathcal{I}(G_p(j\omega)H(j\omega)) < 0$. We see that the point -1 cannot be encircled. We have plotted the "imaginary axis" and its image under $G_p(s)H(s)$ in Figure 5.9. Once again, we find that the system is stable. Here, however, the image does make it into the left half-plane.

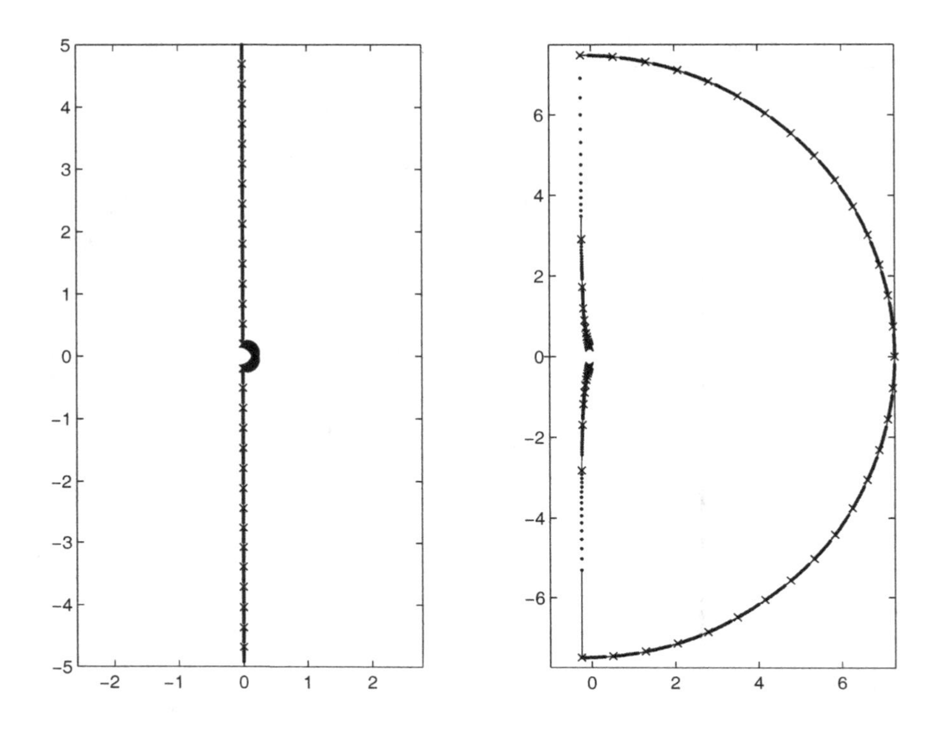

Fig. 5.9 The Imaginary Axis (with Detour) and Its Image under $\frac{s+3}{s(s+2)}$

5.7 MATLAB® and Nyquist Plots

MATLAB has a command to plot Nyquist plots. The command is called `nyquist`. One can find the Nyquist plot of a system `G` by typing:

```
nyquist(G)
```

MATLAB will reply with a plot of

$$G_p(j\omega), \quad |\omega| > \epsilon.$$

In order to determine the stability of the system, one must know how one's system behaves near the origin. MATLAB does not provide that information to you. To deal with this issue, a special set of MATLAB routines is used to plot the Nyquist plots that appear in this book. (This is the only important place where "vanilla" MATLAB is not used in this book.)

5.8 The Nyquist Plot and Delays

From Theorem 7 of Chapter 1, we see that if an element's output is its input delayed by T seconds then the transfer function of the element is e^{-Ts}. On the imaginary axis this is $e^{-j\omega T}$—a phase shift of $-\omega T$ radians. What effect does this have on the stability of a system? Let us consider a system for which $G_p(s) = 1/s$ and $H(s) = e^{-Ts}$. This corresponds to a system for which the plant behaves like an integrator and for which the feedback element—often the measurement device at the output—introduces some "processing delay."

We have seen that if $T = 0$, then the system is stable. In Figure 5.10 we show the Nyquist plot of the system. Since the plot does not enclose $s = -1$ and since $1 + G_p(s)H(s)$ has no poles in the right-half plane other than zero—which is not in our region, we know that the system is stable.

Now consider the same system with $T = 1$. We find that the Nyquist plot is as shown in Figure 5.11. In Figure 5.12 we expand this plot near the origin. Considering these two pictures we see a number of things that are typical of Nyquist plots of systems that include delays.

First of all, we see that near the origin the plot seems to spiral. When is this to be expected? The region near the origin corresponds to $|s| >> 1$. If $s = j\omega$, then we find that $e^{-j\omega T}$ is adding more and more phase as $\omega \to \infty$. On the other hand, since $G_p(s)$ is low-pass, $G_p(j\omega) \to 0$. For low-pass systems with delays, one expects the Nyquist plot to spiral into the origin.

Additionally, we find that even though the plot corresponding to $1/s$ remained in the right half-plane, $\mathcal{R}(s) \geq 0$, we find that after the addition of e^{-Ts} the plot extends into the left half-plane. This too is to be expected. Let us consider $G_p(s)H(s)$ which is a rational function of s and which has a first order pole at zero. That is, $G_p(s)H(s)$ has the form:

$$G_p(s)H(s) = \frac{1}{s}\frac{P(s)}{Q(s)}, \frac{P(0)}{Q(0)} = c > 0.$$

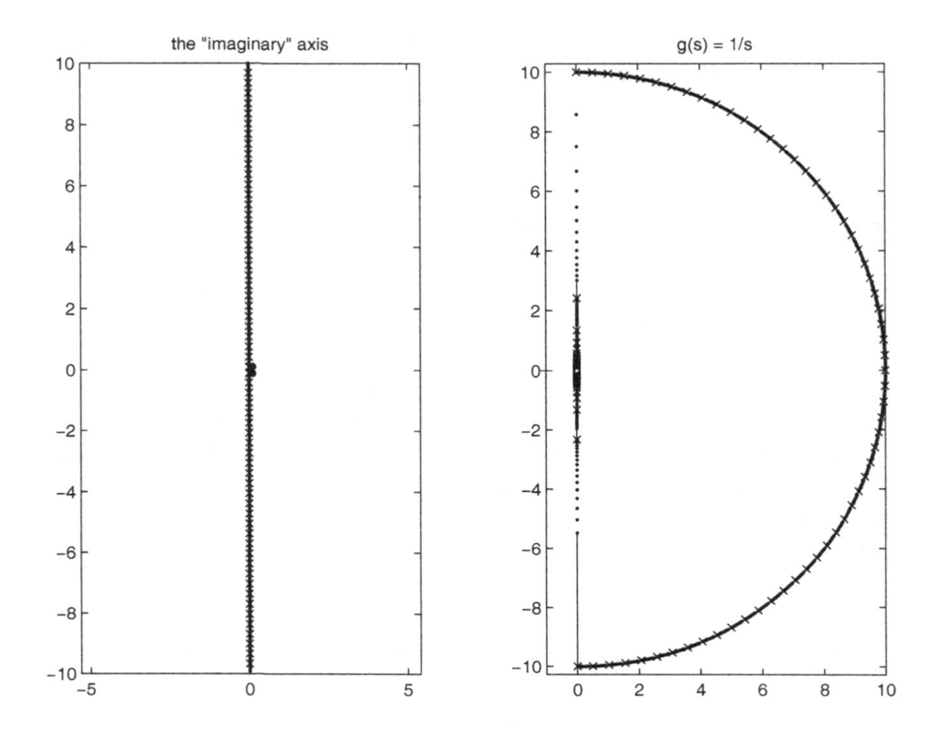

Fig. 5.10 The Imaginary Axis (with Detour) and Its Image under $\frac{1}{s}$

For $|s| << 1$ we find that:

$$\begin{aligned}
\frac{P(s)}{Q(s)} &= \frac{a_m s^m + \cdots + a_1 s + a_0}{b_n s^n + \cdots + b_1 s + b_0} \\
&= \frac{a_0}{b_0} \frac{(a_m/a_0)s^m + \cdots + (a_1/a_0)s + 1}{(b_n/b_0)s^n + \cdots + (b_1/b_0)s + 1} \\
&\approx \frac{a_0}{b_0} \frac{1 + (a_1/a_0)s}{1 + (b_1/b_0)s} \\
&\approx \frac{a_0}{b_0} \left(1 + (a_1/a_0)s\right)\left(1 - (b_1/b_0)s\right) \\
&\approx \frac{a_0}{b_0} \left(1 + \left(\frac{a_1}{a_0} - \frac{b_1}{b_0}\right) s\right) \\
&= \frac{a_0}{b_0} + \left(\frac{a_1}{b_0} - \frac{a_0 b_1}{b_0^2}\right) s
\end{aligned}$$

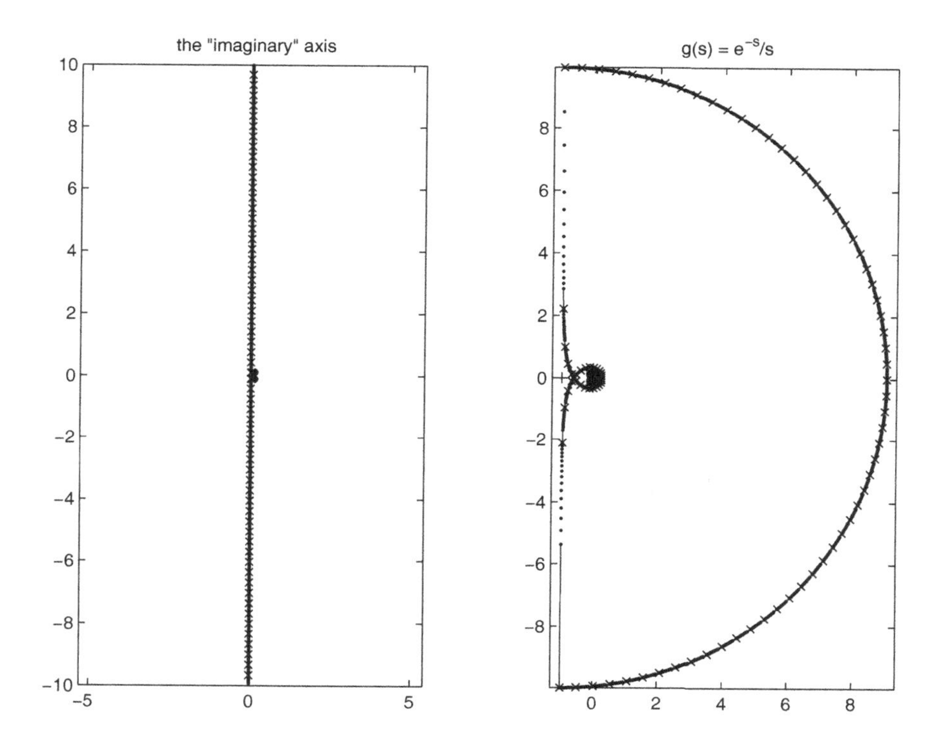

Fig. 5.11 The Imaginary Axis (with Detour) and Its Image under $\frac{e^{-s}}{s}$. (The spiral appear to be finite because the imaginary axis only extends from $-10j$ to $+10j$.)

We see that for small s, $G_p(s)H(s)$ is approximately:

$$G_p(s)H(s) \approx \frac{a_0}{b_0 s} + \left(\frac{a_1}{b_0} - \frac{a_0 b_1}{b_0^2}\right).$$

Thus, as $s = j\omega$ tends to zero, the value of $G_p(s)H(s)$ tends to a point on the line $\mathcal{R}(s) = a_1/b_0 - a_0 b_1/b_0^2$. Recall[4] that for small values of $|Ts|$ we know that:

$$e^{-Ts} \approx 1 - Ts.$$

[4]The Taylor series expansion of e^{-Ts} about zero is:

$$e^{-Ts} = 1 - Ts + \frac{(Ts)^2}{2!} + \cdots + \frac{(-Ts)^n}{n!} + \cdots.$$

For small values of Ts one can—to a first approximation—ignore the second- and higher-order terms.

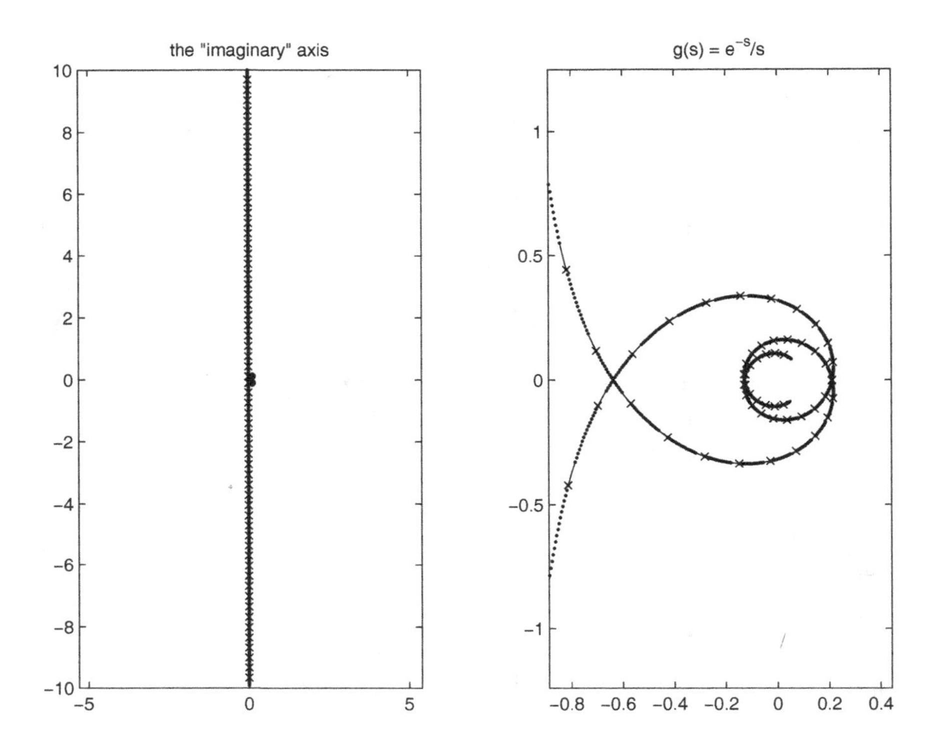

Fig. 5.12 The Imaginary Axis (with Detour) and Its Image under $\frac{e^{-s}}{s}$

We find that the effect of e^{-Ts} on $G_p(s)H(s)$ is, for small s:

$$\begin{aligned}
G_p(s)H(s) &= e^{-Ts}\frac{1}{s}\frac{P(s)}{Q(s)} \\
&\approx (1 - Ts)\left(\frac{a_0}{b_0 s} + \left(\frac{a_1}{b_0} - \frac{a_0 b_1}{b_0^2}\right)\right) \\
&\approx \frac{a_0}{b_0 s} + \left(\frac{a_1}{b_0} - \frac{a_0 b_1}{b_0^2} - T\frac{a_0}{b_0}\right).
\end{aligned}$$

We see that the movement towards the left half-plane is expected and that the amount of the shift of the asymptotes should be $-Ta_0/b_0$ units.

5.9 Delays and the Routh-Hurwitz Criterion

One of the advantages of the Nyquist Plot is that it allows us to deal with any analytic function directly. This does not mean that there is no way to use our previous techniques in such cases. If we can find "good" rational function approximations to our functions, then we can use our other techniques as well.

When do the destabilizing effects of e^{-Ts} become apparent? In low-pass systems they become apparent at low frequencies. At high-frequencies the smallness of $G_p(j\omega)H(j\omega)$ prevents problems from starting. In order to use our polynomial based methods, we need to find a good approximation to e^{-Ts} at low frequencies that has some of the properties of e^{-Ts} at high frequencies.

The "standard" approximation to e^{-Ts} is $1-Ts$—the first two terms in the Taylor series of e^{-Ts} about zero. This function has a rather unfortunate effect—it's absolute value is large for large frequencies—for large values of $j\omega$. This is not the case for e^{-Ts}, and this can be a problem. How else can we approximate e^{-Ts}?

A simple approximation[5] relies on the fact that for small s:

$$\begin{aligned}\frac{1-(T/2)s}{1+(T/2)s} &= (1-(T/2)s)\frac{1}{1+(T/2)s}\\ &= (1-T/2)s)(1-(T/2)s+((T/2)s)^2-\cdots)\\ &\approx (1-(T/2)s)(1-(T/2)s+(T/2)^2s^2)\\ &\approx (1-Ts+(T^2/2)s^2)\\ &\approx e^{-Ts}.\end{aligned}$$

For large values of s this approximation is near -1 and for $s=j\omega$ we find that:

$$\left|\frac{1-(T/2)j\omega}{1+(T/2)j\omega}\right| = \frac{|1-(T/2)j\omega|}{|1+(T/2)j\omega|} = 1.$$

We see that on the imaginary axis this function always has unit magnitude, for small s this function is a good approximation to e^{-Ts}, and for large s the function is very near minus one. As we do not care too much about the precise behavior for large s (as the low-pass nature of $G_p(s)H(s)$ is supposed to guarantee that these values will not affect the stability of the

[5]This approximation is a particular example of a *Padé approximation* [Bor97]—of the approximation of a function by a quotient of polynomials.

system), this function should give us a useful approximation to e^{-Ts}. We consider a simple example.

The Function $G_p(s) = \frac{1}{s(s+1)}$—An Example

Consider a system for which

$$G_p(s) = \frac{1}{s(s+1)}$$

and for which the feedback loop adds a delay of T seconds. We would like to know what the maximum value of T is for the stable operation of our system.

After a number of trials, we found that with $T = 1.1$ the Nyquist plot (see Figure 5.13) does not encircle -1. From the plot, it is clear the if T is much bigger than 1.1, then there will be encirclements. Let us see what value of T our approximation finds for us.

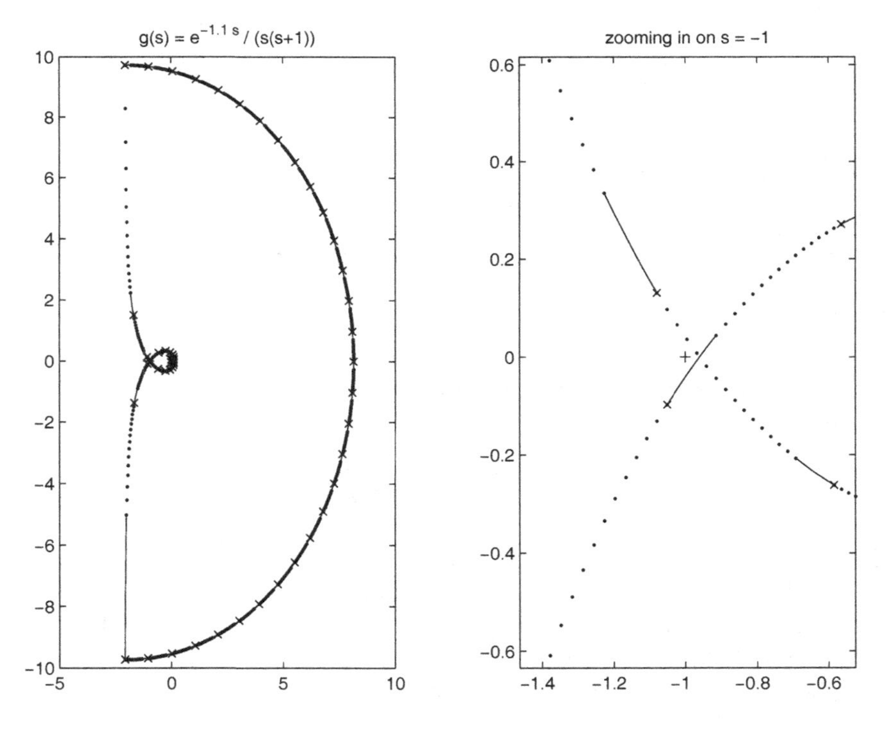

Fig. 5.13 The Nyquist Plot of $e^{-1.1s}\frac{1}{s(s+1)}$ and a Zoom on Part of the Plot

We are interested in determining the values of T for which:

$$1 + G_p(s)e^{-Ts} \approx 1 + \frac{1-(T/2)s}{1+(T/2)s}\frac{1}{s(s+1)}$$

has zeros in the right half-plane. Rewriting this expression, we find that we are interested in the zeros of:

$$\frac{(T/2)s^3 + (1+T/2)s^2 + (1-T/2)s + 1}{s(s+1)(1+(T/2)s)}.$$

These are just the zeros of the polynomial in the numerator. To determine when there are zeros in the right half-plane, we use the Routh-Hurwitz criterion. Building the Routh array, we find that:

$$\begin{array}{ll} T/2 & \frac{2-T}{2} \\ \frac{2+T}{2} & 1 \\ \frac{2-T}{2} - \frac{T}{2+T} & 0 \\ 1 & 0 \end{array}$$

Thus, we need to determine when:

$$\frac{2-T}{2} - \frac{T}{2+T}$$

is negative. This leads to a quadratic equation, and we find that the solution is that there are roots in the right half-plane if $T \geq \sqrt{5} - 1 = 1.23$. The value that we get here is just a little bit too optimistic; the exact value (as a bit of experimenting shows) is between 1.1 and 1.2. (See Problem 10 to see how to calculate the exact value.) For more information about systems with delays, see [Eng08].

5.10 Relative Stability

We have now seen several ways to answer the question, "is a system stable?" We now consider a question whose meaning is somewhat less clear, "*How* stable is the system?"

Why must this question be asked? If a system can be shown to be stable, one would think that that would be the end of the process. However, the mathematical *models* that we use when determining the system's stability are generally inaccurate in a number of ways. First of all, there are often nonlinearities that are not modeled at all. Also there are often parasitic reactances (which often contribute negative phase to the system) that are not taken into account by our models. Furthermore we never know the values of the elements of our system with 100% accuracy. It is, therefore, critical that we be able to show that we have some margin for error. We consider two of the most commonly used measures of relative stability—of how stable the system is—the *gain margin* and the *phase margin*.

Suppose that $G_p(j\omega)H(j\omega)$ is real and negative for some set of values $\omega = \omega_1, ..., \omega_n > 0$. Let $k = \min_i G_p(j\omega_i)H(j\omega_i)$. Then we define the gain margin to be $1/|k|$ in dB—i.e. $-20\ln_{10}|k|$dB. If $|k| < 1$, the gain margin will be positive. What information does the gain margin give us about the stability of our system? Consider Figure 5.14. We have drawn one line passing through -1 and another line passing through k. We see that $1/|k|$ is the amount by which the Nyquist diagram must be multiplied in order for the diagram to first intersect $s = -1$. That is, any gain larger than or equal to $1/|k|$ will cause the system to be unstable. This gain—when measured in decibel—is $-20\log_{10}|k|$. It has been found (empirically) that a gain margin of 8 dB is usually sufficient without being excessive. If $G_p(j\omega)H(j\omega)$ is never negative, then the gain margin is infinite.

The gain margin gives one an idea about how much gain can be added to a system without the system becoming unstable. The phase margin, on the other hand, gives one an idea about how much phase can be "subtracted" from the system without causing the system to become unstable. Suppose that $|G_p(j\omega)H(j\omega)| = 1$ for some set of values $\omega = \omega_1, ..., \omega_n > 0$ and suppose that the phase of the system[6] is never less that -360°. Let Φ be the value of $\angle G_p(j\omega_i)H(j\omega_i)$ that is most negative. Then the phase margin, ϕ, is equal to $+180^o + \Phi$. To understand this idea consider Figure 5.15. We have plotted the unit circle and the Nyquist plot corresponding to:

$$G_p(s)H(s) = \frac{2s+5}{5s^3+5s^2+5s}.$$

Note that only the lower portion of the Nyquist plot comes from positive frequencies. Thus, it is only the intersections in the lower half-plane that count for our purposes. The phase margin is the angle ϕ shown in the figure. This angle is the difference between -180° and the angle of the sole

[6]If the phase of the system can be less than -360°, then the phase margin is defined slightly differently. The idea remains that same—the phase margin is the "amount" of phase that must be subtracted from the system in order to cause the system to become unstable.

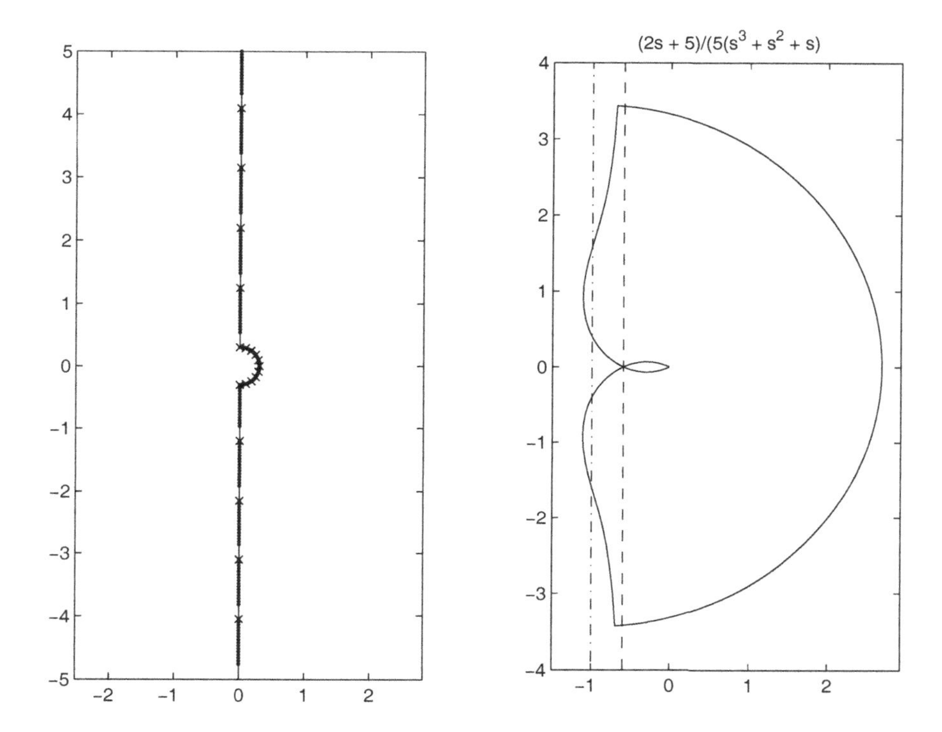

Fig. 5.14 A Graphical Representation of the Gain Margin

relevant point on the Nyquist plot whose magnitude is one. What does the phase margin tell one? It says that if one were to introduce a negative phase shift of ϕ degrees into the system, one would cause the system to become unstable. A phase margin in the neighborhood of 45° is usually sufficient to guarantee the stability and proper operation of the system.

No claim is being made that these two margins fully characterize the stability of the system. It is quite possible to have a system with infinite phase and gain margins; this does not mean that the system cannot be made unstable. It means is that one cannot cause the system to become unstable by the introduction of either a pure gain or a pure phase shift. It is also possible to have a system for which the gain margin is negative but which is still stable. The gain and phase margins have proved to be useful ways of characterizing the extent to which a system is stable. They do not completely characterize it.

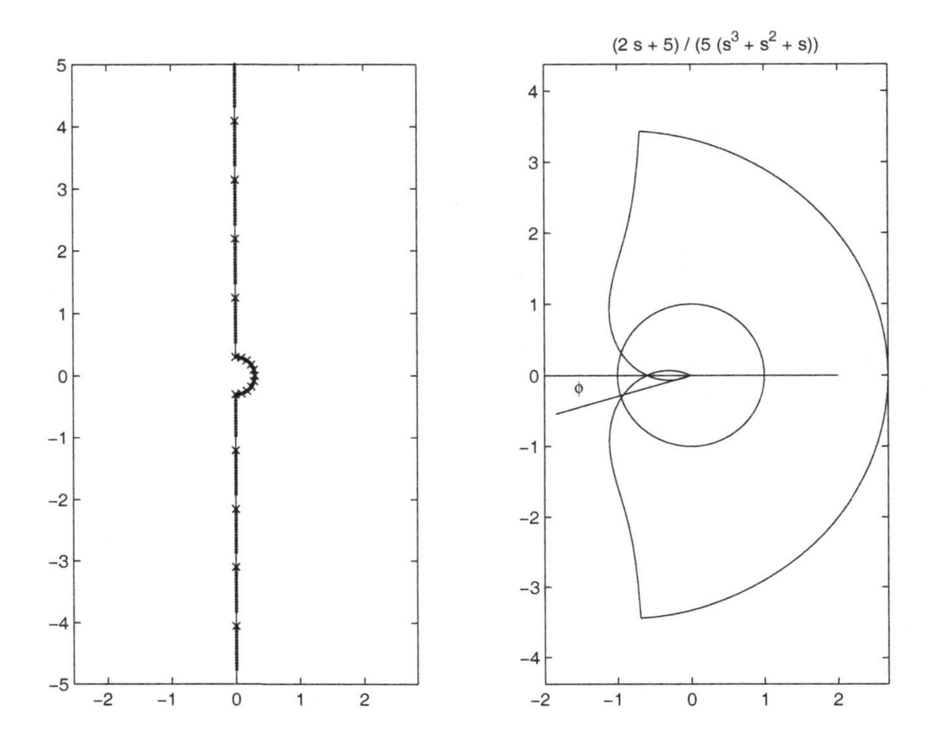

Fig. 5.15 A Graphical Representation of the Phase Margin

A Conditionally Stable System—An Example

Consider the system described by the transfer function:

$$G_p(s)H(s) = \frac{(s+1)(s+2)}{s(s+0.1)(s+0.2)}.$$

We have already seen how to show that for small values of ω this function is approximately:

$$G_p(j\omega)H(j\omega) \approx \frac{100}{j\omega} - 1350.$$

For very large values of ω the transfer function looks like $1/(j\omega)$—that is the Nyquist plot approaches zero through negative imaginary numbers as we see in the (partial) Nyquist plot in Figure 5.16. What is more interesting is that for "intermediate" values of ω the $G_p(j\omega)H(j\omega)$ has positive imaginary part.

(For example, we find that $G_p(.5j)H(.5j) \approx -16+5j$.) Consider Figure 5.16 again. When one considers the stability of the system one finds that as things stand the system is stable—minus one is encircled once in the positive direction (by the semi-circle at infinity) and once in the negative direction—there are zero net encirclements. Such systems are called conditionally stable. If one were to raise or *lower* the gain, the system would become unstable. In this example the gain margin is certainly negative—each of the intersections of the Nyquist plot with the negative real axis occurs for values of $G_p(j\omega)H(j\omega)$ that are larger than one. Here, the system is stable, and the gain margin is not really meaningful.

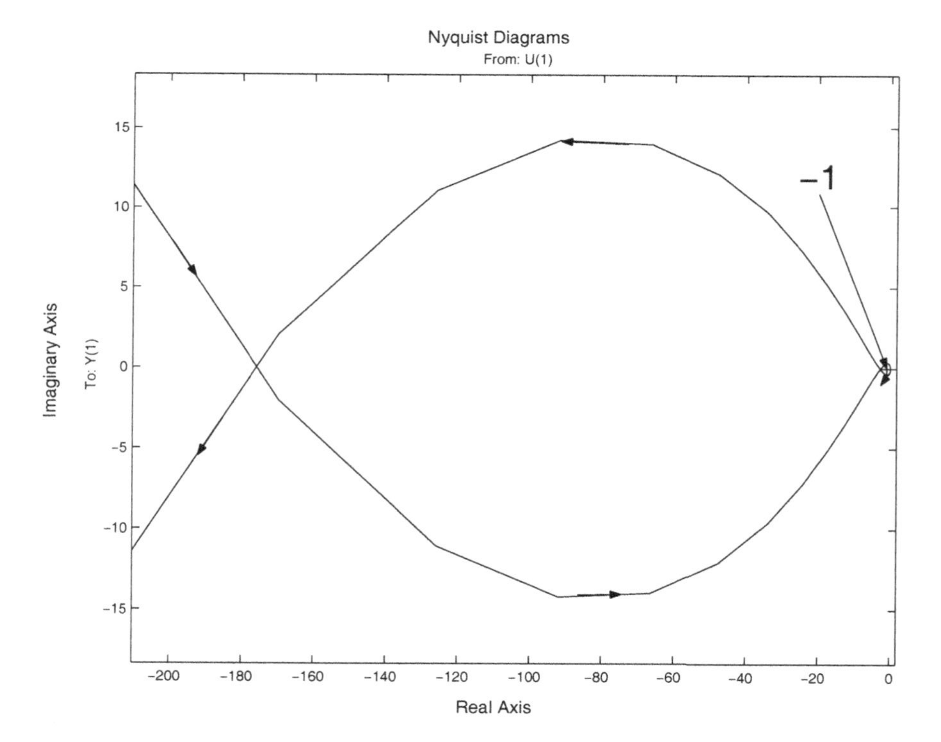

Fig. 5.16 The Nyquist Plot Near the Origin

It is also possible to examine this system using the methods of the previous chapter. The denominator of the transfer function with a gain, K, added before $G_p(s)$ is:

$$1 + K\frac{(s+1)(s+2)}{s(s+0.1)(s+0.2)}.$$

We find that we must identify the values of K for which the polynomial:

$$s(s+0.1)(s+0.2)+K(s+1)(s+2) = s^3+(0.3+K)s^2+(0.02+3K)s+2K$$

has zeros in the right half-plane. We find that the Routh array that corresponds to this polynomial is:

$$\begin{array}{cc} 1 & 0.02+3K \\ 0.3+K & 2K \\ 0.02+3K-\frac{2K}{0.3+K} & 0 \\ 2K & 0 \end{array}.$$

Clearly in order for the sign of the left-hand column to be constant we find that $K > 0$. The only additional condition is that of the third row. As $K > 0$ we find that this condition is equivalent to:

$$(0.02+3K)(0.3+K) - 2K > 0.$$

This is a quadratic inequality. Solving for the zeros of the quadratic, we find that they are:

$$K = 0.172, 0.188.$$

As it is clear that for small K and for large K the condition is met, we find that the system is unstable when K is between 0.172 and 0.188 and is stable otherwise. This information supplements the information gleaned from the Nyquist plot quite nicely.

5.11 The Bode Plots

We have seen how to determine the gain and phase margins of a system using a Nyquist plot. In principle, with a sufficiently precise diagram, one could measure the margins from the diagram. The Nyquist plot is not, however, the easiest way to measure the systems margins. It is easier to measure the margins using the Bode plots. The Bode plots are just plots of the magnitude of $G_p(j\omega)H(j\omega)$ (measured in decibels) against frequency (plotted on a logarithmic scale) and the phase of $G_p(j\omega)H(j\omega)$ (measured in degrees) against frequency (again plotted on a logarithmic scale). The plots are generally presented one above the other and the frequency scales are identical. It is *very easy* to calculate the system's margins using the Bode plots. To calculate the gain margin one locates all points on the phase plot for which the phase is $-180^o + n360^o$. Then one looks up at the gains that correspond to those phases. The additive inverse of the largest of these values is the gain margin. To find the phase margin one looks for all frequencies at which the gain is 0dB. At those frequencies one looks at the phase. The phase closest to $-180^o + n360^o$—the most problematic phase—is the one that fixes the phase margin. Also, MATLAB has a command which allows one to calculate the Bode plots and the margins at one blow–`margin`. This command also shows one where the margins occur.

Let us consider a simple example. Suppose that $G_p(s) = \frac{1}{s(s+1)}$ and $H(s) = 1$. Then one would define $G_p(s)$ by:

```
G = tf([1],[1 1 0])

Transfer Function:
   1
-------
s^2 + s
```

To find the margins, one types:

```
margin(G)
```

MATLAB replies to this with Figure 5.17 which has all of the desired information on it. Note that the phase margin is $180^o + \angle G_p(j\omega)H(j\omega)$ for the ω for which $|G_p(j\omega)H(j\omega)| = 1$.

5.12 An (Approximate) Connection between Frequency Specifications and Time Specification

Let us consider the "typical" second order system of Figure 5.18. The phase margin of the system is determined by the phase of the function $G_p(j\omega_1)$ for the frequency that satisfies $|G_p(j\omega_1)| = 1$. Let us find ω_1.

We must solve the equation:

$$|G_p(j\omega)|^2 = \frac{\omega_n^4}{\omega^4 + 4\zeta\omega_n^2\omega^2} = 1.$$

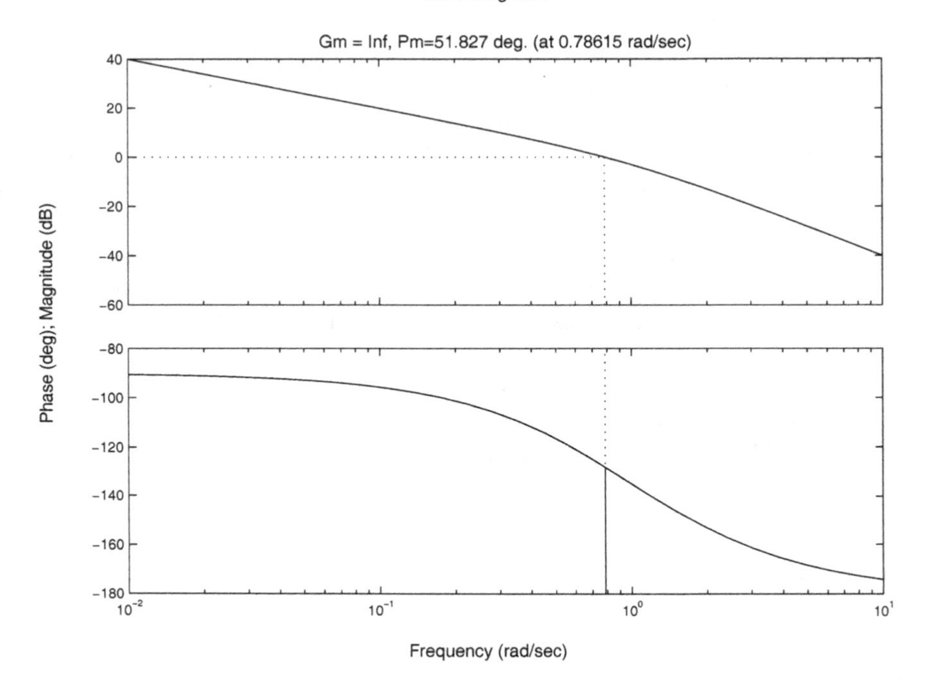

Fig. 5.17 The Bode Plots

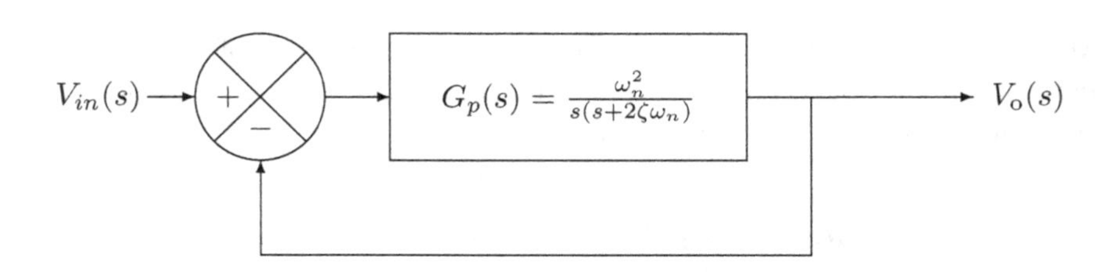

Fig. 5.18 A Simple Feedback Circuit

After a little bit of manipulation, we find that this is equivalent to solving:

$$\omega_v^4 + 4\zeta^2\omega_v^2 - 1 = 0, \quad \omega_v \equiv \omega/\omega_n.$$

This is a quadratic equation in ω_v^2, and the solutions are:

$$\omega_v^2 = -2\zeta^2 \pm \sqrt{4\zeta^4 + 1}.$$

As we are only interested in real positive values of ω_v this leads us to:

$$\omega_v = \sqrt{-2\zeta^2 + \sqrt{4\zeta^4 + 1}},$$

or

$$\omega_1 = \omega_n \sqrt{-2\zeta^2 + \sqrt{4\zeta^4 + 1}}.$$

What is the value of $G_p(j\omega_1)$? Making use of the fact that $\omega_1 = \omega_v \omega_n$ and the fact that $\omega_v^4 + 4\zeta^2\omega_v^2 = 1$, we find that:

$$\begin{aligned} G_p(j\omega_1) &= \frac{\omega_n^2}{-\omega_v^2\omega_n^2 + 2j\zeta\omega_v\omega_n^2} \\ &= \frac{1}{-\omega_v^2 + 2j\zeta\omega_v} \\ &= -\omega_v^2 - 2j\zeta\omega_v \\ &= 1\angle(\arctan(2\zeta/\omega_v) - 180^{\text{o}}). \end{aligned}$$

We see that the phase margin is:

$$\varphi_M = \arctan(2\zeta/\omega_v) = \arctan\left(\frac{2\zeta}{\sqrt{-2\zeta^2 + \sqrt{4\zeta^4 + 1}}}\right).$$

This leads to the approximation, valid for relatively small ζ:

$$\begin{aligned} \varphi_M &= \arctan(2\zeta/\omega_v) \\ &= \arctan\left(\frac{2\zeta}{\sqrt{-2\zeta^2 + \sqrt{4\zeta^4 + 1}}}\right) \\ &\approx \arctan(2\zeta) \\ &\approx 2\zeta \,\text{rad} \\ &\approx 100\zeta^{\text{o}}. \end{aligned}$$

The transfer function of the system with feedback is:

$$T(s) = \frac{\omega_n^2}{s^2 + 2\zeta\omega_n s + \omega_n^2}.$$

We have seen (p. 47) that the time constant of the system is approximately $1/(\omega_n\zeta)$. If one assumes that one's system settles "completely" after four time constants, then one finds that the settling time of this second order system, T_s, is:

$$T_s \approx \frac{4}{\omega_n\zeta}.$$

We have just proved that:

$$\tan(\varphi_M) = \frac{2\zeta}{\omega_v} = \frac{2\zeta}{\omega_1/\omega_n} = \frac{2\zeta\omega_n}{\omega_1}.$$

Combining these results, we find that:

$$T_s \approx \frac{8}{\tan(\varphi_M)\omega_1}. \tag{5.1}$$

Though this relationship between the settling time, T_s, the frequency ω_1 (also called the gain crossover frequency), and the phase margin was derived for a specific second order system, it is sometimes used in a more general context when one knows two of the numbers T_s, φ_M, and ω_1 and wants to find the third. For example, if one is given a value of the phase margin, and one is given the settling time, then this approximation gives one an idea of the angular frequency for which the gain of the system must equal one.

Though (5.1) need not hold exactly for generic systems, it is reasonable that the settling should be inversely proportional to ω_1 and to something like $\tan(\varphi_M)$. The frequency ω_1 is in some sense the system's "bandwidth," and the systems speed *ought* to be inversely proportional to its bandwidth. Also, as $\varphi_M \to 0$ we know that a pole (or poles) is approaching the imaginary axis. As the pole(s) get closer, the system's time constant—which is fixed by the pole closest to the imaginary axis—must get larger, and the system must get slower. Thus, though (5.1) need not hold precisely for generic systems, something like it ought to be true, and we use (5.1) as a first approximation to the exact, but unknown, relationship.

5.13 Some More Examples

An Integrator—An Example

Suppose that $G_p(s)H(s) = 1/s$. How stable is the system? We consider the Bode plots associated with the system. We use the MATLAB `margin` command to draw the plots and find the margins. In Figure 5.19, we see that the gain margin is infinite and the phase margin is ninety degrees. Is this reasonable? Certainly. $G_p(j\omega) = 1/(j\omega)$ is always ninety degrees from the negative real axis, and it is never real. Thus the phase margin is ninety degrees and the gain margin is infinite.

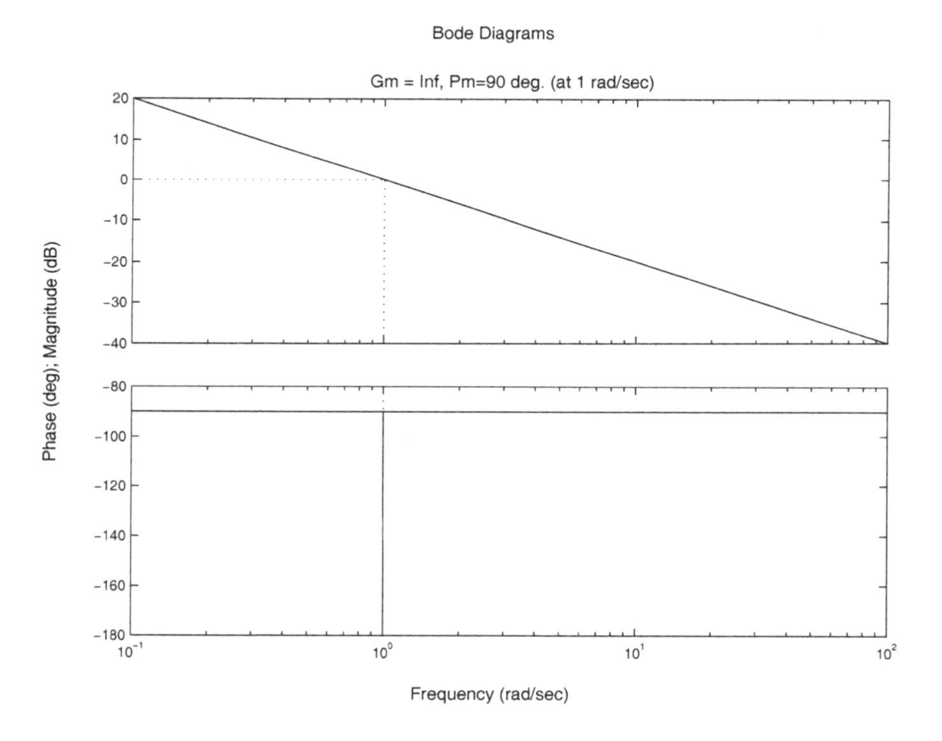

Fig. 5.19 The Bode Plots Pertaining to $G_p(s)H(s) = 1/s$

An Integrator with a "Parasitic Low-Pass Filter"—An Example

Next consider a system for which:

$$G_p(s) = \frac{1}{s}\frac{1}{0.05s + 1}.$$

One can think of the this as an integrator followed by a (parasitic) low-pass filter. Consider the Bode plots that pertain to this system—shown in Figure 5.20. We find that the gain margin is still infinite—and this is as it should be. For the gain margin to be finite the open-loop transfer functions—in this case $G_p(s)$—must be negative for some value of ω. However, the phase of $G_p(s)$ only tends towards -180° as $\omega \to \infty$. As $\omega \to \infty$, $G_p(j\omega) \to 0$. The phase margin of the new system is 87.14^o. This too is reasonable. The phase contribution of a low pass filter is always negative. Thus, the parasitic filter tends to force the Nyquist plot of the integrator towards the negative real axis and, consequently, to make the closed-loop system less stable.

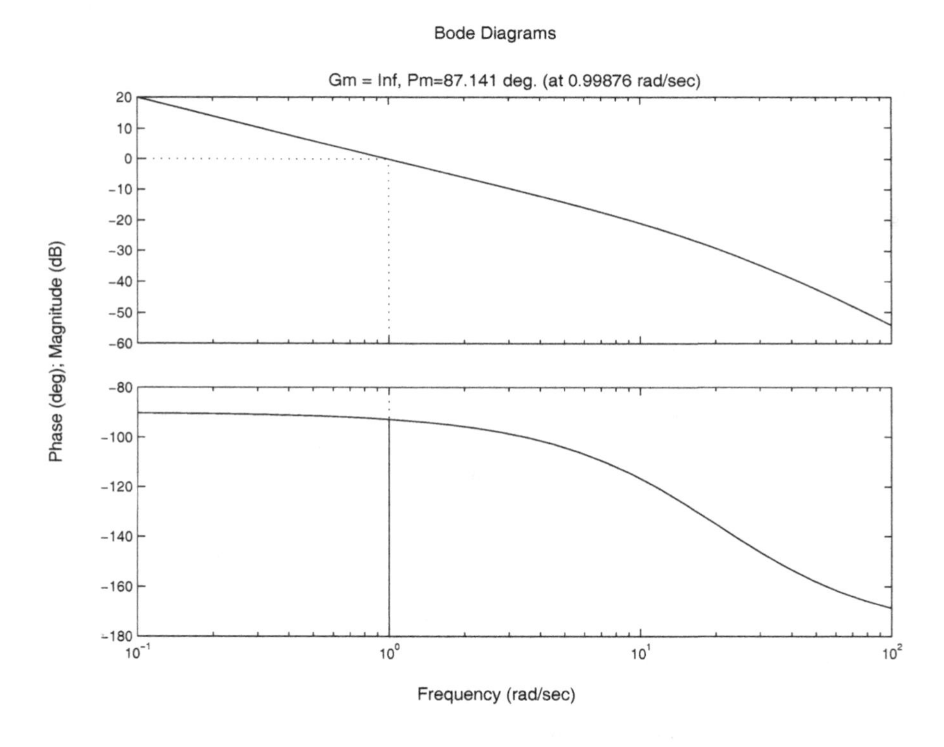

Fig. 5.20 The Bode Plots Pertaining to $G_p(s)H(s) = (1/s)(1/(0.05s + 1))$

A Motor with a "Parasitic Low-Pass Filter"—An Example
In Section 5.11 we saw that if

$$G_p(s) = \frac{1}{s(s+1)}, H(s) = 1,$$

then the gain margin of the system is infinite and the phase margin is 51.8°. Suppose that the motor is preceded by a (parasitic) low-pass filter whose transfer function is $1/(0.05s+1)$. What will the effect of the filter be on the system's margins? In Figure 5.21 we find that the new gain margin is not infinite—it is 26.4dB. This too is reasonable. From the form of the final open-loop transfer function of the system:

$$\frac{1}{s(s+1)(0.05s+1)}$$

we see that as $s = j\omega$ increases through the imaginary numbers the phase of the denominator:

$$\angle(j\omega)(j\omega+1)(0.05j\omega+1) = 90^\circ + \tan^{-1}(\omega)\tan^{-1}(0.05\omega)$$

must increase from ninety degrees to 270°. Thus, the phase of the system—which is just minus the phase of the denominator in this case—passes through -180° once. It is easy to determine the frequency at which the phase is equal to -180°. At this frequency we find that:

$$\begin{aligned}\Im(j\omega(j\omega+1)(0.05j\omega+1)) &= \Im(-j0.05\omega^3 - 1.05\omega^2 + j\omega)\\ &= -0.05\omega^3 + \omega\\ &= 0.\end{aligned}$$

Thus, we find that $\omega = 0, \sqrt{20}, -\sqrt{20}$. As we are interested in positive values of ω, the value of interest to us is $\omega = \sqrt{20}$. At this frequency, we find that:

$$G_p(j\sqrt{20})H(j\sqrt{20}) = \frac{1}{-1.05 \cdot 20}.$$

As this is almost exactly $-1/20$ it must correspond to a gain margin of approximately $20\log 20 \approx 26$ dB.

Additionally we find that the phase margin of the new system is 49.6°. This decrease in the phase margin is also reasonable—

after all the (parasitic) low-pass filter always contributes negative phase and helps push the phase closer to -180°.

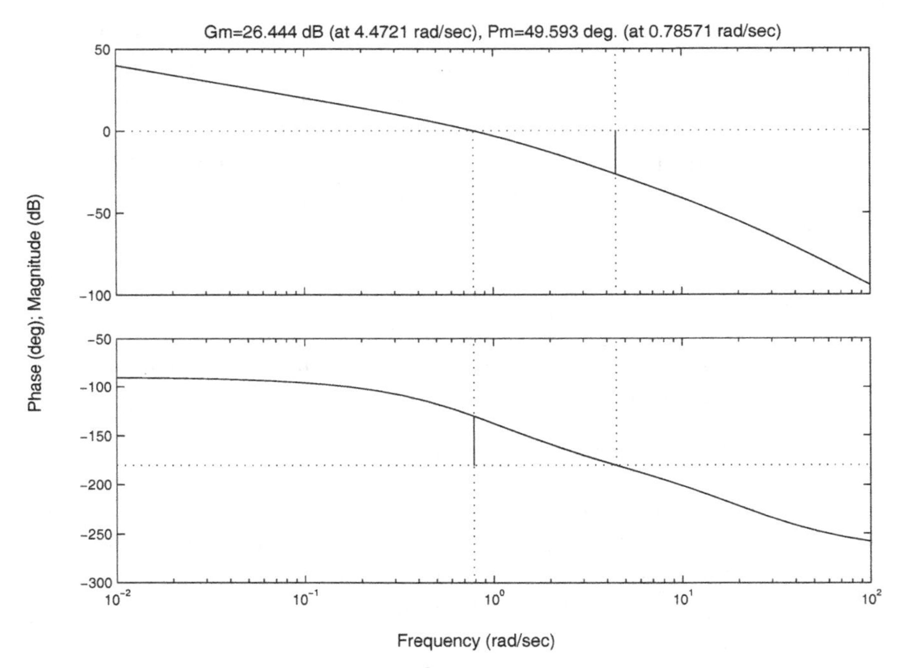

Fig. 5.21 The Bode Plots Pertaining to $G_p(s)H(s) = 1/(s(s+1)) \cdot 1/(0.05s+1)$

A Fourth-Order System—An Example

Consider a unity-feedback system for which:

$$G_p(s) = \frac{1}{s(s+1)(s+2)(s+3)} = \frac{1}{s(s^3+6s^2+11s+6)}$$

(and, of course, $G_c(s) = 1$). We examine the stability of the system and its gain and phase margins by considering the Nyquist plot.

Clearly, for $\omega > 0$, $|G_p(j\omega)|$ is monotonically decreasing. Additionally, for $\omega > 0$:

$$\angle G_p(j\omega) = -90^o - \tan^{-1}(\omega) - \tan^{-1}(\omega/2) - \tan^{-1}(\omega/3). \quad (5.2)$$

This function decreases from -90° to -360° as ω goes from zero to infinity. Thus, we expect low positive frequencies to lead to an asymptote in the third quadrant, we expect the Nyquist plot to approach zero through the first quadrant for large positive frequencies, and for positive frequencies, we expect the plot to intersect the real axis at a single point.

We now calculate the point of intersection. Shortly, we will determine the asymptote's location. To find the point of intersection, one must find a point $s = j\omega$ for which $G_p(j\omega) \in \mathcal{R}$. As the numerator of $G_p(j\omega)$ is real, one must find a point at which the denominator is real. One must look for the positive zeros of the imaginary part of the denominator—for the positive solutions of $6(j\omega)^2 + 6 = 0$. Clearly, the solution is $\omega = 1$. We find that the point of intersection is $G_p(j) = -1/10$.

Now we move on to the asymptote. We know that we are interested in small values of s, so we anticipate making use of the approximation:

$$\frac{1}{1+s} \approx 1 - s, \qquad |s| \ll 1.$$

Considering $G_p(s)$ for small values of s we find that:

$$\begin{aligned} G_p(s) &= \frac{1}{s(s^3 + 6s^2 + 11s + 6)} \\ &\approx \frac{1}{s(11s + 6)} \\ &= \frac{1}{6s} \frac{1}{(1 + (11/6)s)} \\ &\approx \frac{1}{6s}(1 - (11/6)s) \\ &= \frac{1}{6s} - 11/36. \end{aligned}$$

For small values of $s = j\omega, \quad \omega > 0$, the imaginary part of $G_p(j\omega)$ tends to $-\infty$ while the real part tends to the constant value $-11/36$.

As the behavior of the $G_p(s)$ for small s is like that of $1/(6s)$, the Nyquist plot has a large semicircle that is drawn counterclockwise. See Figures 5.22 and 5.23.

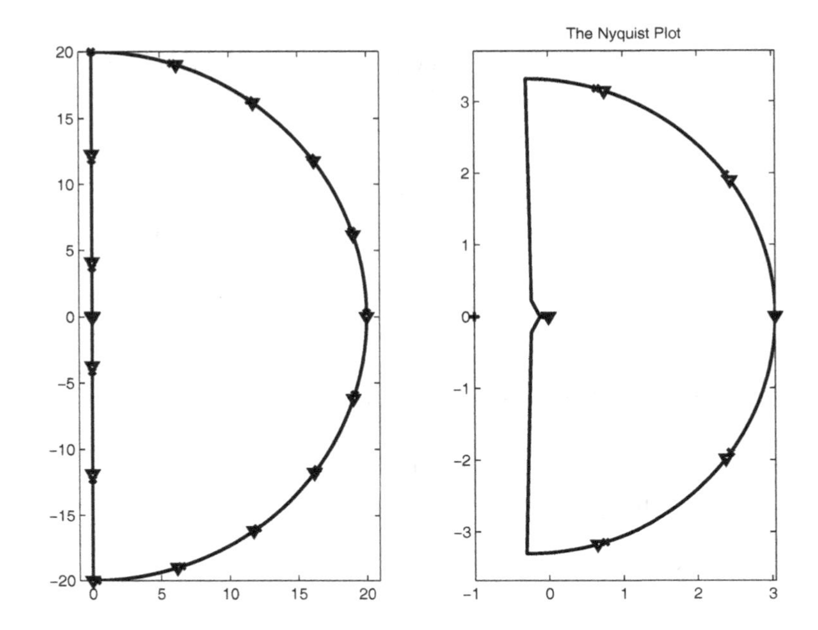

Fig. 5.22 The Region $\Omega_{\rho,R}$ Where $\rho = 0.05$ and $R = 20$ and the Nyquist Plot

As -1 is not encircled by the plot and as $P = 0$, the closed-loop system is stable. Clearly, the gain margin of the closed-loop system is $20\log_{10}(10) = 20\,\text{dB}$. (This result can, of course, be checked by using the Routh array to determine the largest value of K for which a unity feedback system with the same $G_p(s)$ and with $G_c(s) = K$ is stable.)

Considering the form of $G_p(s)$, it should be reasonably easy to estimate the system's phase margin as well. For small $s = j\omega$, $|G_p(j\omega)| \approx 1/(6|\omega|)$. As $s = j/6$ makes this estimate equal one, we expect the frequency $\omega = 1/6\,\text{rad/s}$ to be near the frequency that determines the phase margin. A first estimate of the phase margin would then be 90°. One can improve this estimate by plugging the estimated frequency, $\omega = 1/6\,\text{rad/s}$, into (5.2). This calculation leads to an estimated phase of -107° and an estimated phase margin of 73°. The plots generated by the MATLAB `margin` command are given in Figure 5.24. The gain margin given is precisely what we calculated, and the phase margin is quite close to our estimate.

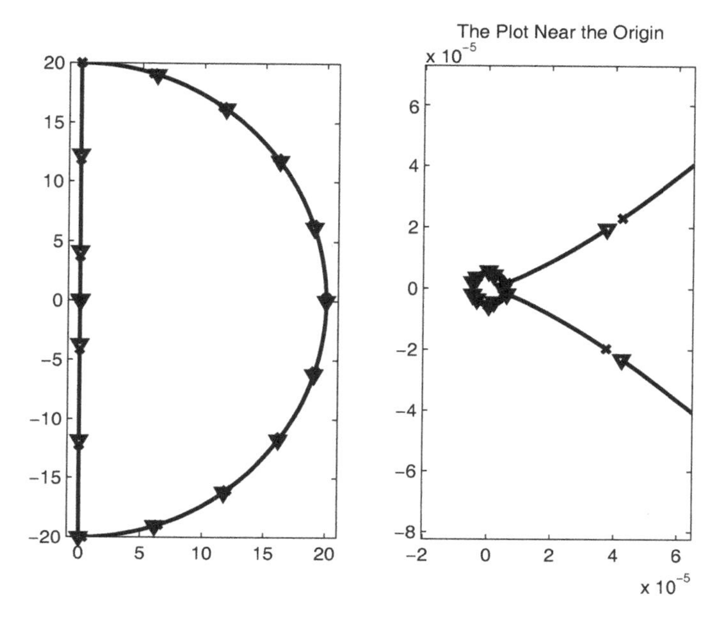

Fig. 5.23 A zoom of the region near the origin. Note that the plot leaves the origin via the first quadrant. (The two very small circles very near the origin are the image of the very large semi-circle that is a border of the region of interest.)

For values of ω that are reasonably small, it is not hard to show that our estimate for the absolute value of $G_p(j\omega)$ is quite good. Thus, we *know* that $\omega \approx 1/6\,\text{rad/s}$ is a *good* estimate of the frequency that determines the phase margin—of the gain crossover frequency.

5.14 Exercises

(1) Using the principle of the argument, determine the number of poles and zeros of the following functions in the region $\{z \,\big|\, |z| < 1\}$. (Consider using MATLAB to help plot the functions.)

(a) $f(z) = \frac{1}{z^2}$
(b) $f(z) = \frac{1}{z(z-2)}$
(c) $f(z) = \frac{z}{(z^2-4)}$

(2) Use the Nyquist plot to show that the system of Figure 5.25 is stable for all $K > 0$.

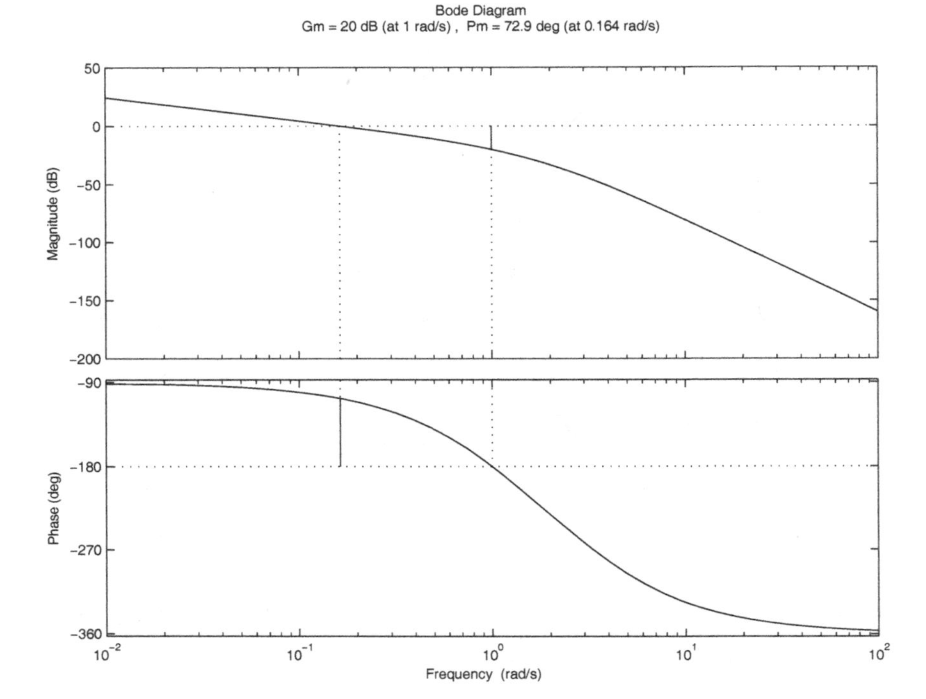

Fig. 5.24 The Bode Plots Pertaining to $G_p(s) = 1/(s(s+1)(s+2)(s+3))$

(3) Using the Nyquist plot, show that the system of Figure 5.25 is unstable for all $K > 0$ when:

$$G_p(s) = \frac{K}{s^2(s+1)}.$$

(4) Using the Nyquist plot, show that the system of Figure 5.25 is stable for all $K > 0$ when:

$$G_p(s) = \frac{K}{s(s+2)}.$$

(This corresponds to using proportional feedback on a DC motor. See §7.6.)

(5) Show that for $K > 0$ the system of Figure 5.26 has poles in the right half-plane for all $T > 0$. Note that when $T = 0$ the system has two poles on the imaginary axis—the system is on the (wrong side of

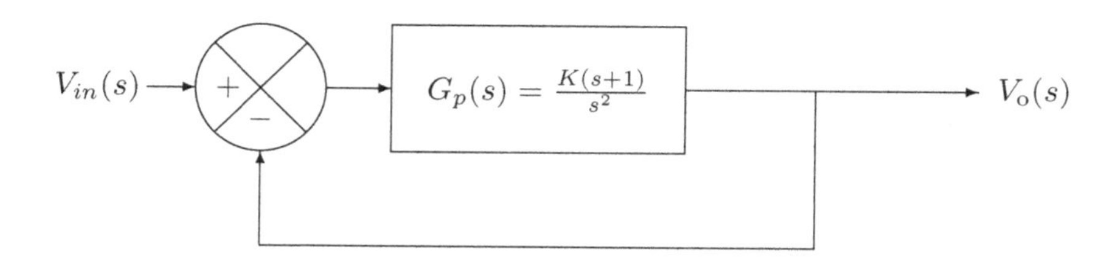

Fig. 5.25 Control of a Simple Satellite Revisited

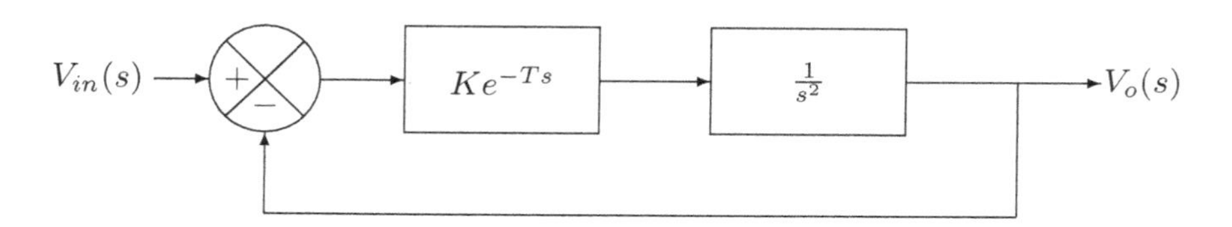

Fig. 5.26 A Controller for a Simple Satellite

the) border of stability; any delay causes the system to become wildly unstable.

(6) (a) Find the Bode plots that correspond to the system of Figure 5.25 when $K = 1$.
(b) What are the gain and phase margins of the system?
(c) What do these margins indicate about the stability of the system?
(d) Would it be wise to use this system in practice? Explain.

(7) (a) Find the Bode plots that correspond to the system of Figure 5.26 when $K = 1, T = 0.01$.
(b) What are the gain and phase margins of the system?
(c) What do these margins indicate about the stability of the system?
(d) Would it be wise to use this system in practice? Explain.

(8) Consider Figure 5.27.

(a) Using the Nyquist plot and some simple calculations, find the exact value of T for which the system becomes unstable.
(b) Now use the approximation:

$$e^{-Ts} \approx \frac{1 - (T/2)s}{1 + (T/2)s}$$

and the Routh-Hurwitz technique to approximate the range of T

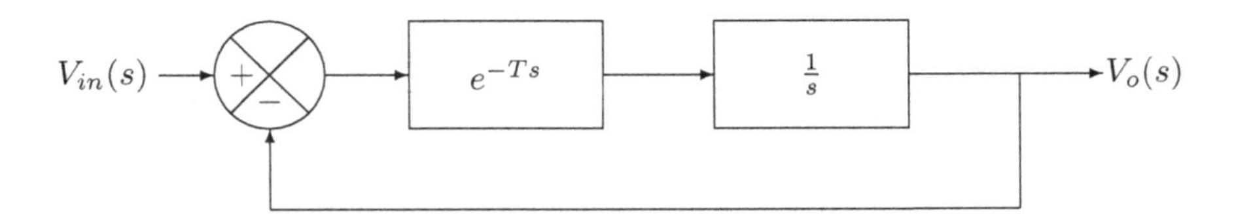

Fig. 5.27 A Simple Integrator with Delay

for which the system is stable. Is the value one finds reasonable in light of the solution to the first part of this question?

(9) One of the reasons that the Bode plots are useful is that the information required to generate them can often be measured with relative ease. In order to measure the frequency response of $G_p(s)H(s)$ one "opens the loop" of one's system as done in Figure 5.6, inputs sine waves at a variety of frequencies, and measures their amplitude and phase at $\hat{v}_o(t)$. Assuming that the system is low-pass (as most physical systems are) one need not worry about the semi-circle at infinity. However, the system may have a pole or poles at zero. Explain how one can use the Bode plots to determine the number of poles at zero. Note that:

$$20\log_{10}\left(\frac{K}{\omega^n}\right) = -20n\log_{10}(\omega) + 20\log_{10}(K).$$

(10) Consider the example of p. 126 again. Calculate the exact value of T for which the system becomes unstable by:

(a) Finding the value of ω for which $|G_p(j\omega)| = 1$.
(b) Finding the phase of $G_p(j\omega)$ at that frequency.
(c) Finding the value of T that makes the phase of the system exactly $-180°$.

(11) Consider Figure 5.27 once again.

(a) Use the approximation:

$$e^{-Ts} \approx 1 - Ts$$

to determine the range of T for which the system is stable.

(b) Use the approximation:

$$e^{-Ts} \approx 1 - Ts + (T^2/2)s^2$$

to determine the range of T for which the system is stable.

(c) Use the approximation:

$$e^{-Ts} \approx 1 - Ts + (T^2/2)s^2 - (T^3/6)s^3$$

to determine the range of T for which the system is stable.

(d) Use the Nyquist plot to examine the results of the previous part. (Consider particularly the image of the *large* semi-circle.)

(12) Consider a system for which:

$$G_p(s)H(s) = \frac{40(s+1)}{s(s+2)(s+20)}.$$

(a) Find the Bode plots that correspond to the system.
(b) What are the gain and phase margins of the system?
(c) What do the margins imply about the stability of the system?

(13) Consider a unity feedback system for which:

$$G_c(s)G_p(s) = \frac{s+1}{s(s+3)(s+5)}.$$

(a) Find the system's Nyquist plot. Pay particular attention to the asymptotes.
(b) Find the Bode plots that correspond to the system.
(c) What are the gain and phase margins of the system.
(d) What do the margins imply about the stability of the system?

(14) Consider a unity feedback system for which:

$$G_c(s) = K > 0, \qquad G_p(s) = \frac{1}{s(s+1)(s-1)}.$$

(a) Find the system's Nyquist plot.
(b) For which values of K is the system stable?
(c) When the system is unstable, how many closed-loop poles does the system have in the right half-plane?
(d) Use the Routh array to check your answer to exercise 14c.

(15) Consider a unity feedback system for which $G_c(s) = 1$ and:

$$G_p(s) = \frac{1}{s(s+1)^6}.$$

(a) Sketch the system's Nyquist plot. (MATLAB or another program may be used to find the roots of high-order polynomials.)
(b) Is the closed-loop system stable?

(c) Let $G_c(s) = K$. For which values of K is the closed-loop system stable?

(16) Consider a unity feedback system for which $G_c(s) = 1$ and:

$$G_p(s) = \frac{s+a}{s(s+1)}, \qquad a > 0.$$

For which values of a does the Nyquist plot have asymptotes in the right half-plane? Explain.

(17) Consider a unity feedback system for which $G_c(s) = 1$ and:

$$G_p(s) = \frac{s^2+a^2}{s(s+1)^2}, \qquad a > 0.$$

(a) Plot the Nyquist plot of the system when $a = 0.6$. What does the plot tell you about the stability of the closed-loop system?
(b) Plot the Nyquist plot of the system when $a = 1$. What does the plot tell you about the stability of the closed-loop system?
(c) Plot the Nyquist plot of the system when $a = \sqrt{2}$. What does the plot tell you about the stability of the closed-loop system?
(d) Use the Routh array to determine the gain margin of the system as a function of a. Explain how the results of this section serve as a partial confirmation of the results of the previous three sections.

(18) Consider a unity feedback system for which $G_c(s) = 1$ and:

$$G_p(s) = \frac{s+4}{s(s+1)(s+2)}.$$

(a) Use the MATLAB `bode` command to plot the Bode plots of $G_p(s)$.
(b) By considering the Bode plots, determine the phase and gain margins of the system.
(c) Use the MATLAB `margin` command to check the answers to the preceding section.
(d) Is the closed-loop system sufficiently stable that a responsible engineer (and there should be no other type!) would a allow the system to be built and used? Explain.

(19) Consider a unity feedback system for which $G_c(s) = 1$ and:

$$G_p(s) = \frac{s}{(s+1)(s+2)(s+3)}.$$

(a) Plot the system's Nyquist plot, and use the plot to determine the system's gain margin.

(b) Use the Routh array to check the answer to the previous section.

(20) Consider a unity feedback system for which:

$$G_c(s) = K > 0, \qquad G_p(s) = \frac{s-1}{s+1}.$$

(a) Plot the system's Nyquist plot.
(b) Use the Nyquist plot to show that the system is stable for small gains and unstable for large gains.
(c) Determine the range of K for which the system is stable.
(d) Use the Routh array to check the answer to the preceding section.

(21) Consider a unity feedback system for which:

$$G_c(s) = 1, \qquad G_p(s) = \frac{1}{s(s+2)(s+4)}.$$

(a) Plot the system's Nyquist plot.
(b) Use the Nyquist plot to calculate the system's gain margin.
(c) Use the Routh array to check the answer to the preceding section.
(d) Use the Nyquist plot to *estimate* the system's phase margin.
(e) Is this system sufficiently stable to use? Explain.

(22) In this exercise, we consider three unity feedback systems for which $G_c(s) = 1$.

(a) Let:

$$G_p(s) = \frac{1}{s(s+3)(s+5)}.$$

i. Plot the system's Nyquist plot.
ii. Use the Nyquist plot to determine whether or not the system is stable.

(b) Let:

$$G_p(s) = \frac{1}{2}\frac{s+1}{s(s-1)}.$$

i. Plot the system's Nyquist plot.
ii. Use the Nyquist plot to determine whether or not the system is stable.

(c) Let:

$$G_p(s) = 2\frac{s+1}{s(s-1)}.$$

i. Plot the system's Nyquist plot.
ii. Use the Nyquist plot to determine whether or not the system is stable.

Be particularly careful when answering parts 22b and 22c; they are *interesting* problems.

(23) In the following two sections, use the Nyquist plot to determine whether or not unity feedback systems for which $G_c(s) = 1$ and with the given values of $G_p(s)$ are stable. In each section, provide a second proof that your answer is correct.

(a)
$$G_p(s) = \frac{s}{(s+1)(s-1)}.$$

(b)
$$G_p(s) = \frac{s+2}{(s+1)(s-1)}.$$

(24) Consider a unity feedback system for which:
$$G_c(s) = 1, \qquad G_p(s) = \frac{10}{s(s+1)(s+2)}.$$

(a) Plot the Bode plots corresponding to $G_p(s)$. (You may use the MATLAB `bode` command to produce the plots.)
(b) Use the Bode plots to calculate the gain and phase margins of the closed-loop system.
(c) Use the Nyquist plot to check that gain margin you measured is correct.
(d) Would a responsible engineer use a system with these gain and phase margins? Explain.

Chapter 6

The Root Locus Diagram

6.1 The Root Locus—An Introduction

In this chapter we consider a third method of analyzing a system with feedback. We examine systems of the form shown in Figure 6.1. We consider the blocks $G_p(s)$ and $H(s)$ to be given and the gain, K, to be variable. Our goal is to plot the position of the roots of the denominator of the transfer function of the system:

$$T(s) = \frac{KG_p(s)}{1 + KG_p(s)H(s)}$$

with K as an implicit parameter. This is the plot of the system's poles and is called the root locus diagram[1] of the system. It gives us an idea of the effect of gain on the system's stability and performance. Often the gain, K, is not present in the physical system being analyzed. The virtual gain block is added to the system in order to understand the effect that additional gain would have on the system's stability and performance.

The System $G_p(s) = 1/s^2, H(s) = 1$—An Example

As we have seen $G_p(s) = 1/s^2$ is the transfer function of a simple satellite. We now analyze a system to control the angle of the satellite. We find that the transfer function of the system is:

$$T(s) = \frac{K/s^2}{1 + K/s^2}.$$

We need to determine all of the roots of $1 + K/s^2$. The solutions

[1] This technique was invented by Walter R. Evans (1920-1999) in 1948 while he was at Autonetics, a division of North American Aviation (now Rockwell International) [Ano00].

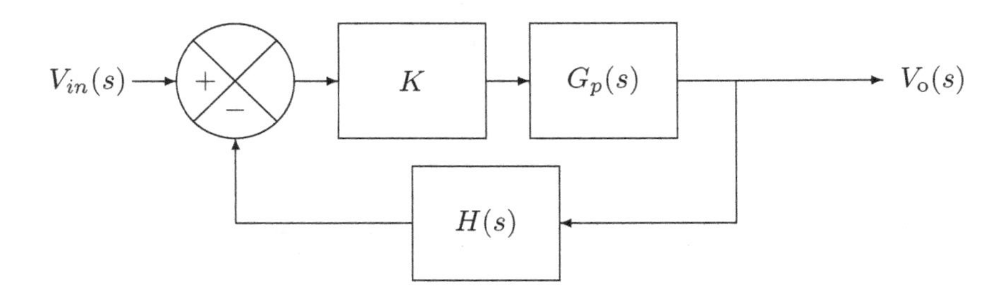

Fig. 6.1 The Typical System Analyzed by Means of Root Locus Techniques

are $s = \pm j\sqrt{K}$. Thus, when $K = 0$ the roots are both at the origin, and as K increases the two roots "climb" the imaginary axis—one "branch" heading towards $+j\infty$ and the other towards $-j\infty$. We see that this control strategy leads to a system that is unstable (marginally stable) at any gain. Figure 6.2 shows the poles "climbing" the imaginary axis.

Figure 6.2 was drawn using the MATLAB `rlocus` command. To cause MATLAB to produce a root-locus plot, one defines $G_p(s)H(s)$; then one uses the `rlocus` command. In this example, the plot was generated by the sequence of commands (with the responses produced by MATLAB):

```
Gp = tf([1],[1 0 0])

Transfer function:
 1
---
s^2

rlocus(Gp)
```

to which MATLAB responds by generating Figure 6.2. The command `rlocus(Gp)` generates a plot of the solutions of $1 + KG_p(s)$ for $K \geq 0$ with K acting as a (implicit) parameter. Poles of $G_p(s)H(s)$ are marked with x's and zeros are marked with o's. This is a general convention in root locus diagrams.

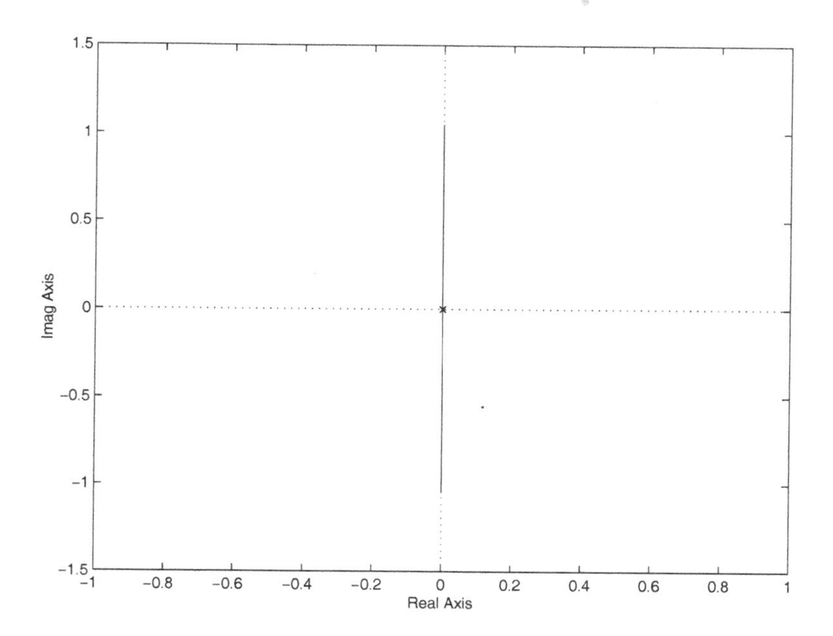

Fig. 6.2 The Root Locus Diagram for $G_p(s) = 1/s^2, H(s) = 1$

6.2 Rules for Plotting the Root Locus

We now develop a set of rules to be used when plotting the root locus diagram that corresponds to $G_p(s)H(s)$. We assume that $G_p(s)H(s)$ is a rational function of s with real coefficients—that $G_p(s)H(s) = P(s)/Q(s)$ where $P(s)$ and $Q(s)$ are polynomials with real coefficients. We are looking for all of the solutions of the equation:

$$1 + KG_p(s)H(s) = 1 + KP(s)/Q(s) = 0.$$

This is equivalent to looking for all of the solutions of the equation:

$$Q(s) + KP(s) = 0.$$

This is a *polynomial* with *real* coefficients.

6.2.1 *The Symmetry of the Root Locus*

The first rule is:

Rule 1 The root locus diagram is symmetric about the real axis.

This follows from the fact that the roots of a polynomial with real coefficients are always either real—leading to the points on the root locus being on the real axis—or they appear in complex-conjugate pairs—leading to branches symmetric with respect to the real axis. Figure 6.2 shows this symmetry nicely.

6.2.2 *Branches on the Real Axis*

Next we consider the question, "Which points on the real axis solve the equation $1 + KP(s)/Q(s) = 0$ for some value of K?" We make an assumption about the form of $P(s)/Q(s)$—we assume that $P(s) = s^n + \cdots$ and $Q(s) = B(s^m + \cdots)$ where $B > 0$. All rational functions with real coefficients can, by a simple division, be reduced to this form save that some will have $B < 0$. We exclude such polynomials from our consideration; they do not occur too frequently in practice. (When the need arises, it is simple to recast the rules we develop to handle the case $B < 0$.) We find that:

Rule 2 Points on the real axis to the left of an odd number of real zeros and poles of $G_p(s)H(s)$ are on the root locus. Points to the left of an even number of real zeros and poles of $G_p(s)H(s)$ are not on the root locus.

> PROOF: Polynomials of the form $R(s) = s^n + a_1 s^{n-1} + \cdots + a_n$ can be written as $R(s) = \Pi_{i=0}^{n}(s - r_i)$. If the polynomial has real coefficients, then all of the r_i are either real or they appear along with their complex conjugates. Let us start by assuming that $P(s) = \Pi_{i=0}^{n}(s - (r_P)_i)$, $Q(s) = B\Pi_{i=0}^{m}(s - (r_Q)_i)$, that all of the roots are real, and that B is positive. It is clear that for large positive values of s both $P(s)$ and $Q(s)$ are positive. As $K > 0$, it is clear that for s large and positive $1 + KP(s)/Q(s)$ cannot be zero—all the terms are positive. Suppose now that s is smaller than the largest of the roots of $Q(s)$ or $P(s)$ but is larger than all of the other roots. Then one of the factors of either $P(s)$ or $Q(s)$ is negative, but all of the other factors are positive. For such a value of s we find $P(s)/Q(s)$ is negative and, consequently, that $1+KP(s)/Q(s)$ must equal zero for some positive value of K. After crossing another one of the roots of $P(s)$ or $Q(s)$ there will be two negative factors in $P(s)/Q(s)$—thus $P(s)/Q(s) > 0$ and once again there will be no positive K

for which $1 + KP(s)/Q(s) = 0$. Such values of s do not belong to the root locus. We see that if all of the roots of $P(s)$ and $Q(s)$ are real then Rule 2 holds.

One can remove the condition that the roots must all be real. Consider a factor of the form $(s - (r_P)_i)$ where $(r_P)_i = c$ is complex. Then we know that there exists a j such that $(r_P)_j = \overline{(r_P)_i} = \overline{c}$. Now consider the product of the factors that correspond to these two roots. We find that when s is real the product of the two factors is:

$$(s - c)(s - (\overline{c})) \stackrel{s \in \mathcal{R}}{=} (s - c)(\overline{s - c}) = |s - c|^2 > 0.$$

Thus such factors cannot contribute a minus sign for points s on the real axis—they are totally irrelevant from our perspective.

Let us now consider some simple examples.

$G_p(s) = \frac{1}{1+s^2}, H(s) = 1$—An Example

In this example there are two complex poles and no real poles or zeros. Thus there should not be any points of the root locus on the real axis. MATLAB generates the root locus diagram of Figure 6.3 for this example. We see that this bears out rule 2.

$G_p(s) = 1/(s(s + 1)), H(s) = 1$—An Example

In this example there are two poles—$s = 0, -1$. When s is to the left of one of them—when s is between 0 and -1—s should on the root locus. Using MATLAB we find that the root locus is as shown in Figure 6.4. This too bears out rule 2. (Note that rule 2 gives no information about complex values of s.)

6.2.3 *The Asymptotic Behavior of the Branches*

Next we consider the question "what do the roots of $Q(s)+KP(s) = 0$ look like for small values of K?" We assume that $G_p(s)H(s)$ is low-pass[2]—that $\deg(Q(s)) \geq \deg(P(s))$. As $K \to 0$ we find that the equation approaches $Q(s) = 0$. As K approaches 0, we see that the points on the root locus approach the roots of $Q(s) = 0$—the poles of $G_p(s)H(s)$.

[2]Note that this definition differs slightly from the definition of low-pass that was given when we were considering Nyquist plots. There we required that the degree of the denominator be greater than that of the numerator—here we only require that it be greater than or equal to the degree of the numerator.

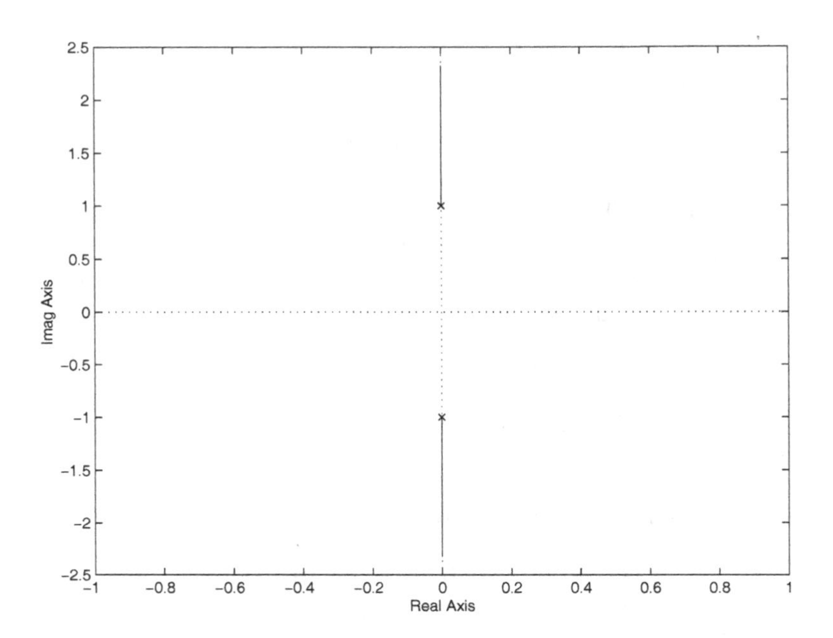

Fig. 6.3 The Root Locus Diagram for $G_p(s) = 1/(1+s^2), H(s) = 1$

The trickier question is "what happens as $K \to \infty$?" First, let us suppose that $\deg(P(s)) = \deg(Q(s))$. Then as $K \to \infty$ we find that:

$$Q(s) + KP(s) = 0 \Leftrightarrow P(s) + Q(s)/K = 0 \Rightarrow P(s) = 0.$$

That is we find that the solutions of the equation tend to the roots of $P(s)$—the zeros of $G_p(s)H(s)$. In fact, we can now state:

Rule 3 Each "branch" of the root locus diagram starts at a pole $G_p(s)H(s)$ and ends on a zero of $G_p(s)H(s)$.

What happens when there are more poles of $G_p(s)H(s)$ than there are zeros? Consider the polynomial whose roots we would like to find:

$$Q(s)/K + P(s) = B(s^m + \cdots)/K + (s^n + \cdots) = 0.$$

This is an mth order polynomial and, therefore, has m roots. In Section 6.4 we prove that n of the roots indeed converge to the roots of $P(s)$—as they should. The other $m - n$ roots all tend to infinity. We note that one can reasonably say that a function whose denominator is of higher degree than its numerator has zeros at infinity—after all as $|s| \to \infty$ the function tends

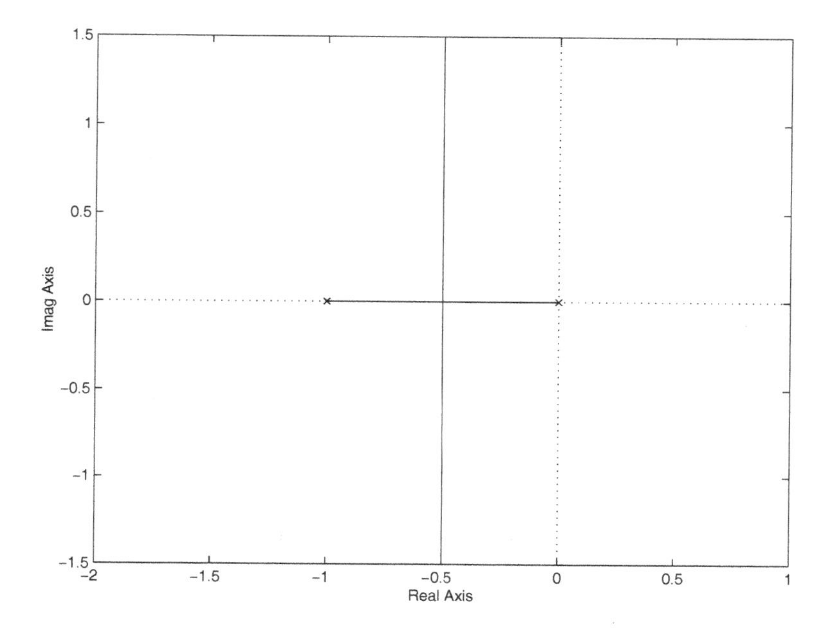

Fig. 6.4 The Root Locus Diagram for $G_p(s) = 1/(s(s+1))$, $H(s) = 1$

to zero. *Defining* the number of zeros at infinity as the difference between the degree of the denominator and the numerator, we see that with this definition rule 3 always applies.

We consider *how* the $m - n$ roots that tend to infinity actually progress towards infinity as $K \to \infty$. In Section 6.4 we show that the $m - n$ roots are approximately $s = -(K/B)^{\frac{1}{m-n}} e^{\frac{2\pi ji}{m-n}}, i = 1, ..., m-n$. This is fairly easy to see (though a bit of technical work is required for a proof). We know that any m^{th} order equation—no matter how small the coefficient of s^m—must have m roots. As the coefficient of s^m gets smaller the effect of the term will be felt only for very large values of $|s|$. For large values of $|s|$, the two important terms in the equation $Q(s)/K + P(s) = 0$ are Bs^m/K and s^n. Thus the equation that we need to solve is (for large s) approximately $Bs^m/K + s^n = 0$. The first n solutions of this equation are zero—they correspond to the n solutions that stay finite—the n solutions of $P(s) = 0$. The other $m - n$ solutions are the ones listed above. In fact, we can prove more (and we do in Section 6.4). We can show that the $m - n$

roots tend towards:

$$s = (K/B)^{\frac{1}{m-n}} e^{\frac{j\pi + 2j\pi i}{m-n}} + \frac{\sum \text{finite poles} - \sum \text{finite zeros}}{m-n}, i = 1, ..., m-n.$$

These are rays that go off to infinity and start at the point:

$$\frac{\sum \text{finite poles} - \sum \text{finite zeros}}{m-n}.$$

Summing up we find that:

Rule 4 If the degree of the denominator exceeds the degree of the numerator by d, then d branches of the root locus tend to infinity. Asymptotically these branches are straight lines, their angles are $\frac{180^\circ + 360^\circ i}{d}$, $i = 0, ..., d-1$, and they have as their common x-intercept:

$$\frac{\sum \text{finite poles} - \sum \text{finite zeros}}{m-n}.$$

Figures 6.2 and 6.4 Revisited—An Example

Figure 6.2 corresponds to a system with two poles at zero and no finite zeros. According to rule 2 there should be no branches of the root locus on the real axis, and this is indeed the case. According to rule 3 there should be two branches that tend to infinity—and there are. According to rule 4 the branches should head off with angles $180^\circ/2 = 90^\circ$ and $540^\circ/2 = 270^\circ$—and they do. Moreover the asymptotes should have as their common point of origin $(0-0)/2 = 0$ and this is indeed the case.

In Figure 6.4 we consider a system with no finite zeros and whose poles are 0 and -1. According to rule 3 there should be two branches heading off to zeros at infinity—and this is indeed so. Also, as before, the angles that they head off at should be 90° and 270°—and they are. Finally the common point of origin should be $((0+-1)-0)/2 = -1/2$, and this is indeed the case.

6.2.4 *Departure of Branches from the Real Axis*

Rules 2 and 3 have a very nice corollary. Suppose that one has a pole in position $2n+1$ and another pole in position $2n+2$. From rule 2 one knows that all of the point between the first pole and the second pole must be part of the root locus. From rule 3 one knows that branches of the root locus start on poles and must end on zeros—either finite zeros or zeros at infinity. Combining these two rules, we find that branches must begin on each of the two poles and *cannot* end on either of them. This leaves us with only one possibility. The region between the two poles must consist of at least two branches and the root locus must leave the real axis at some point or points between the two poles. Similarly if there are zeros at positions $2n + 1$ and $2n + 2$ then at some point (or points) between the two zeros other branches enter the real axis. We find that:

Rule 5 If one has two poles (or two zeros) at locations $2n+1$ and $2n+2$, then the root locus leaves (or enters) the real axis at some point (or points) between the two poles (or zeros).

Rule 5 states that under certain conditions branches of the root locus leave or enter the real axis. We would like to answer the question "where do the branches leave the real axis?" Define the functions $s_j(K)$ as the location of the points on the jth branch of the root locus as a function of K. We know that while the jth branch remains on the real axis:

$$\frac{d}{dK}s_j(K) \in \mathcal{R}$$

for if $s_j(K)$ is real, its derivative must be real as well. To calculate the derivative of $s_j(K)$, we differentiate the equation:

$$Q(s_j(K)) + KP(s_j(K)) = 0$$

with respect to K. Differentiating implicitly we find that:

$$\frac{d}{ds}Q(s_j(K))\frac{ds_j(K)}{dK} + P(s_j(K)) + K\frac{d}{ds}P(s_j(K))\frac{ds_j(K)}{dK} = 0.$$

We see that:

$$\frac{ds_j(K)}{dK} = -\frac{P(s_j(K))}{Q'(s_j(K)) + KP'(s_j(K))}$$

where the $'$ denotes differentiation with respect to s. We assume that $P(s)$ and $Q(s)$ are polynomials with no common factors. This implies that the two polynomials do not share any zeros. Thus, if we know the s is a point at which $Q(s) = -KP(s)$, then we know that at that point *neither* of the polynomials can have a zero. Thus, for s on the root locus we have:

$$\frac{d}{ds}(P(s)/Q(s)) = \frac{P'Q - Q'P}{Q^2} \overset{s \text{ on the root locus}}{=} \frac{-KPP' - PQ'}{K^2P^2} = -\frac{KP' + Q'}{K^2P}.$$

We find that for points s on the root locus we can write:

$$\frac{d}{dK}s_j(K) = \left.\frac{1}{K^2(P(s)/Q(s))'}\right|_{s=s_j(K)}. \tag{6.1}$$

We would like to answer the question, "When does $s_j(K)$ leave (or enter) the real axis?" We need to find a point on the real axis at which $\frac{d}{dK}s_j(K)$ is not real. Looking at (6.1) we find that on the real axis *that cannot happen.* As all our polynomials have real coefficients, when evaluated at real values of s the derivative cannot be complex. How can branches of the root locus leave (or enter) the real axis?

Branches of the root locus can leave (or enter) the real axis in one way—if the derivative in (6.1) fails to exist at some point. If the derivative fails to exist at some point on the root locus diagram, s_*, then $1+KP(s_*)/Q(s_*) = 0$ and

$$\frac{d}{ds}(1 + KP(s)/Q(s)|_{s=s_*} = K\frac{d}{ds}(P(s)/Q(s)|_{s=s_*} = 0.$$

That is, $1 + KP(s)/Q(s)$ has a second- or higher-order zero at s_*, and at least two branches of the root locus coalesce at this point. As the zeros of the $(P(s)/Q(s))'$ are isolated, so are the points at which several branches coalesce. Near s_*, branches must be leaving (or entering) the root locus. We find that:

Rule 6 Branches of the root locus leave or enter the real axis at all points on the root locus that satisfy the condition:

$$\frac{d}{ds}(P(s)/Q(s)) = \frac{d}{ds}(G_p(s)H(s)) = 0.$$

$G_p(s)H(s) = 1/(s(s+1))$ Revisited—An Example

We can now determine where the root locus corresponding to this system leaves the real axis. Let us differentiate $G_p(s)H(s)$ with respect to s. We find that:

$$\frac{d}{ds}\frac{1}{s(s+1)} = \frac{-2s-1}{s^2(s+1)^2}.$$

This expression equals zero precisely when $s = -1/2$. That is where the root locus leaves the real axis.

$G_p(s)H(s) = \frac{1+10s}{1+70s}\frac{10}{s(s+1)}$—An Example

This system has three poles—$0, -1/70$, and -1—and one zero—$-1/10$. From rule 2 we find that branches of the root

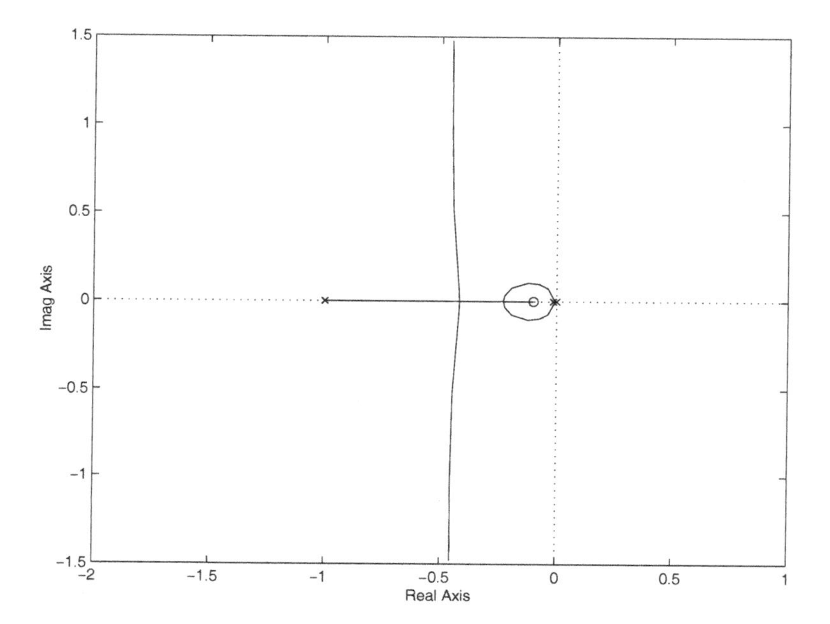

Fig. 6.5 The Root Locus Diagram for $G_p(s)H(s) = 10(10s + 1)/(s(s + 1)(70s + 1))$

locus are located between 0 and $-1/70$ and between $-1/10$ and -1. From rule 5 we find that the root locus must leave the axis at some point between 0 and $-1/70$. Let us use rule 6 to determine where. We find that:

$$(G_p(s)H(s))' = \frac{-14000s^3 - 9200s^2 - 1420s - 10}{((1 + 70s)(s(s + 1)))^2}.$$

We find (numerically) that the roots of this function—the roots of the numerator—are -0.4193, -0.2304, and -0.0074. The last root is between 0 and $-1/70$. The other two roots, however, are between $-1/10$ and -1. We find that the branches that leave at -0.0074 reenter at -0.2304 and leave again at -0.4193. Making use of rule 4 we find that there are two asymptotes that leave at $\pm 90^{\circ}$ and that have as their common x-intercept:

$$x\text{-intercept} = \frac{0 - 1/70 - 1 - (-1/10)}{2} = -0.4571.$$

This root locus diagram is shown in Figure 6.5. We see that

though branches of the root locus *must* enter or leave the real axis between two poles or zero that appear in the positions $2n$ and $2n+1$, they *may* enter or leave the real axis under other conditions as well.

Consider the behavior of $ds_j(K)/dK$ just after the point $s_j(\hat{K})$ at which the branches of the root locus leave the real axis. Assuming that two branches of the root locus coalesce when $K = \hat{K}$, we find that $s_j(\hat{K})$ must be a double zero of $Q(s) + \hat{K}P(s) = 0$. That is, we can write:

$$\begin{aligned} Q(s) + KP(s) &= Q(s) + \hat{K}P(s) + (K - \hat{K})P(s) \\ &= (s - s_j(\hat{K}))^2 R(s) + (K - \hat{K})P(s), \end{aligned}$$

where $R(s_j(\hat{K})) \neq 0$. We find that $s_j(K)$ must solve:

$$(s_j(K) - s_j(\hat{K}))^2 R(s_j(K)) + (K - \hat{K})P(s_j(K)) = 0.$$

Let $\Delta s_j \equiv s_j(K) - s_j(\hat{K})$. Differentiating implicitly with respect to K and solving for:

$$\frac{d}{dK} s_j(K),$$

we find that:

$$\frac{d}{dK} s_j(K) = -\frac{P(s_j(K))}{2\Delta s_j R(s_j(K)) + (\Delta s_j)^2 R'(s_j(K)) + (K - \hat{K})P'(s_j(K))}. \tag{6.2}$$

We have already seen that $|ds_j(K)/dK| \to \infty$ as $K \to \hat{K}$. Clearly $\Delta s_j \to 0$ as well. Thus, as K tends towards $\hat{K}$ it is clear that $\Delta s_j/(K - \hat{K}) \to \infty$. We see that in the denominator in (6.2) the leftmost term will be much larger than the other two terms for values of K near $\hat{K}$. Thus we see that:

$$\frac{d}{dK} s_j(K) \approx -\frac{P(s_j(K))}{2\Delta s_j R(s_j(K))}.$$

For K near $\hat{K}$ it is clear that $s_j(K)$ will be very nearly real. Thus, the angle of the derivative will be fixed by the angle of:

$$\frac{1}{\Delta s_j}.$$

For two nearby points both Δs_j and $ds_j(K)/dK$ point in the same direction. That is, there exists $\beta > 0$ such that:

$$\Delta s_j \approx \beta \frac{d}{dK} s_j(K) \approx -\beta \frac{P(s_j(K))}{2\Delta s_j R(s_j(K))} \approx \gamma \frac{1}{\Delta s_j}, \gamma \in \mathcal{R}.$$

This approximation is increasingly good as $K \to \hat{K}$. From this approximation we find that $(s_j(K) - s_j(\hat{K}))^2 \approx \gamma \in \mathcal{R}$. This happens if $s_j(K) - s_j(\hat{K}) \approx \sqrt{\gamma}$. As γ is real, this square root is either pure real or pure imaginary. It cannot be pure real—we know that the jth branch has left the real axis. Thus, the change in the position of the root locus must be imaginary. This leads us to:

Rule 7 When *two* branches of the root locus coalesce and then (enter or) leave the real axis, they (enter or) leave at right angles to the real axis.

6.2.5 *A "Conservation Law"*

We have already seen that the poles of our system are the roots of the equation:

$$Q(s) + KP(s) = 0.$$

We also know (see item 1 on p. 89) that for any polynomial of the form $s^n + a_1 s^{n-1} + \cdots$ the coefficient a_1 is equal to:

$$a_1 = -r_1 - \cdots - r_n.$$

where the r_i are the roots of the polynomial. If the degree of $Q(s)$ is greater than the degree of $P(s)$ by two or more, then the coefficient of s^{n-1} does not depend on $KP(s)$ at all, and, therefore, the sum of the roots of the polynomial is fixed. We see that:

Rule 8 If the degree of $Q(s)$ is greater than the degree of $P(s)$ by two or more, then the sum of the poles of the system is constant.

Thus, the sum of the poles for any value of K is equal to the sum of the poles of $G_p(s)H(s)$—which are the roots of the equation when $K = 0$. This rule forces a certain "balance" on the root locus when the degree of the denominator of $G_p(s)H(s)$ is greater than the degree of the numerator by two or more.

$G_p(s)H(s) = 1/(s(s+1))$ Revisited—An Example

We have seen previously that the root locus diagram that corresponds to this function has two branches and that the branches must head off to $\pm j\infty$ as $K \to \infty$. We now use Rule 1 and Rule 8 to show that when they leave the real axis their real part is always -0.5. From rule 1, we see that from the point at which the branches leave the real axis the points on the root locus are a and $\overline{a}$. From rule 8 we know that the sum of the poles must equal the sum of the poles when $K = 0$—which is just $0 + (-1) = -1$. Thus, we find that $a + \overline{a} = 2\mathcal{R}(a) = -1$. Thus the real part of any point on the root locus that is off the real axis is -0.5. This is borne out by Figure 6.4.

6.2.6 *The Behavior of Branches as They Leave Finite Poles or Enter Finite Zeros*

We have seen how the branches of the root locus behave as they tend to infinity in those cases where they tend to infinity. We would now like to know how the branches behave as they tend to finite zeros or poles. We know that branches depart from finite poles and end on finite zeros. We would like to know the angle of departure and the angle of entry. Once again, we start by assuming that $P(s) = \Pi_{i=0}^{n}(s-(r_P)_i)$, $Q(s) = B\Pi_{i=0}^{m}(s-(r_Q)_i)$, and that B is positive. Suppose that we are near the pole $(r_Q)_I$. Then $s = (r_Q)_I + \xi$ and ξ is relatively small. The expression $s - (r_Q)_i \approx (r_Q)_I - (r_Q)_i$ except when $i = I$. We find that ξ must make the expression:

$$\left.\frac{P(s)}{Q(s)}\right|_{s=(r_Q)_i+\xi} = \left.\frac{\Pi_{i=0}^{n}(s-(r_P)_i)}{B\Pi_{i=0}^{m}(s-(r_Q)_i)}\right|_{s=(r_Q)_i+\xi}$$

$$\approx \frac{\Pi_{i=0}^{n}((r_Q)_I-(r_P)_i)}{B\Pi_{i=0,i\neq I}^{m}(r_Q)_I-(r_Q)_i)}\frac{1}{\xi} \tag{6.3}$$

negative. Consider the "meaning" of this statement. Each term of the form $(r_Q)_I - (r_P)_i$ can be thought of as the vector going from $(r_P)_i$—which is either a pole or a root—to $(r_Q)_I$—which is the point of interest to us. To say that $P(s)/Q(s)$ must be negative is to say that its angle must be $-180^\circ + n360^\circ$. If all that we want is information about angles, then (6.3) can be rewritten as:

$$\angle \frac{P(s)}{Q(s)}\bigg|_{s=(r_Q)_i+\xi} \approx \angle \left\{ \frac{\Pi_{i=0}^{n}((r_Q)_I - (r_P)_i)}{B\Pi_{i=0,i\neq I}^{m}((r_Q)_I - (r_Q)_i)} \frac{1}{\xi} \right\}$$
$$= \sum_{i=0}^{n} \angle((r_Q)_I - (r_P)_i) - \sum_{i=0,i\neq I}^{m} \angle((r_Q)_I - (r_Q)_i) - \angle\xi$$
$$= -180^{\circ} + n360^{\circ}$$

with the approximation tending to equality as $|\xi|$ tends to zero. This leads us to:

Rule 9 The angle of departure from the pole $(r_Q)_I$, $\angle\xi$, satisfies:

$$\angle\xi = \sum_{i=0}^{n} \angle((r_Q)_I - (r_P)_i) - \sum_{i=0,i\neq I}^{m} \angle((r_Q)_I - (r_Q)_i) + 180^{\circ} - n360^{\circ}.$$

Similarly the angle at which the root locus approaches a zero $(r_P)_I$, $\angle\xi$, satisfies:

$$\angle\xi = -\sum_{i=0,i\neq I}^{n} \angle((r_P)_I - (r_P)_i) + \sum_{i=0}^{m} \angle((r_P)_I - (r_Q)_i) - 180^{\circ} + n360^{\circ}.$$

$G_p(s)H(s) = 1/(s(s+1)(s^2+1))$—An Example

In this example we find that there are four more finite poles than finite zeros. According to rule 4 that there are four branches going off to infinity and that they tend to infinity at angles of $45^{\circ}, 135^{\circ}, 225^{\circ}$, and 315°. Thus it is clear that for large K the system is unstable. From rule 2 we know that there is a branch connecting the pole at 0 to the pole at -1. Near $K = 0$ this branch cannot cause trouble. From rule 3, we know that the branches leave from each pole $G_p(s)H(s)$. The question remains what happens at the branches that leave from $\pm j$? Let us use rule 9 to answer this question. We find that the angle at which the branch must leave the pole at $s = j$ is:

$$\angle\xi = -\angle(j-0) - \angle(j-(-1)) - \angle(j-(-j)) + 180^{\circ} - n360^{\circ}$$
$$= -90^{\circ} - 45^{\circ} - 90^{\circ} + 180^{\circ} - 0 \cdot 360^{\circ}$$
$$= -45^{\circ}.$$

According to rule 1—symmetry—the branch departing from $-j$ must leave at 45°. We see that Figures 6.6 and 6.7 bear out these statements. (In Figure 6.7 we see that near $s = j$ the branch is indeed departing at $-45°$.) We see that the system is unstable at any gain.

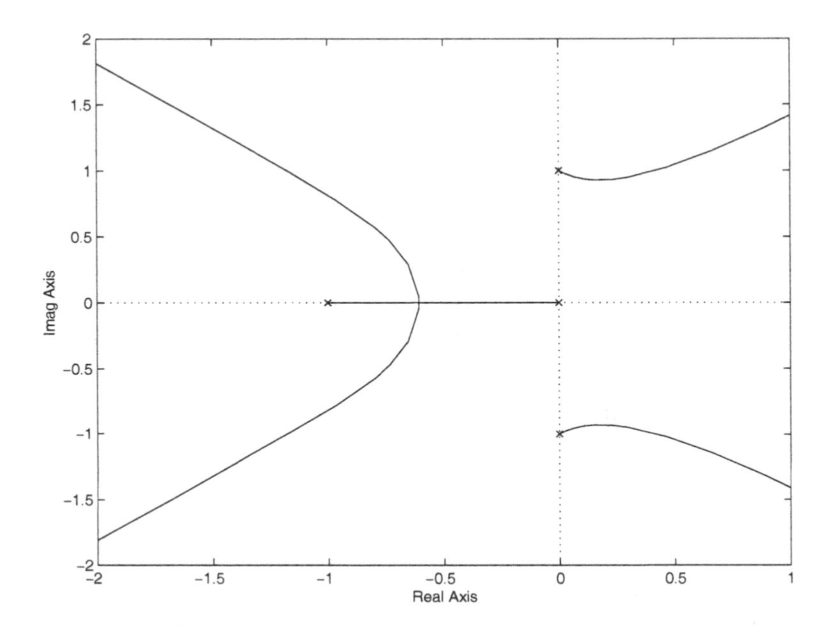

Fig. 6.6 The Root Locus Diagram for $G_p(s)H(s) = 1/(s(s+1)(s^2+1))$

6.2.7 *A Group of Poles and Zeros Near the Origin*

Making use of the ideas of the last section, we develop one more rule. Suppose that one has one group of poles near the origin, other groups of poles farther from the origin, and no poles on the positive real axis. We show that while one is near the origin one can—to a first approximation—ignore the poles far from the origin. The idea is quite simple. In order for a point, s, to be on the root locus, it is necessary that:

$$\angle\left\{\frac{P(s)}{Q(s)}\right\} = -180° + n360°.$$

Now suppose that r is a root of either $P(s)$ or $Q(s)$ that is far from the origin and s is relatively small. If r is real and negative, then $\angle(s - r) \approx$

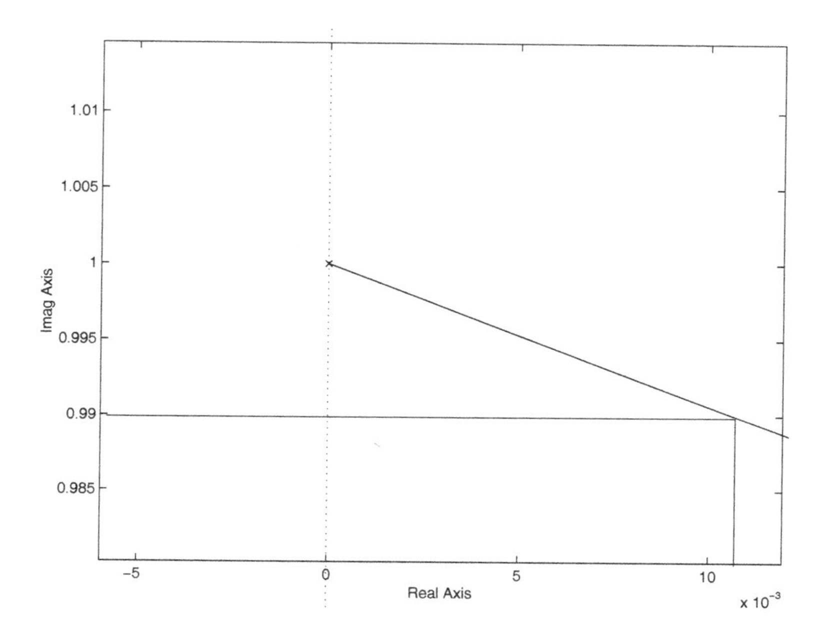

Fig. 6.7 A Zoom of the Root Locus Diagram for $G_p(s)H(s) = 1/(s(s+1)(s^2+1))$

$\angle(-r) = 0^\circ$ because s is very small compared to r. If r is complex, then $\overline{r}$ is also a pole/zero. Thus, we can consider the overall contribution of the two poles/zeros. This contribution is

$$\angle\{(s-r)(s-\overline{r})\} = \angle(s^2 - 2s\mathcal{R}(r) + |r^2|) \approx \angle|r|^2 = 0^\circ.$$

We have shown that:

Rule 10 In a system which has no poles on the positive real axis, poles and zeros far from the origin have little effect on the root locus diagram near the origin.

In other words, for "moderate gains" (and systems with no poles in the right half-plane) the poles near the origin fix the behavior of the system. This is one reason why the effects of poles far from the origin are often ignored in a first order approximation.

$G_p(s)H(s) = 10(10s + 1)/(s(s + 1)(70s + 1)$ Revisited—An Example

We note that in this example one pole—$s = -1$—is quite a lot farther from the origin than the rest of the poles. Let us see how the system behaves without this pole. Using MATLAB we find that the root locus diagram of this new system is as shown in Figure 6.8. Note how similar this plot is to Figure 6.5 for points near the origin.

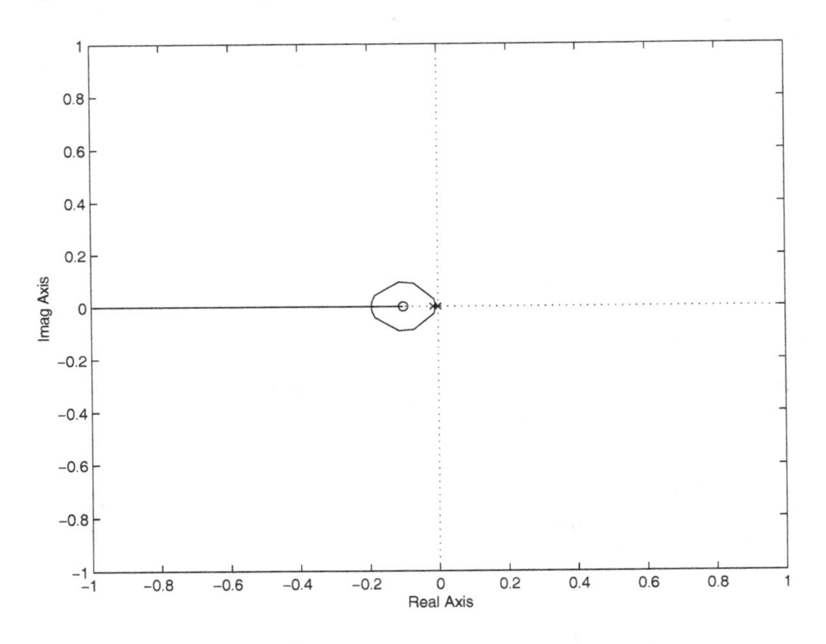

Fig. 6.8 The Root Locus Diagram for $G_p(s)H(s) = 10(10s + 1)/(s(70s + 1)$

6.3 Some (Semi-)Practical Examples

6.3.1 *The Effect of Zeros in the Right Half-Plane*

Suppose one has a transfer function with a zero in the right half-plane. What implications does this zero have for system stability? From Rule 3 we see that for sufficiently large gain a branch of the root locus crosses into the right half-plane and approaches the right half-plane zero.

We see that a right half-plane *pole* indicates an unstable component in the system and indicates that at the very least at low gains the system will be unstable. A right half-plane *zero* indicates that the system will be unstable for sufficiently high gain. In other words, *a system with a zero in the right half-plane has finite gain margin.* This is one reason why elements with zeros in the right half-plane are considered somewhat problematic. (See §2.6 and exercises 6 and 7 of §4.4 for other ways to see that zeros in the right half-plane can be problematic.)

6.3.2 *The Effect of Three Poles at the Origin*

Suppose that one has a system all of whose poles are in the left half-plane except for three poles at the origin. Then one can use rule 10 because the three poles near the origin constitute a "group" near the origin. We see that the branches leave the origin just like the poles of the system described by $G_p(s)H(s) = 1/s^3$.

Let us consider the root locus diagram that corresponds to the the system described by $G_p(s)H(s) = 1/s^3$. The poles of the system are the solutions of:

$$1 + K\frac{1}{s^3} = 0 \Leftrightarrow s^3 = -K.$$

The solutions of this equation are three lines emanating from the origin. One leaves the origin at 60°, one at -60°, and one at 180°. We see that two of the branches leave the origin and move into the right half-plane. Thus the system is unstable for any value of K greater than zero.

Now consider the system $G_p(s)H(s) = (s + 1)^3/s^3$. From rule 10, we see that for gains near zero this system is not stable—two of the poles are located in the right half-plane. On the other hand, from rule 3 we know that as $K \to \infty$ the poles of the system tend to -1. The root locus diagram corresponding to $G_p(s)H(s) = (s + 1)^3/s^3$ is plotted in Figure 6.9. This system is an example of a conditionally stable system.

6.3.3 *The Effect of Two Poles at the Origin*

In Figure 6.2, we see that a system with precisely two poles at the origin is on the border of stability. Let us see what effect adding various types of "compensation" has on the system's stability. Consider the system

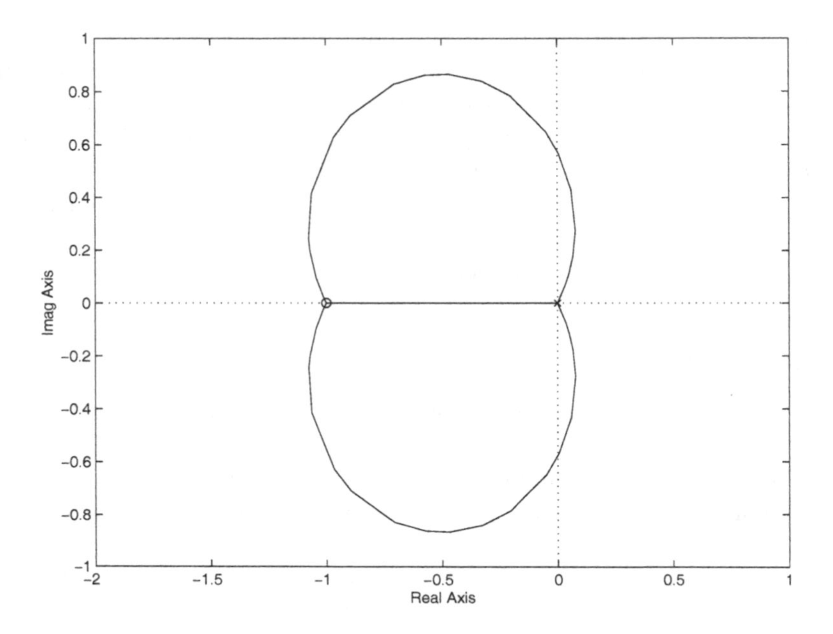

Fig. 6.9 The Root Locus Diagram for $G_p(s)H(s) = (s+1)^3/s^3$

described by:

$$G_p(s)H(s) = \frac{1 + s/10}{1 + s}\frac{1}{s^2}.$$

Here the additional term reduces the gain at high frequencies. Let us consider the root locus diagram that corresponds to the new system. The order of the numerator here is two less than the order of the denominator. Thus the sum of the poles is conserved. There is a branch between the pole at -1 and the zero at -10. Also there are two branches that go off to $\pm j\infty$. Let us assume that the two branches that go to infinity start at the two poles at the origin. (This can be checked by using rule 6.) Then the two complex points on the root locus must be complex conjugates. As the sum of the roots must be constant, since the one real root is moving to the left, the two complex roots must be moving to the right. Thus, the system will be unstable at any gain. Figure 6.10 shows the root locus diagram that corresponds to this system.

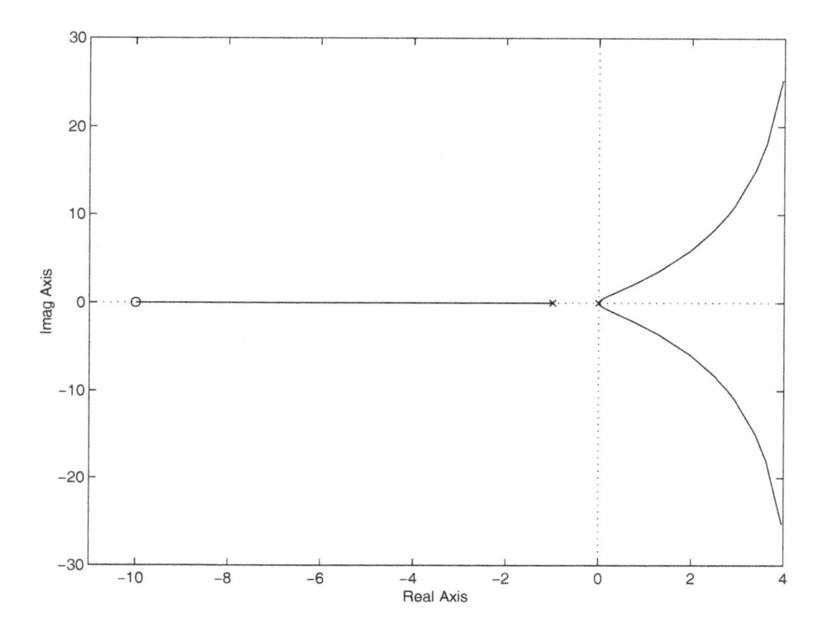

Fig. 6.10 The Root Locus Diagram for $G_p(s)H(s) = (1 + s/10)/((1 + s)s^2)$

On the other hand, consider the system:

$$G_p(s)H(s) = \frac{1+s}{1+s/10}\frac{1}{s^2}.$$

It is easy to see that this system is stable at any gain. When there are two poles at the origin the problem is *more complicated* then when there are three poles at the origin. When there are two poles at the origin some types of compensation lead to unconditionally stable systems; others lead to unstable systems.

6.3.4 *A System that Tracks* $t\sin(t)$

Consider a unity-feedback system for which

$$G_c(s)G_p(s) = \frac{4s(s^2+s+1)}{(s^2+1)^2} = \frac{(s+1)^4 - (s^2+1)^2}{(s^2+1)^2}.$$

As $G_c(s)G_p(s)$ has poles at $\pm j$, if the root-locus diagram indicates that the closed loops system is stable for some gain, K, the closed loop system certainly track $\sin(t)$. To check that this is so, consider the transfer function of the closed loop system, $T(s)$. We find that:

$$T(s)|_{s=j} = \left.\frac{KG_c(s)G_p(s)}{1 + KG_c(s)G_p(s)}\right|_{s=j} = 1.$$

Because the poles of $G_c(s)G_p(s)$ are of *second order*, we expect the system to track $t\sin(t)$. (See exercise 18 for a proof that the system indeed tracks $t\sin(t)$.)

The open loop transfer function has zeros at 0 and $-1/2 \pm j\sqrt{3}/4$, and branches of the root locus must tend to these zeros for large K. It is clear that the negative real axis is part of the root locus as well. Additionally, as $G_c(s)G_p(s)$ has zeros at 0 and at infinity, at a minimum, branches must enter the negative real axis. To find out where where they enter and, perhaps, leave the axis, we differentiate $G_c(s)G_p(s)$ and set the derivative to zero. The derivative is:

$$\begin{aligned}\frac{d}{ds}G_c(s)G_p(s) &= \frac{4(s+1)^3(s^2+1)^2 - 4s(s^2+1)(s+1)^4}{(s^2+1)^4} \\ &= \frac{(s+1)^3(4(s^2+1) - 4s(s+1))}{(s^2+1)^3} \\ &= \frac{4(s+1)^3(1-s)}{(s^2+1)^3}.\end{aligned}$$

The only real numbers for which the derivative is zero are 1 and -1. The only value that is located on the root locus is -1.

Let us examine the behavior of the root locus near $s = -1$. We consider $s = -1 + \epsilon$ and find the values of ϵ for which s is on the root locus.

For $s = -1 + \epsilon$, we find that:

$$G_c(s)G_p(s) = \frac{\epsilon^4}{(s^2+1)^2} - 1.$$

For small values of ϵ, the denominator is, essentially, $2^2 \in \mathcal{R}$. In order for $G_c(s)G_p(s)$ to be real and negative, we need ϵ^4 to be real. For a given value of $a = |\epsilon|$, $\epsilon = a\exp(n\pi/4)$, $n = 0, \ldots, 7$ gives the eight different solutions of the two equations $\epsilon^4 = a^4$ and $\epsilon^4 = -a^4$. Thus *eight branches* leave or depart from $s = -1$.

Looking at all the information we have, we find that the root locus must:

- Have four branches that start from $\pm j$—with two branches starting at each pole.

- The four branches must all enter the negative real axis at -1.
- After entering the negative real axis, two of the branches leave to head off towards $-1/2 \pm j\sqrt{3}/4$.
- One of the other two branches heads off to 0, and the last branch heads off to $-\infty$.

The root locus plot is given in Figure 6.11. We find that the system is stable for all positive gains.

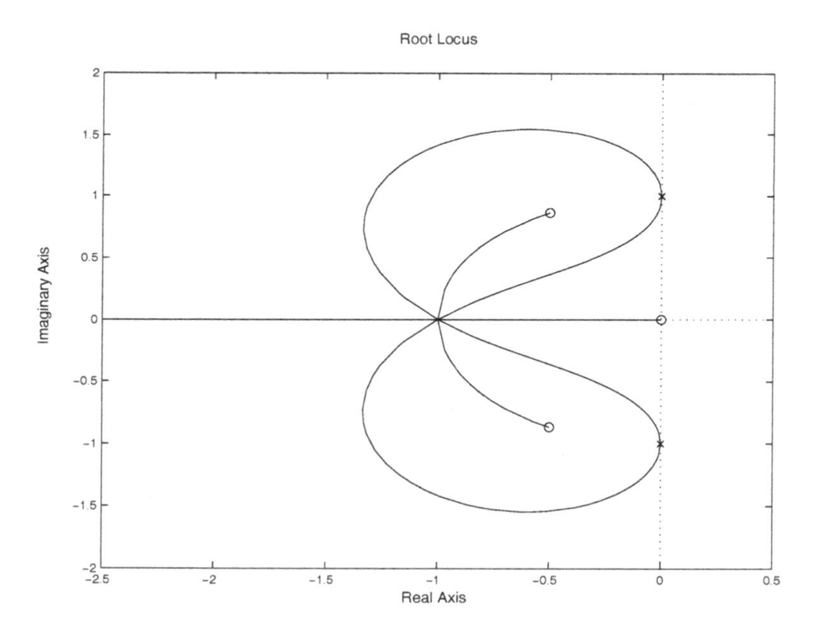

Fig. 6.11 The Root Locus Diagram Corresponding to $G_c(s)G_p(s) = 4s(s^2+s+1)/(s^2+1)^2$

Let's consider the system's output when its input is $t\sin(t)$ and $K = 1$. In this case, the closed loop transfer function of the system is:

$$T(s) = \frac{(s+1)^4 - (s^2+1)^2}{(s+1)^4}.$$

For $x(t) = t\sin(t)u(t)$, we find that:

$$X(s) = -\frac{d}{ds}\frac{1}{s^2+1} = \frac{2s}{(s^2+1)^2}.$$

The output of the system to this input, $y(t)$, satisfies:

$$Y(s) = T(s)X(s) = \frac{2s}{(s^2+1)^2} - \frac{2s}{(s+1)^4} = \frac{2s}{(s^2+1)^2} - \frac{2(s+1)}{(s+1)^4} + \frac{2}{(s+1)^4}.$$

Thus, $y(t) = (t\sin(t) - (t^2 - t^3/3)e^{-t})u(t)$. Clearly, as $t \to \infty$, $y(t) - t\sin(t) \to 0$.

6.3.5 *Variations on Our Theme*

All of the rules that we have considered are rules for systems whose poles are the solutions of an equation of the type $1 + KP(s)/Q(s)$ where both $P(s)$ and $Q(s)$ are polynomials with real coefficients, where the degree of $Q(s)$ is at least as large as the degree of $P(s)$, and where the the coefficients of the highest powers of s have the same sign. We can use our techniques for any system that can put in this form. Consider a system for which K is fixed at one but for which $G_p(s)H(s) = \frac{1}{s(\tau s+1)}$—this might describe a motor whose shaft angle is being controlled and whose time constant τ is a "free" parameter. We would like to understand the effect of the motor's time constant on the location of the poles of the system.

To find the poles of the closed-loop system, we must solve the equation:

$$1 + \frac{1}{s(\tau s + 1)} = 0.$$

Rearranging terms, we find that this is equivalent to considering how changes in τ effect the solutions of:

$$1 + \tau\frac{s^2}{s+1} = 0.$$

This is now in the correct form for our rules—with τ assuming the role of the gain, K—except that the degree of the numerator is greater than the degree of the denominator (i.e. the system is not low-pass). Considering our rules (and their proofs), we find that the only rules that depended in an unobvious way on the relation of the degree of numerator and the denominator are rules 3 and 4. It is easy to see that if one is willing to allow poles at infinity then rule 3 is unchanged. As we will be able to determine the asymptotic behavior of the poles without recourse to rule 4, we do not consider it further.

From rule 2 we find that the real axis to the left of -1 is part of the root locus diagram. From rule 3 we see that two branches end on the double zero at the origin. From rule 5, we see that there is one point on the negative real axis at which the branches leave the axis. This point is found (using rule 6) by solving:

$$\frac{d}{ds}\frac{s^2}{s+1} = \frac{2s(s+1)-s^2}{(s+1)^2} = \frac{s^2+2s}{(s+1)^2} = 0.$$

The solutions are $s = 0, -2$. The root locus has been plotted in Figure 6.12.

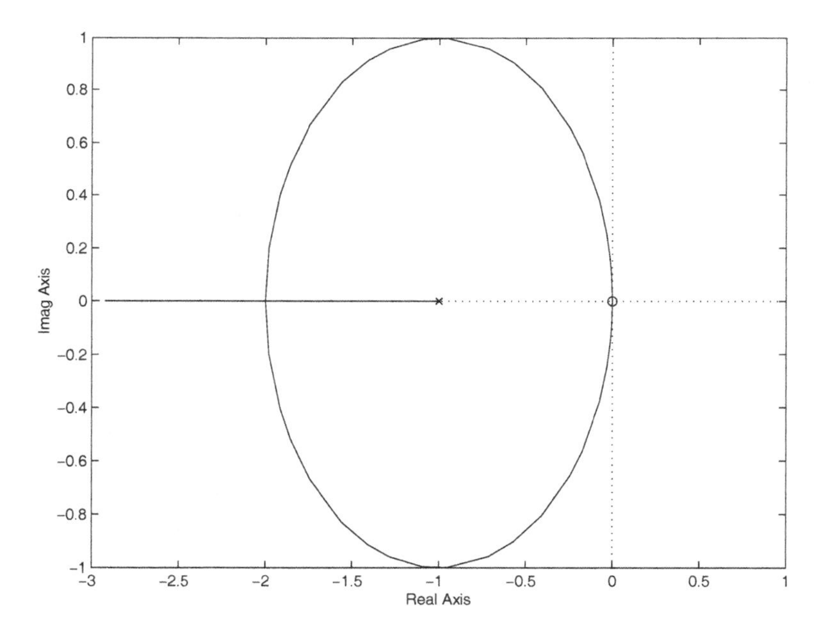

Fig. 6.12 The Root Locus Diagram for $G_p(s)H(s) = s^2/(s+1)$

What value of τ gives us the fastest system for which there are no oscillations? We see that the fastest system without oscillation is the system with both of its poles at -2. At $s = -2$, we find that we must solve $1+\tau 2^2/(-2+1) = 0$ for τ. We find that $\tau = 1/4$ gives the "optimal" system. In practice τ is not generally something we can choose—nonetheless we see that the optimal value for τ is neither zero nor infinity.

6.3.6 The Effect of a Delay on the Root Locus Plot

Consider the system defined by $G_p(s)H(s) = e^{-Ts}/s$. When $T = 0$, we know that the system is stable for any gain. Let us consider the root locus diagram when $T > 0$. We make use of the approximation:

$$e^{-Ts} \approx \frac{1 - (T/2)s}{1 + (T/2)s}$$

in order to find a transfer function that is a rational function of s. We must analyze the system described by:

$$G_p(s)H(s) = \frac{1 - (T/2)s}{1 + (T/2)s} \frac{1}{s}.$$

The denominator of $G_p(s)/(1 + G_p(s)H(s))$ is:

$$1 + \frac{1 - (T/2)s}{1 + (T/2)s} \frac{1}{s}.$$

Rearranging terms, we find that we must find the solutions of:

$$1 + (T/2)\left(\frac{s(s-1)}{s+1}\right) = 0.$$

Thus our "$G_p(s)H(s)$" is:

$$\frac{s(s-1)}{2(s+1)}.$$

We are in the same position we were in in the last example. Here we find that there are branches of the root locus between 1 and 0 and between -1 and $-\infty$. Because there is a pole at infinity, we know that the root locus must leave the real axis at some point to the left of -1. Because the branches must end on the two finite zeros, we know that the branches that leave the real axis must reenter it between 0 and 1. To find these points we solve the equation:

$$\frac{d}{ds}\frac{s(s-1)}{s+1} = \frac{(2s-1)(s+1) - s(s-1)}{(s+1)^2} = \frac{s^2 + 2s - 1}{(s+1)^2} = 0.$$

The solutions of this equation are $-1 \pm \sqrt{2} = -2.41, 0.41$. For small values of T, our system is very nicely behaved. As T is increased, the root locus shows the onset of sinusoidal solutions and, finally, for sufficiently large T the system is unstable. The root locus diagram is shown in Figure 6.13.

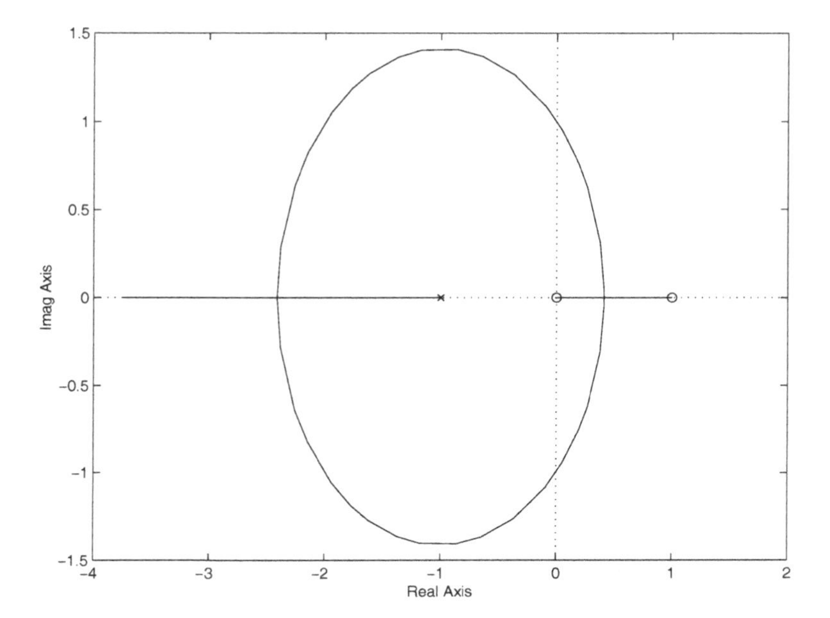

Fig. 6.13 The Root Locus Diagram for $G_p(s)H(s) = s(s-1)/(s+1)$

There is no reason to expect the preceding analysis to be correct for relatively large values of Ts. Our approximation of e^{-Ts} is only effective for small values of Ts. At what point will our system *really* first become unstable? Let us consider the Nyquist plot corresponding to our system—see Figure 6.14 for the Nyquist plot and Figure 6.15 for a blowup of Nyquist plot near the region near $s = 0$. The Nyquist plot crosses the negative real axis for the first time—and gives us the largest negative value—when $e^{-Tj\omega} = -90^o$. At that point, we find that $G_p(s)H(s) = -1/\omega$. The smallest value of ω for which this crossing leads to instability is $\omega = 1$, and we find that the smallest value of T for instability is $\pi/2$.

Using our approximations to find the smallest value of T that causes instability, we find that we must find the smallest value of T for which $1 + (T/2)(s(s-1)/(s+1))$ has zeros in the right half-plane. Rearranging terms we find that we must determine the values of T for which:

$$2(s+1) + T(s^2 - s) = Ts^2 + (2-T)s + 2$$

has zeros in the right half-plane. Clearly, this occurs when $T \geq 2$. We see

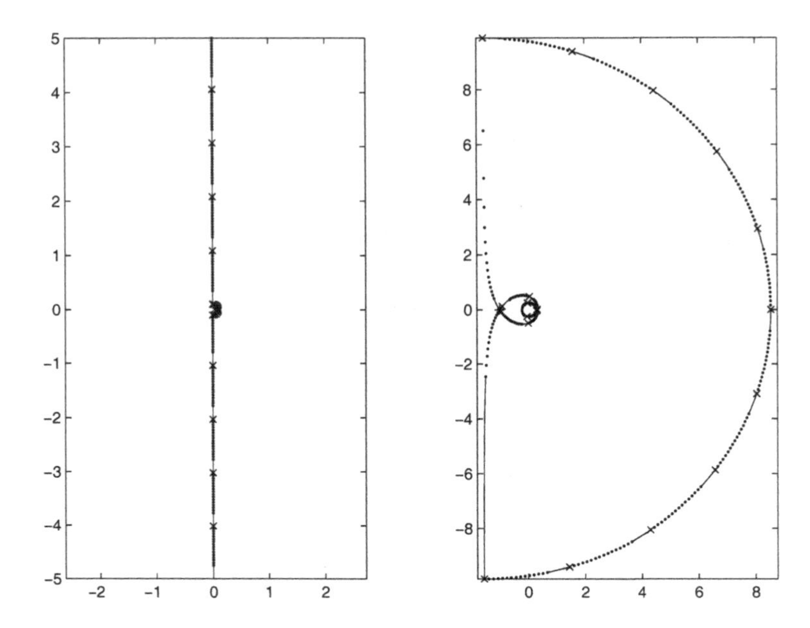

Fig. 6.14 The Nyquist Plot of $G_p(s)H(s) = e^{-Ts}/s, T = \pi/2$

that our approximation mistakenly predicts a stable system for values of T between $\pi/2$ and 2. Our approximation gives us a feel for the effect of T on the location of the roots for *small value of* Ts. It does not work well for large values of Ts.

In fact, there is another related problem—there are actually other solutions of $1 + G_p(s)H(s) = 0$. If we take as our domain:

$$\Omega = \{s||s-a| \leq R, \mathcal{R}(s) \leq -a\}$$

we find that we must consider the how $G_p(s)H(s)$ maps the line $s = -a+j\omega$ and our (more or less) usual semi-circle. On the line, we find that:

$$G_p(s)H(s) = \frac{e^{a-j\omega}}{-a+j\omega} \overset{\omega>>1}{\approx} e^a \frac{e^{-j\omega}}{\omega},$$

and on the semi-circle $G_p(s)H(s) \approx 0$. For large value of ω, $G_p(s)H(s)$ maps the boundary of our new region into (approximately) e^a times the map of the boundary of our standard region. In Figure 6.15, we see that this figure spirals in to the origin for large values of ω. We see that in the new region this spiral gets larger—and will encircle -1 a number of

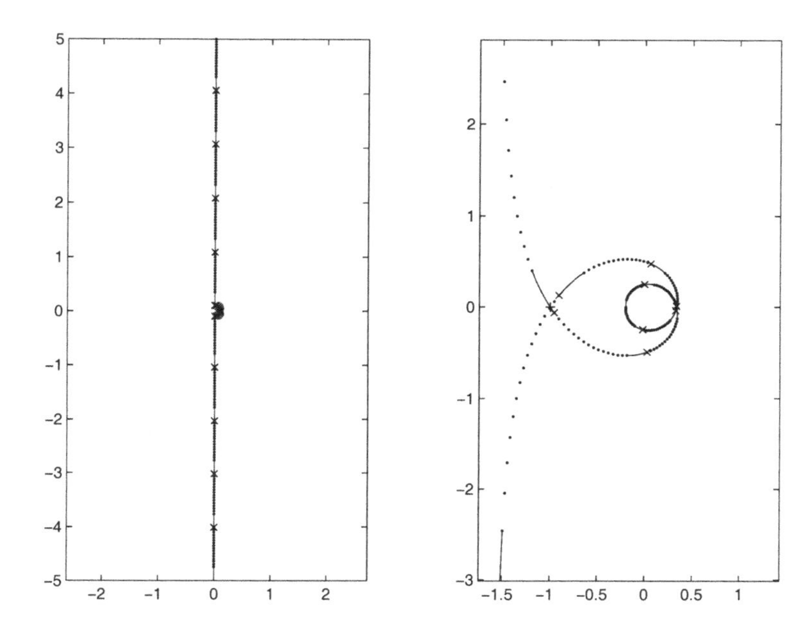

Fig. 6.15 A Blowup of the Nyquist Plot of $G_p(s)H(s) = e^{-Ts}/s, T = \pi/2$

times. This means that there will be a number of roots with relatively large negative real part. These roots may well be complex—there is no reason to expect that our system will not have oscillations in its output even for relatively small T. The root locus diagram tells us (approximately) how the pole nearest the origin behaves. (See [Eng08].)

6.3.7 *The Phase-lock Loop*

We now turn our attention to the phase-lock loop (PLL)—a device that finds the phase and frequency of an input sinusoid and outputs a sinusoid of the same frequency whose phase differs from the phase of the input by a constant. If the input to a PLL is $\alpha \sin(wt + \Delta\omega t + \varphi_1)$, then the output will be $\cos(wt + \Delta\omega t + \varphi_2)$—a scaled and phase-shifted version of the input. PLLs are useful because if they are fed a signal with a sinusoid and other terms they tend to "find" the sinusoid and lock onto its phase and frequency. In Figure 6.16 we show the block diagram of a PLL.

The components of this phase-lock loop are a multiplier, a low-pass filter (LPF), and a voltage controlled oscillator (VCO). We assume that

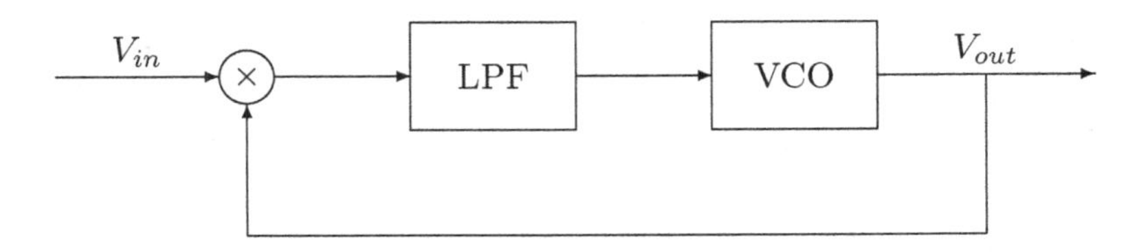

Fig. 6.16 The Block Diagram of a Phase-Lock Loop

the input is of the form $\alpha \sin(\omega t + \Delta\omega t + \varphi)$. The transfer function of the LPF is assumed to be $1/(\tau s + 1)$, and the output of the VCO is $\cos(\omega t + \beta \int_0^t V_{vco-in}(\xi)\, d\xi)$.

Consider the relationship between the inputs to the multiplier and its output. The output of the multiplier—the input to the LPF—is:

$$\begin{aligned} &\alpha \sin(\omega t + \Delta\omega t + \varphi) \cos(\omega t + \beta \int_0^t V_{vco-in}(\xi)\, d\xi) \\ &= \frac{\alpha}{2} \left(\sin(\Delta\omega t + \varphi - \beta \int_0^t V_{vco-in}(\xi)\, d\xi) + \right. \\ &\qquad \left. \sin(2\omega t + \Delta\omega t + \varphi + \beta \int_0^t V_{vco-in}(\xi)\, d\xi) \right) . \end{aligned}$$

If $\Delta\omega$ is not too big, it is reasonable to hope that the difference in the first term of the equation above is small. Also, if one picks a reasonable low-pass filter, it is reasonable to ignore the effect of the the second term altogether—it is a high frequency term and is removed by the LPF. Making use of the the fact that for small $|x|$ we know that $\sin(x) \approx x$, we see that the output of the multiplier can be taken to be:

$$\text{output} \approx \frac{\alpha}{2} \left(\Delta\omega t + \varphi - \beta \int_0^t V_{vco-in}(\xi)\, d\xi \right) .$$

This is $\alpha/2$ times the difference in the phases of the input and the output. If we use this approximation and we consider the phases of the input and the output relative to ωt rather than the input and the output themselves, we find that we can model our system by the block diagram of Figure 6.17. (Note that the VCO—when we consider only phases relative to ωt—acts by integrating its input and multiplying the result by β.)

We see that we can now apply what we know about control theory to

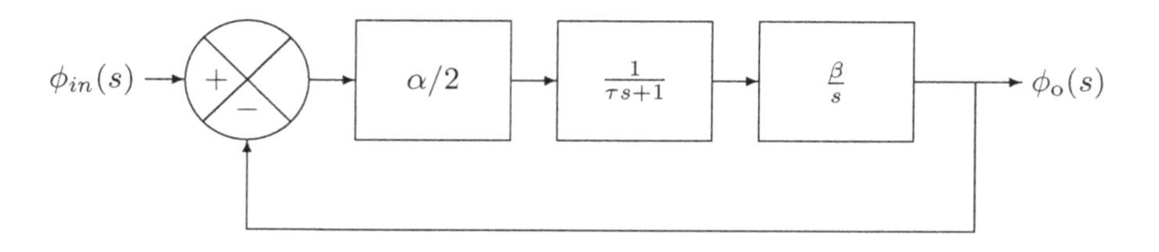

Fig. 6.17 The Phase-Lock Loop—with Respect to Phase

the PLL. First, we note that:

$$G_p(s)H(s) = \frac{\alpha\beta}{2}\frac{1}{s(\tau s + 1)}.$$

We can consider the term $\alpha\beta/2$ to be our K—our gain. We find that the system has two poles—one at 0 and one at $-1/\tau$. Thus, the root locus diagram has the same form as Figure 6.4. We see that the system is stable for all "gains". For low gains—while $\alpha\beta/2$ is small—the system has two real poles. Thus the phase of the output tends to converge to the phase of the input without oscillations. As the gain gets larger the two roots coalesce. At this point the system is as fast as it can get without oscillations in the output. As the gain passes this point the system gets no faster, but the phase of the output approaches the phase of the input in an oscillatory fashion.

As $G_p(s)H(s)$ has one pole at zero, we saw in §3.5 that if the input is a step function—if the phase of the input and the VCO initially differ by a constant—then the output tracks the input perfectly. This shows that a PLL tracks phase differences perfectly. If, however, the input is a ramp—if there is a difference in frequency as well—then the frequency of the output tracks that of the input, but there will be a fixed phase difference between the two signals. The larger the system gain, the smaller this difference in phase will be. For this reason, one often finds that the gain in the PLL is sufficiently large to cause the phase of the output to tend towards the phase of the input in a "mildly" oscillatory fashion. Such a gain causes the input and output phase to be more nearly equal in the steady-state.

6.3.8 *Sounding a Cautionary Note—Pole-Zero Cancellation*

Consider the system of Figure 6.18. Here we are trying to control the angle of a "simple satellite" by feeding back $s\Theta(s)$—in other words by feeding back the angular velocity. It does not seem too reasonable that this should work—or even be stable. Let us analyze our system using root locus techniques. We find that $G_p(s)H(s) = 1/s$. The root locus of this system is given in Figure 6.19. We see that this system is stable at any gain! Moreover because $G_p(s)H(s)$ has one pole at the origin, this system should track its input perfectly in the steady-state too. This seems to be an ideal system, but it is too good to be true.

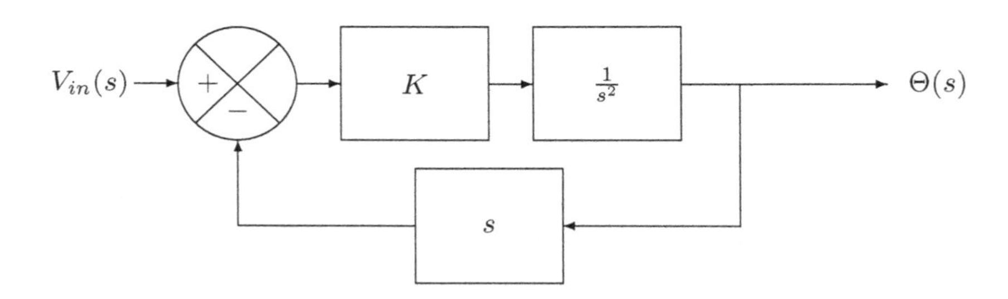

Fig. 6.18 An Example of Pole-Zero Cancellation

What is wrong with this picture? Let us consider the ideas that underlie the root locus technique. We assume that all of the poles of the transfer function:

$$T(s) = \frac{KG_p(s)}{1 + KG_p(s)H(s)}$$

come from the roots of the denominator. We have left out a possibility—perhaps some of the poles come from the *poles of the numerator*. Generally speaking this is not a problem. A pole in the numerator will be effectively canceled by a pole in the denominator. At any point at which the numerator tends to blow up, the denominator—which also contains $G_p(s)$ will blow up. However, if a pole of $G_p(s)$ is canceled by a zero of $H(s)$, then the numerator will blow up at a place where the denominator does not blow up. That is, there will be poles of the system at points other than the zeros

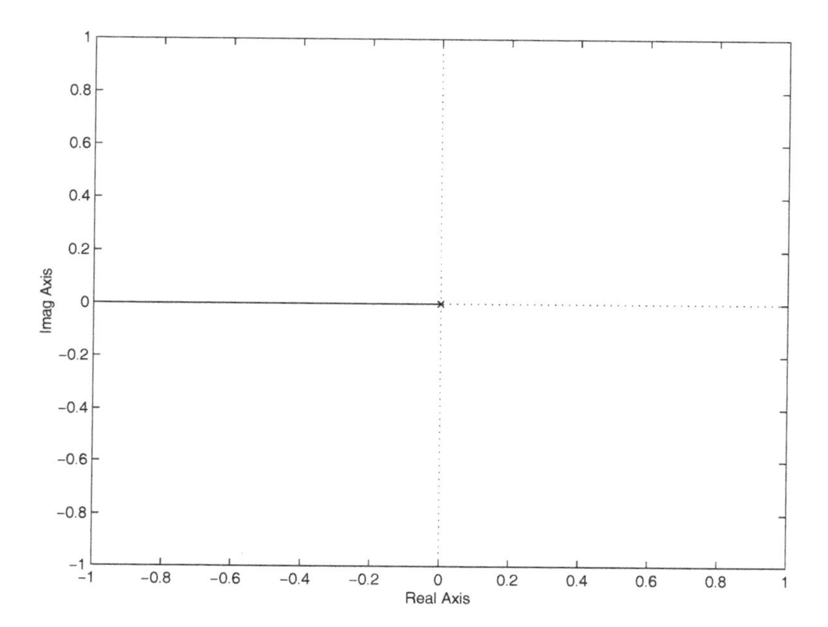

Fig. 6.19 The Root Locus Plot Corresponding to a System for which $G_p(s)H(s) = 1/2$

of the denominator of $T(s)$.

In our example, we see that:

$$T(s) = \frac{KG_p(s)}{1 + KG_p(s)H(s)} = \frac{K/s^2}{1 + K/s} = \frac{K}{s(s+K)}.$$

We see that our system has one pole at $s = -K$—and this is the pole that the root locus diagram shows—and one pole at $s = 0$ which is not shown in the root locus diagram. The pole at zero means that this system has one pole in the right half-plane and is *unstable at any gain.*

This example shows how when using techniques that only consider the roots of the denominator of the system's transfer function one must be careful that the zeros of $H(s)$ not cancel poles of $G_p(s)$. As we saw in §3.7, using a right half-plane zero to cancel a right half-plane pole produces a system that is not internally stable. Here we see that sometimes such a cancellation leads to a system that is not even BIBO stable.

6.4 More on the Behavior of the Roots of $Q(s)/K + P(s) = 0$

We now continue the discussion of §6.2.3 and prove that n of the roots of the equation:

$$Q(s)/K + P(s) = 0$$

tend to the roots of $P(s)$ and that the remaining roots tend to infinity. To do this, we make use of Rouché's theorem [CB90] (which is a corollary of the principle of the argument). Rouché's theorem states that if one has two analytic functions, $f(s)$ and $g(s)$, and if $|f(s)| > |g(s)|$ on some simple closed contour, then the number of zeros of $f(s) + g(s)$ inside the contour is equal to the number of zeros of $f(s)$ inside the contour.

We now show that our polynomial has n roots inside a fixed circle and has $m - n$ roots in the region:

$$(1-\epsilon)(K/B)^{\frac{1}{m-n}} < |s| < (1+\epsilon)(K/B)^{\frac{1}{m-n}}.$$

We start by picking $f(s) = s^n$ and our contour to be $|s| = C$ where C is a large, fixed, number. For large enough C it is clear that $|s^n|$ is bigger than the absolute value of the remaining terms in $P(s)$. Moreover, for all K larger than some number it is clear that $|s^n|$ is larger than $|Q(s)/K|$ on this contour. In fact, it is clear that for sufficiently large C we will find that $|s^n|$ is larger than all the remaining terms in the polynomial. Thus, the number of zeros of the polynomial that are located in the disk $|s| \leq C$ is the same as the number of zeros of s^n in that region. That is, there are n zeros of the polynomial inside $|s| \leq C$.

Next, we consider the same function, $f(s)$, but on the contour $|s| = (1-\epsilon)(K/B)^{\frac{1}{m-n}}$. Clearly here $|s^n|$ will be much larger than the remaining terms in $P(s)$—it always includes at least a factor of $(1-\epsilon)(K/B)^{\frac{1}{m-n}}$ more than they do. In fact :

$$|s^n| = (1-\epsilon)^n (K/B)^{\frac{n}{m-n}}.$$

What about the terms corresponding to $Q(s)/K$? Let us consider the largest term there—Bs^m/K. We find that:

$$B|s^m|/K = (1-\epsilon)^m (K/B)^{\frac{m}{m-n}} (B/K) = (1-\epsilon)^m (K/B)^{\frac{n}{m-n}} < |s|^n$$

because $(1-\epsilon) < 1$ and $m > n$. Clearly all of the rest of the terms will add only a relatively insignificant amount. We see that our function, $f(s)$, is in fact larger than all the other terms on the circle. Thus there are only n zeros inside this new larger circle, and there are no zeros of our polynomial in the region:

$$C < |s| < (1-\epsilon)(K/B)^{\frac{1}{m-n}}.$$

Now we let $f(s) = Bs^m/K$, and we consider the contour $|s| = (1+\epsilon)(K/B)^{\frac{1}{m-n}}$. We find that:

$$B|s^m|/K = (1+\epsilon)^m(K/B)^{\frac{m}{m-n}}(B/K) = (1+\epsilon)^m(K/B)^{\frac{n}{m-n}}.$$

Furthermore, we find that:

$$|s^n| = (1+\epsilon)^n(K/B)^{\frac{n}{m-n}} < B|s^m|/K.$$

Clearly all of the rest of the terms become insignificant as $K \to \infty$. We find that for large K the number of zeros of the mth order polynomial that are inside the circle of radius $(1+\epsilon)(K/B)^{\frac{1}{m-n}}$ is m. That is there are $m-n$ zeros located in the region:

$$\Omega = \left\{s|(1-\epsilon)(K/B)^{\frac{1}{m-n}} < |s| < (1+\epsilon)(K/B)^{\frac{1}{m-n}}\right\}.$$

We have now pinned the zeros down to two regions. What more can we say about their positions? Let us consider the zeros located inside the disk $|s| < C$. It is clear that as $K \to \infty$ that the contribution of $Q(s)/K \to 0$. Thus, the zeros inside this disk must tend to the zeros of the remaining term—$P(s)$. What about the zeros in the annulus Ω? We know that the zeros are the solutions of the equation $Q(s)/K + P(s) = 0$. When s is in Ω we have already seen that Bs^m/K and s^n are of the same order. All the rest of the terms are much smaller. Thus, for s to be a solution of the equation it must be an approximate solution of $Bs^m/K + s^n = 0$ and it must be large. The solutions of this equation are:

(1) n zeros at 0.
(2) The $m-n$ numbers $s = (K/B)^{\frac{1}{m-n}}e^{\frac{j\pi+j2\pi i}{m-n}}, i = 1, ..., m-n$.

The first set of solutions are irrelevant to us—they are not large. The second set of solutions are the approximate solutions of our equation. We note that their size is just right—they are in the annulus. We now know—approximately—what the solutions are.

Let us define $\mathcal{A}_i = (K/B)^{\frac{1}{m-n}} e^{\frac{j\pi + j2\pi i}{m-n}}$. Let us plug $s = \mathcal{A}_i + c_i$ into $Q(s)/K + P(s)$. We find that:

$$\begin{aligned} Q(s)/K + P(s) &= (B/K)\left((\mathcal{A}_i + c_i)^m + b_1(\mathcal{A}_i + c_i)^{m-1} + \cdots\right) \\ &\quad + (\mathcal{A}_i + c_i)^n + a_1(\mathcal{A}_i + c_i)^{n-1} + \cdots \\ &= (B/K)\left(\mathcal{A}_i^m + mc_i\mathcal{A}_i^{m-1} + \cdots + b_1\mathcal{A}_i^{m-1} + \cdots\right) \\ &\quad + \mathcal{A}_i^n + nc_i\mathcal{A}_i^{n-1} + \cdots + a_1\mathcal{A}_i^{n-1} + \cdots \\ &= 0 \end{aligned}$$

We have already seen that $(B/K)\mathcal{A}_i^m + \mathcal{A}_i^n = 0$. The next set of terms in the equation—the terms that correspond to $(B/K)\mathcal{A}_i^{m-1}$ and $\mathcal{A}_i^{n-1}$ are all listed above. In order for these "second order" terms to cancel we find that:

$$(B/K)\left(mc_i\mathcal{A}_i^{m-1} + b_1\mathcal{A}_i^{m-1}\right) + nc_i\mathcal{A}_i^{n-1} + a_1\mathcal{A}_i^{n-1} = 0.$$

As $-(B/K)\mathcal{A}_i^m = \mathcal{A}_i^n$, we see that this leads to the condition:

$$mc_i + b_1 - nc_i - a_1 = 0 \Leftrightarrow c_i = \frac{a_1 - b_1}{m - n}.$$

We have seen previously (p. 89) that $a_1 = -\sum(r_P)_i$ and $b_1 = -\sum(r_Q)_i$. Thus c_i is equal to the sum of the poles of $G_p(s)H(s)$ less the sum of the zeros of $G_p(s)H(s)$ divided by the difference in the degrees of the denominator and the numerator. Note that all the c_i are identical.

To sum up, we have found that "to second order" the $m - n$ large roots of $Q(s)/K + P(s)$ are:

$$s = (K/B)^{\frac{1}{m-n}} e^{\frac{j\pi + j2\pi i}{m-n}} + \frac{\sum \text{finite poles} - \sum \text{finite zeros}}{m - n}, i = 1, ..., m - n.$$

6.5 Exercises

(1) Sketch the root locus diagram that corresponds to the system of Figure 6.20 when:

(a) $G_p(s) = \frac{s}{s+1}$.
(b) $G_p(s) = \frac{s+1}{s(s+2)}$.
(c) $G_p(s) = \frac{s^2+1}{s^3}$.

Include all relevant calculations (i.e. the points at which the branches leave the real axis, etc.).

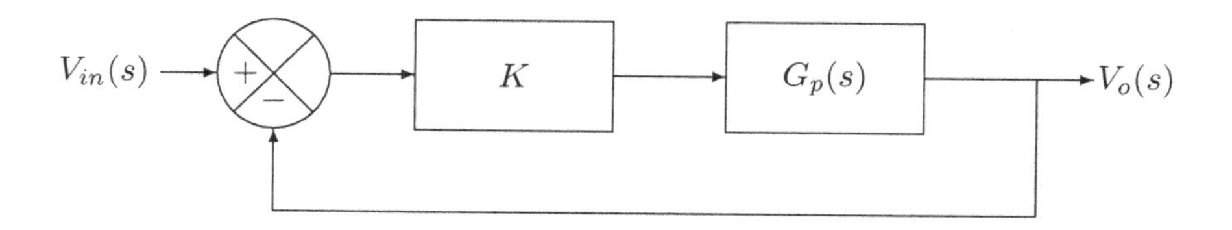

Fig. 6.20 A Generic Unity-Gain Feedback System

(2) Sketch the root locus diagram that corresponds to the system of Figure 6.20 when:

(a) $G_p(s) = \frac{s(s+2)}{s+1}$.
(b) $G_p(s) = \frac{s+1}{(s+2)(s+3)(s+4)}$.
(c) $G_p(s) = \frac{s-1}{s^2+5s+6}$.

Include all relevant calculations (i.e. the points at which the branches leave the real axis, etc.).

(3) Sketch the root locus diagram that corresponds to:

$$G_1(s) = \frac{s+0.1}{s(s+0.2)(s+10)(s+20)}.$$

Compare this with the root locus diagram that corresponds to:

$$G_2(s) = \frac{s+0.1}{s(s+0.2)}.$$

Why are the plots so similar?

(4) Use Rule 3 and Rule 10 to design a conditionally stable system. (I.e. design a system that is unstable at low gains by using Rule 10 but stable at high gains because of Rule 3.)

(5) Let

$$G_p(s) = \frac{s+5}{s(s+1)(s+2)}$$

in the system of Figure 6.20. For which values of K will the system have complex poles?

(6) Let K in Figure 6.20 be replaced by Ke^{-Ts} and let $G_p(s) = 1/(s(s+2))$. Use the approximation:

$$e^{-Ts} \approx \frac{1-(T/2)s}{1+(T/2)s}$$

in order to approximate the root locus diagram of the system. Should one expect the system to be stable for all values of K? Explain.

(7) Let

$$G_p(s) = \frac{s+4}{(s+5)(s^2+1)}$$

in the system of Figure 6.20. Using the rules that we have developed, prove that this system is stable for all $K > 0$. (Rule 8 may be helpful.)

(8) Sketch the root locus of the system of Figure 6.20 when:

$$G_p(s) = \frac{1}{(s^2+1)(s^2-1)}.$$

Make use of the fact that the root locus corresponding to this system is the "square root" of the root locus corresponding to the system for which:

$$G_p(s) = \frac{1}{(s+1)(s-1)}.$$

(9) Consider a system like that of Figure 6.20 where:

$$G_p(s) = \frac{s^2+1}{s(s+1)}.$$

(a) Plot the root locus diagram of the system.
(b) What are the zeros of the system?
(c) What is the transfer function of the system?
(d) What does the frequency response of the system look like when K is large?
(e) What is the net effect of this system?

(10) Consider a system like that of Problem 9 but where both the gain term and $G_p(s)$ have been moved into the feedback. That is, in the new system $G_p(s) = 1$ and:

$$H(s) = K\frac{s^2+1}{s(s+1)}.$$

(a) Plot the root locus diagram of the system.
(b) What are the zeros of the system?
(c) What is the transfer function of the system?
(d) What does the frequency response of the system look like when K is large?
(e) What is the net effect of this system?

(11) Consider a unity-feedback system for which

$$G_c(s)G_p(s) = \frac{(s+1)^3 - s^3}{s^3}.$$

(a) Plot the root locus of the system.
(b) By considering the system with an additional gain, K, use the Routh array to calculate the values of K for which the closed loop system is stable.
(c) Are the results of the preceding two sections consistent with one another?
(d) Is it reasonable to speak of this system's gain margin? Explain.
(e) Calculate the output of the closed loop system (for $K = 1$) when the system's input is $t^2u(t)$. Does the output track the input?

(12) Find the root locus diagram associated with the unity-feedback system for which

$$G_c(s)G_p(s) = \frac{s-1}{(s+1)(s^2+a^2)}, \qquad a \geq 0.$$

(a) Plot the root locus diagram when $a = 0$.
(b) Now consider the diagram for $a > 0$. Pay close attention to the points at which branches of the root locus diagram enter and leave the real axis. Show that the behavior of the root locus changes when $a = \sqrt{5/27}$.
(c) Plot the root locus diagram for the three cases $0 < a < \sqrt{5/27}$, $a = \sqrt{5/27}$, and $a > \sqrt{5/27}$.

(13) Consider a unity feedback system for which:

$$G_p(s) = \frac{1}{s(s+1)^6}, \qquad G_c(s) = 1.$$

(a) Plot the root locus diagram associated with the system.
(b) Is the system's gain margin finite? Explain.

(14) Consider a unity feedback system for which:

$$G_c(s)G_p(s) = \frac{s+1}{s(s+3)(s+5)}.$$

(a) Plot the root locus diagram associated with the system.
(b) Make use of the root locus diagram to determine the system's gain margin.
(c) Check your answer using the Routh array.

(d) Check you answer again by using the Nyquist plot.

(15) Consider a unity feedback system for which:

$$G_c(s)G_p(s) = \frac{1}{s(s^2+2s+2)}$$

(a) Plot the root locus diagram of the system.
(b) Consider the root locus diagram. What can you conclude about the system's gain margin?
(c) Use the Routh array to calculate the system's gain margin. Do the results agree with those of the previous section?
(d) Does the output of the closed-loop system track a step input? Does it track a ramp input? Explain.

(16) Consider a unity feedback system for which:

$$G_c(s)G_p(s) = \frac{s-1}{s(s+1)}.$$

(a) Examine the system's stability by making use of the root locus diagram.
(b) Now examine the system's stability using the Nyquist plot.
(c) Next, examine the system's stability using the Routh Array.
(d) Summarize what you have learned about the system's stability.

(17) Consider a unity feedback system for which:

$$G_c(s)G_p(s) = \frac{s^2+a^2}{s(s+1)(s+2)}.$$

(a) Use the Routh array to determine the closed-loop system's gain margin as a function of a.
(b) For which values of a is the gain margin infinite?
(c) Plot the root locus diagram associated with this system. Make sure to calculate the angle at which branches approach the imaginary axis as $K \to \infty$.

(18) Prove that a stable unity-feedback system for which:

$$G_c(s)G_p(s) = \frac{P(s)}{Q(s)}\frac{1}{(s^2+1)^2}, \qquad P(\pm j) \neq 0, Q(\pm j) \neq 0$$

tracks the signal $t\sin(t)u(t)$. You may find it convenient to make use of the input-to-error transfer function:

$$T_{V_{in}(s)\to E(s)} = \frac{1}{1 + G_c(s)G_p(s)}.$$

(For more about this transfer function, see p. 77.)

Chapter 7

Compensation

7.1 Compensation—An Introduction

We have developed many tools for the *analysis* of linear feedback systems. It is time to work on the problem of how best to *design* such systems. Often, we are given a plant to control—for example a motor—and we are given tools with which the output of the plant is to measured—say a tachometer (a device for measuring a motor's speed). We can then add other parts to the system in order to meet the requirements imposed on the system. One requirement is, of course, that the system be stable. Another requirement might be that the system be over-damped—that there be no overshoot, that the system's settling time be 1 s, or that the system's phase margin be 60°. In this chapter we examine a number of the standard techniques for *compensating* a system—for adding elements to the system to improve the system's performance. This part of control theory is something of an art, and as is typical of engineering problems there is often more than one way to design a reasonable system.

7.2 The Attenuator

Consider the system of Figure 7.1. The transfer function of our plant is

$$G_p(s) = \frac{\sqrt{2}}{s(s+1)},$$

$H(s) = 1$, and we are to design $G_c(s)$. This is a reasonable model for a DC motor with position feedback, for example. Suppose that we are asked to design a compensator, $G_c(s)$, such that the final system's phase margin is 60°.

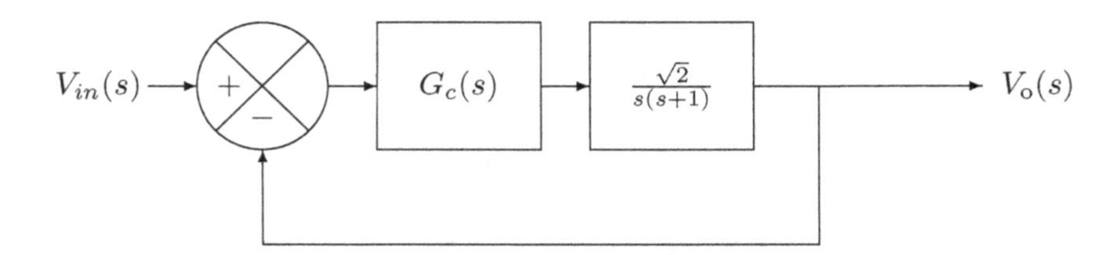

Fig. 7.1 A Generic Feedback Loop with Compensator, $G_c(s)$

Using MATLAB to plot the Bode plots for the system with $G_c(s) = 1$, for the uncompensated system, we find that the phase margin of the system is 45°. (See Figure 7.2.) Looking at the Bode plots it is clear that if we want to achieve a better phase margin—if we want the phase to be further from −180° when the gain is 0 dB—then one method is to simply "add" some attenuation into the system. Reading the Bode plot[1], we see that the system's phase is −120° when $\omega \approx 0.55$ rad/sec. At that point the amplitude is about 7dB or 2.25. If we make $G_c(s) = 1/2.25$, then we find that the phase margin is indeed about 60°. (See Figure 7.3.)

7.3 Phase-Lag Compensation

Unfortunately, though the addition of an attenuator often improves a system's stability, it generally harms system performance in important ways. As we have seen, the greater the forward gain of a system, $G_c(s)G_p(s)$, at a particular frequency, the more nearly the system's output will track its input at that frequency in the steady state. In the preceding example we needed to lower the gain at a *particular* frequency, $\omega \approx 0.55$ rad/sec, and instead we lowered the gain at *all* frequencies. We now consider a strategy that leaves the gain at low frequencies alone while reducing the gain

[1]Here it is relatively easy to find the exact values necessary for the design of the compensator without using the Bode plots. Note that at the point at which $G_p(j\omega)$ is 60° from −180°:

$$\angle G_p(j\omega) = \angle \frac{\sqrt{2}}{j\omega(j\omega+1)} = -90^\circ - \tan^{-1}(\omega) = -120^\circ.$$

We find that $\omega = 1/\sqrt{3} = 0.5774$. We find that for this value of omega:

$$|G_p(j/\sqrt{3})| = 2.12.$$

Thus the exact value of the attenuator is $G_c(s) = 1/2.12$.

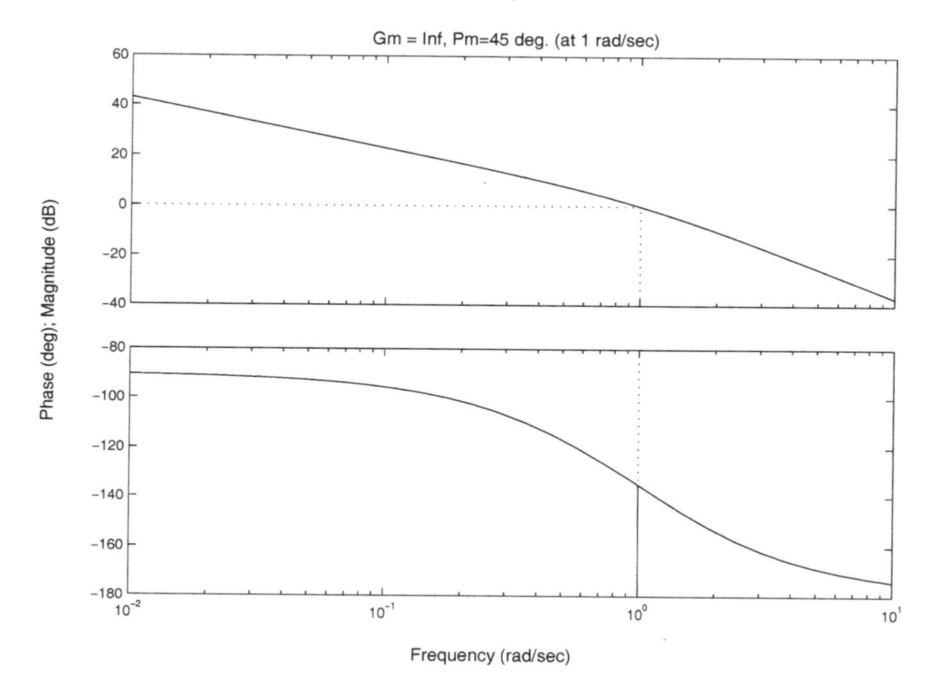

Fig. 7.2 The Bode Plots for $G_p(s) = \sqrt{2}/(s(s+1)), H(s) = 1, G_c(s) = 1$

both where necessary and at higher frequencies than necessary. This allows us to improve the system's phase margin without harming the system's performance at DC.

Consider a compensator $G_c(s)$ of the form:

$$G_c(s) = \frac{s/\omega_o + 1}{s/\omega_p + 1}, \quad \omega_p < \omega_o.$$

The Bode plots of such a system (when $\omega_p = 1, \omega_o = 2$) are shown in Figure 7.4. We see that the gain of this system is always less than one (0 dB), but it approaches one for low frequencies. For large s—at very high frequencies—$G_c(s) \approx \omega_p/\omega_o < 1$. This compensator is generally used to reduce the gain of a system at high frequencies. It is called a phase-lag compensator because its phase is always negative.

The phase-lag compensator is a little bit more complicated than a simple

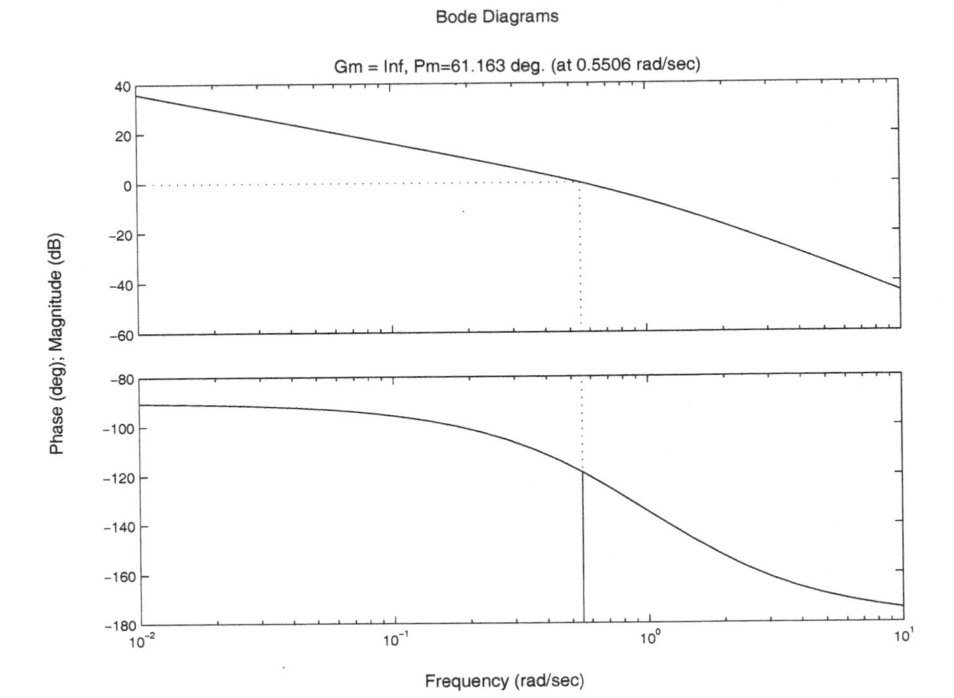

Fig. 7.3 The Bode Plots for $G_p(s) = \sqrt{2}/(s(s+1)), H(s) = 1, G_c(s) = 1/2.25$

attenuator and generally gives somewhat better results. The main advantage of a phase-lag compensator is that it does not significantly alter the low frequency gain of the system. It lowers the gain at high frequencies—something that though not an unmitigated blessing is sometimes useful. (Lowering the gain at high frequencies will tend to reduce the system's sensitivity to high frequency inputs. If the only high frequency "input" one expects is noise, then this reduction is a good thing.)

Designing a phase-lag compensator is relatively simple. One uses the phase-lag compensator to lower the gain of the system to one at the point, ω_1, at which the phase is $-180^\circ + \varphi + 5^\circ$ where φ is the desired phase margin. The main idea is that when s is near infinity the compensator's gain is ω_p/ω_o. Thus, one makes certain that ω_p and ω_o are small enough that $G_c(j\omega_1) \approx \omega_p/\omega_o$—one makes sure that ω_1 is effectively "at infinity" relative to ω_p and ω_o. The general method is:

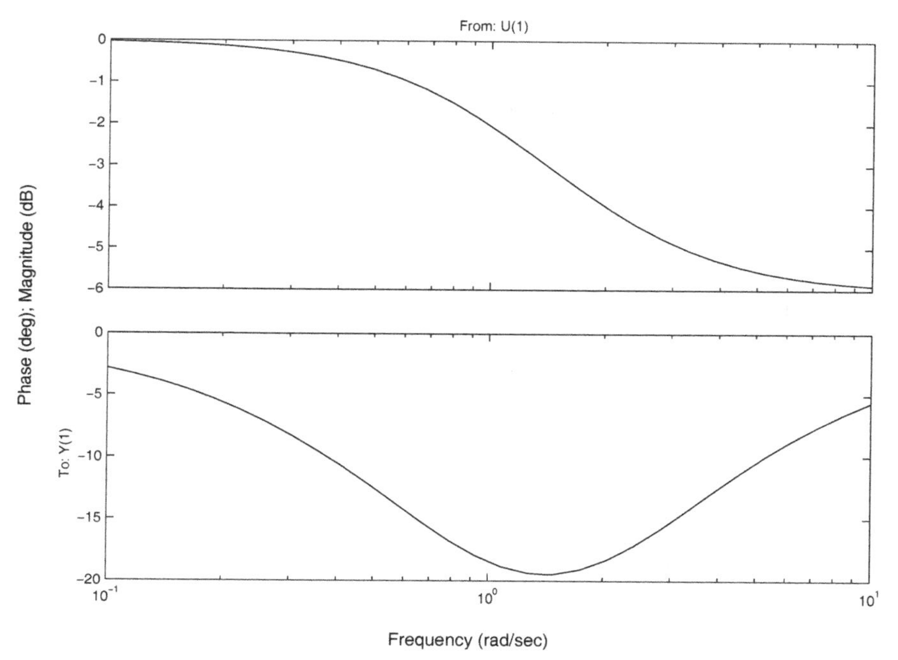

Fig. 7.4 The Bode Plots of a Typical Phase-Lag Compensator

(1) Find ω_1—the frequency at which $\angle G_p(j\omega_1) = -180^o + \varphi + 5^o$—and calculate $|G_p(j\omega_1)|$.
(2) Let $\omega_o = 0.1 \cdot \omega_1$.
(3) Let $\omega_p/\omega_o = 1/|G_p(j\omega_1)| < 1$.
(4) Finally, set:

$$G_c(s) = \frac{s/\omega_o + 1}{s/\omega_p + 1}.$$

$G_p(s) = \frac{\sqrt{2}}{s(s+1)}$—An Example

Once again we would like to design a compensator that will give us a system whose phase margin is 60°. After carefully considering Figure 7.2 (and performing a couple of calculations to add accuracy) we find that $w_1 = 0.47$ and $|G_p(j\omega_1)| = 2.7$.

Thus, $w_o = 0.047$, and $\omega_p = 0.047/2.7 = 0.017$. We see that:

$$G_c(s) = \frac{s/0.047 + 1}{s/0.017 + 1} = \frac{21.3s + 1}{57.4s + 1}$$

The Bode plots of the compensated system are given in Figure 7.5. Note that the phase margin is almost exactly 60°.

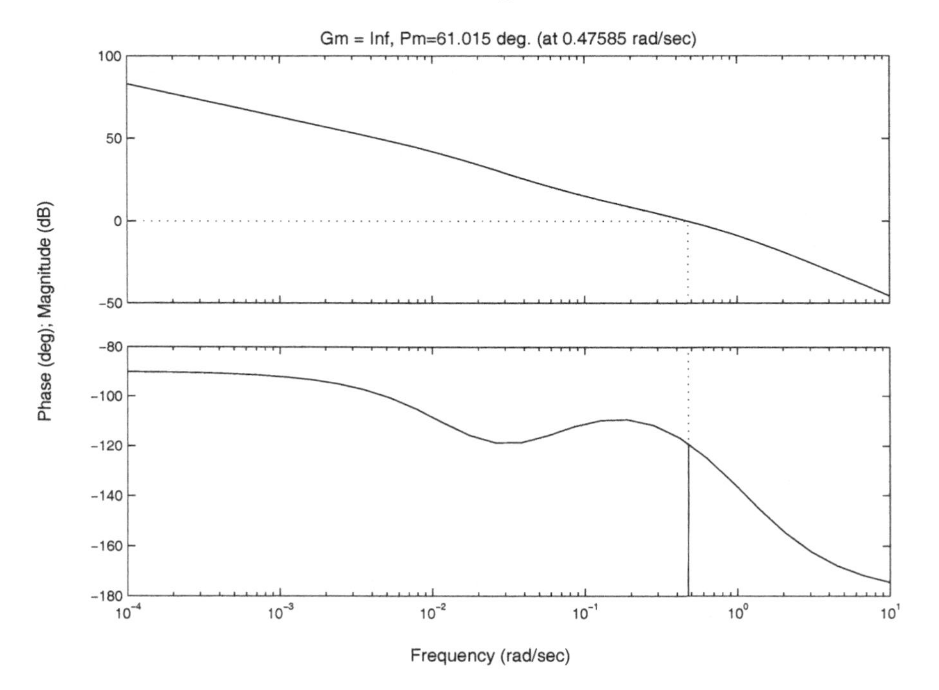

Fig. 7.5 The Bode Plots for $G_p(s) = \sqrt{2}/(s(s+1)), H(s) = 1, G_c(s) = (21.3s + 1)/(57.4s + 1)$

What differences do the two types of compensation make to the final system? Consider Figure 7.6 in which the step responses of the two systems are compared. We find that the response of the system with the attenuator alone is better (both faster and has less overshoot) than the response of the system that uses phase-lag compensation. Is that reasonable? Yes it is. The attenuation of the attenuator is only 2.25, while the attenuation of the phase-lag compensator at high frequencies is 2.7. As the attenuation at high frequencies is larger, it is reasonable that the system's response is slower. Overshoot is sufficiently complicated that it is hard to say how reasonable or unreasonable our results are.

Considering what we have just seen, why prefer phase-lag compensation? Because $G_p(s)$ has one pole at the origin, we know that its steady-state response to a unit step is perfect—and we see this in both of the step responses is Figure 7.6. However, since $G_p(s)$ has only one pole at the origin, the response to a ramp should not be perfect. There should be some steady-state error. We compare the unit ramp responses of the two systems in Figure 7.7. We see that the response of the system that uses phase-lag compensation is in fact closer to a unit ramp, $tu(t)$, than the response of the system whose compensation is performed by means of a simple attenuator. We see that the one thing that we can be certain of is that phase-lag compensation gives better steady-state performance than compensation using a simple attenuator. As that is really all that phase-lag compensation is designed to add, it should not come as a big surprise that (at times) that is all that it does add.

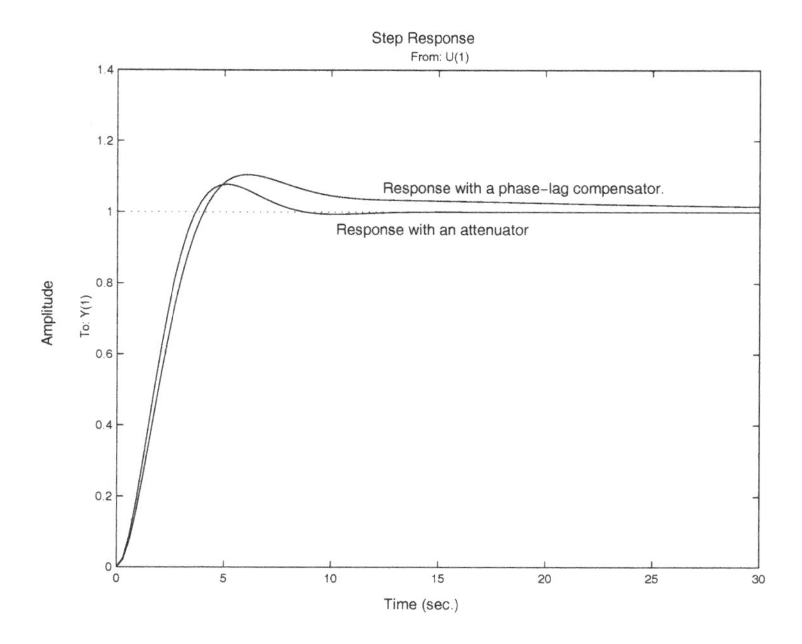

Fig. 7.6 A Comparison of the Unit Step Responses of the Two Systems

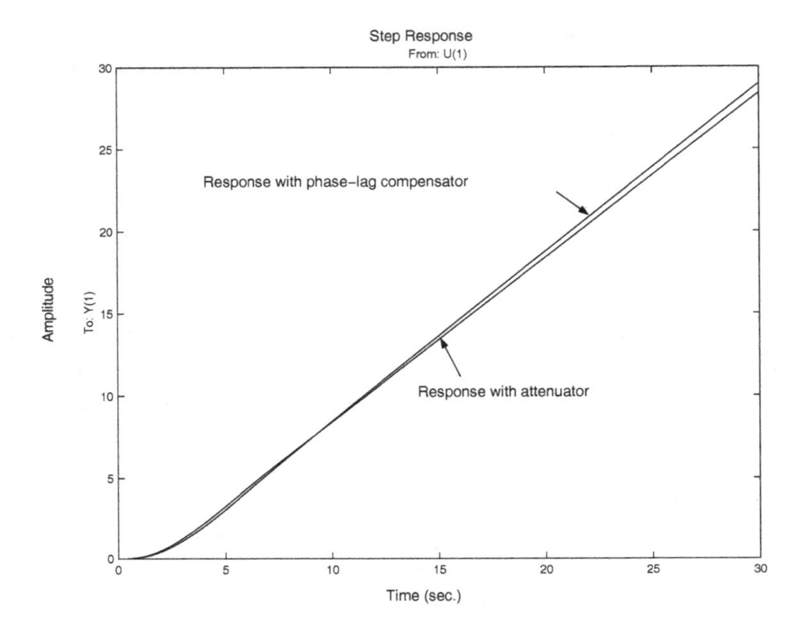

Fig. 7.7 A Comparison of the Unit Ramp Responses of the Two Systems

In the rules for designing a phase-lag compensator, we find that we are interested in ω_1—the frequency at which the phase of $G_p(j\omega)$ is 5° farther from -180° than the phase margin seems to need. Why must we add five degrees? Let us consider our compensator at ω_1. We find that:

$$\angle G_c(j\omega_1) = \angle \frac{j\omega_1/\omega_o + 1}{j\omega_1/\omega_p + 1} = \angle(j10 + 1) - \angle(j\overbrace{\omega_1/\omega_p}^{>10} + 1).$$

It is clear that this angle is between 0° and $\angle(j10 + 1) - 90^\circ \approx -5.7^\circ$. For this reason we add an additional five degrees—the compensator itself may subtract five degrees from the phase at the frequency of interest.

Considering that the compensator can contribute a relatively substantial phase, why don't we worry that the attenuation is not as large as the intended attenuation? Let us perform a quick calculation:

$$|G_c(j\omega_1)| = \left|\frac{10j + 1}{j(\omega_1/\omega_p) + 1}\right| = \sqrt{101}/\sqrt{100(\omega_0/\omega_p)^2 + 1}.$$

It is easy to show[2] that $\sqrt{1+\epsilon} \approx 1 + \epsilon/2$. Thus we find that:

$$|G_c(j\omega_1)| \approx \frac{10(1+.01/2)}{10(\omega_p/\omega_o)\left(1+\frac{1}{200(\omega_o/\omega_p)^2}\right)}$$
$$= \frac{1.005}{1+\frac{1}{200(\omega_o/\omega_p)^2}}\frac{\omega_p}{\omega_o}.$$

This number is *very* close to ω_p/ω_o—it is a much closer match than the angles were. The magnitude of the compensator is almost exactly correct at ω_1 though the phase may be off by almost six degrees.

7.4 Phase-Lead Compensation

Phase-lag compensation is generally used because it gives frequency dependent attenuation. Phase-lead compensation is used because it adds phase to $G_p(s)H(s)$. It improves the phase margin by "main force."

The transfer function of a phase-lead compensator is:

$$G_c(s) = \frac{s/\omega_o + 1}{s/\omega_p + 1}, \quad \omega_o < \omega_p.$$

This is essentially the same as a phase-lag compensator except that here $\omega_o < \omega_p$. The effect of this change is to make the numerator "start adding" phase before the denominator "starts subtracting" it. Another consequence of this difference is that for large s we find that:

$$\lim_{s\to\infty} T(s) = \frac{\omega_p}{\omega_o} > 1.$$

I.e. at high frequencies the phase-lead compensator acts as an amplifier.

As our goal when a adding a phase-lead compensator is to add phase, we ought to see where the phase is added. Let us consider the phase of the compensator as a function of ω. We find that:

$$\angle G_c(j\omega) = \tan^{-1}(\omega/\omega_o) - \tan^{-1}(\omega/\omega_p).$$

Making use of the identity:

$$\tan(a-b) = \frac{\tan(a) - \tan(b)}{1 + \tan(a)\tan(b)},$$

[2]The first two terms in the Taylor series expansion of $\sqrt{1+x}$ are 1 and $x/2$. Thus for small $|x|$ we find that $\sqrt{1+x} \approx 1 + x/2$.

we find that:

$$\tan(\angle(G_c(j\omega))) = \frac{\omega/\omega_o - \omega/\omega_p}{1 + \omega^2/(\omega_o\omega_p)} = \frac{\omega\omega_p - \omega\omega_o}{\omega_o\omega_p + \omega^2}. \tag{7.1}$$

As the phase of the compensator is always between 0° and 90° and as $\tan(x)$ is increasing in this range, to find the value of ω for which the phase of the compensator is maximum it is sufficient to find the maximum of (7.1). Differentiating the right hand side of the equation with respect to ω and setting the result to zero, we see that we must find the roots of:

$$\frac{(\omega_p - \omega_o)(\omega_o\omega_p + \omega^2) - 2\omega^2(\omega_p - \omega_o)}{(\omega_o\omega_p + \omega^2)^2} = 0$$

A little bit of algebra shows that this is solved by $\omega^2 = \omega_o\omega_p$. As we are only interested in positive values of ω, we find that $\omega = +\sqrt{\omega_o\omega_p}$. In order to find the maximum of (7.1), we must find the maximum value that the function takes at its critical points—the points at which the derivative of the function is zero—and at the endpoints of the interval. The range of ω is from 0 to infinity. Because the phase tends to zero as $\omega \to \infty$, and as the phase is 0 at $\omega = 0$, it is clear that the maximum of the function occurs at the critical point

$$\omega_{max} = \sqrt{\omega_o\omega_p}. \tag{7.2}$$

We find that:

$$\text{maximum phase} = \tan^{-1}\left\{\frac{1}{2}\left(\sqrt{\frac{\omega_p}{\omega_o}} - \sqrt{\frac{\omega_o}{\omega_p}}\right)\right\}. \tag{7.3}$$

What is the amplification that the phase-lead compensator contributes at this frequency? We find that:

$$|G_c(j\omega_{max})| = \sqrt{\frac{1 + \omega_p/\omega_o}{1 + \omega_o/\omega_p}} = \sqrt{\frac{\omega_o\omega_p + \omega_p^2}{\omega_o\omega_p + \omega_o^2}} = \sqrt{\frac{\omega_p}{\omega_o}}.$$

We find that both the maximum phase of the compensator and its gain at the point of maximum phase are functions of ω_p/ω_o alone. For the record, all of our conclusions hold for phase-lag compensators too–with *max* replaced by *min*, of course.

$G_p(s)H(s) = \sqrt{2}/(s(s+1))$ Revisited—An Example

As we saw in Figure 7.2, when $G_c(s) = 1$ the phase margin of this system is 45° and is achieved when $\omega = 1$. Let us try to add another 15° at this point by using a phase-lead compensator. We know that $\omega_{max} = \sqrt{\omega_p \omega_o}$. Let us try putting the maximum phase at $\omega = 1$. For this to happen, we find that:

$$\sqrt{\omega_p \omega_o} = 1 \Rightarrow \omega_p = 1/\omega_o.$$

A phase-lead compensator adds gain as well as phase. This gain will tend to change the point at which the phase margin is fixed. At ω_{max} the phase-lead compensator contributes a gain of $\sqrt{\omega_p/\omega_o}$. From Figure 7.2 we see that if one increases the gain, one tends to push the point at which $|G_c(j\omega)G_p(j\omega)H(j\omega)| = 1$ to the right. As one moves to the right, one finds that the phase of system gets more negative. Thus, it makes sense to choose the maximum phase of the phase-lead compensator to be a bit more than is strictly necessary. Let us try 20°. A bit of experimenting with (7.3)—or a simple calculation—leads us to the conclusion that if $\omega_p/\omega_o = (1.5)^2$ then the maximum phase of the phase-lead compensator is approximately 22°. We find that $\omega_p = 1.5, \omega_o = 1/1.5$. Thus, our compensator is:

$$G_c(s) = \frac{1.5s + 1}{s/1.5 + 1}$$

The Bode plots of the system with this compensator added are shown in Figure 7.8. After adding the compensator the phase margin increased from 45° to 57° and the frequency at which it is achieved moved from $\omega = 1$ rightward (to $\omega = 1.4$) as we expected.

In Figure 7.9 we show the system's unit step response. Comparing this result with Figure 7.6, we find that the step response is faster than the response with either of the other compensation schemes. This is reasonable; the phase-lead compensator amplifies the high-frequency components of the control signal—the components that allow a system to respond quickly.

In Figure 7.10 we show the response of the compensator to a unit step input to the system. We see that initially—after the step—the output is quite large. Intuitively this happens because the step has some very high frequency components which are amplified by the phase-lead compensator. It is not hard to prove that this must happen. The transfer function of a system with

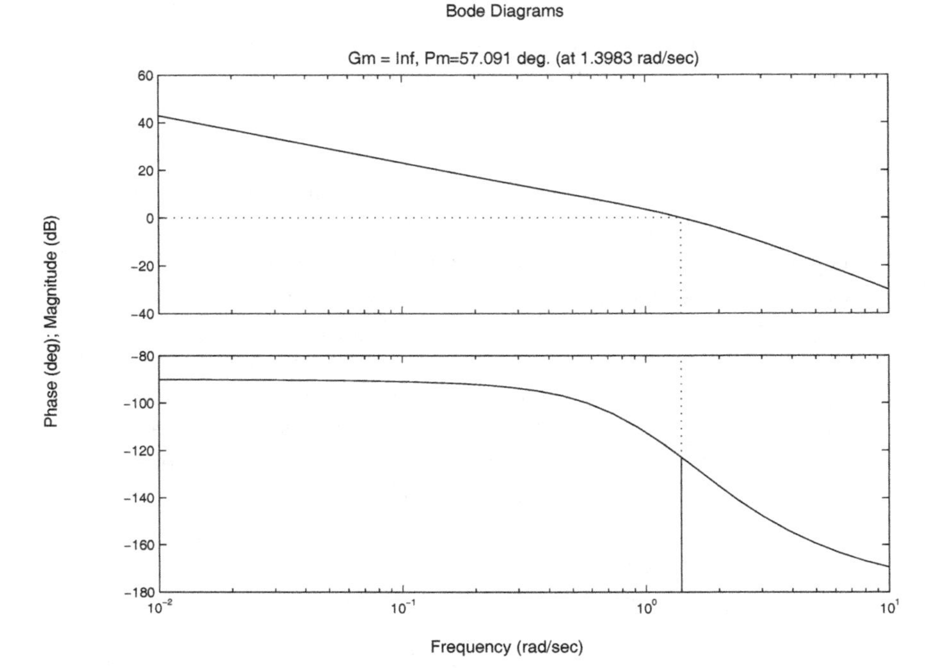

Fig. 7.8 The Bode Plots of the Phase-Lead Compensated System

unity feedback is:

$$T(s) = \frac{G_c(s)G_p(s)}{1 + G_c(s)G_p(s)}.$$

As the output of the compensator is the input to the plant, it is easy to see that the Laplace transform of the output of the compensator, $V_{comp}(s)$, is given by:

$$V_{comp}(s) = \frac{G_c(s)}{1 + G_c(s)G_p(s)} V_{in}(s).$$

Assuming that the plant's transfer function is low-pass—that $\lim_{s\to\infty} G_p(s) = 0$—and making use of the initial value theorem, we find that when the input to the system is a unit step, then the initial value of the output of the compensator satisfies:

$$\begin{aligned} v_{comp}(0^+) &= \lim_{\Re(s)\to\infty} s\frac{G_c(s)}{1+G_c(s)G_p(s)}\frac{1}{s} \\ &= \lim_{\Re(s)\to\infty} \frac{G_c(s)}{1+G_c(s)G_p(s)} \\ &= \frac{G_c(\infty)}{1+G_c(\infty)G_p(\infty)} \\ &= \frac{\omega_p}{\omega_o}. \end{aligned}$$

As in our case $\omega_p/\omega_o = 2.25$, the initial value of the output must be 2.25. This large initial output could be a real problem if the next component cannot handle a relatively large voltage. One must be careful that the output of the phase-lead compensator not be too large for the next element in the loop.

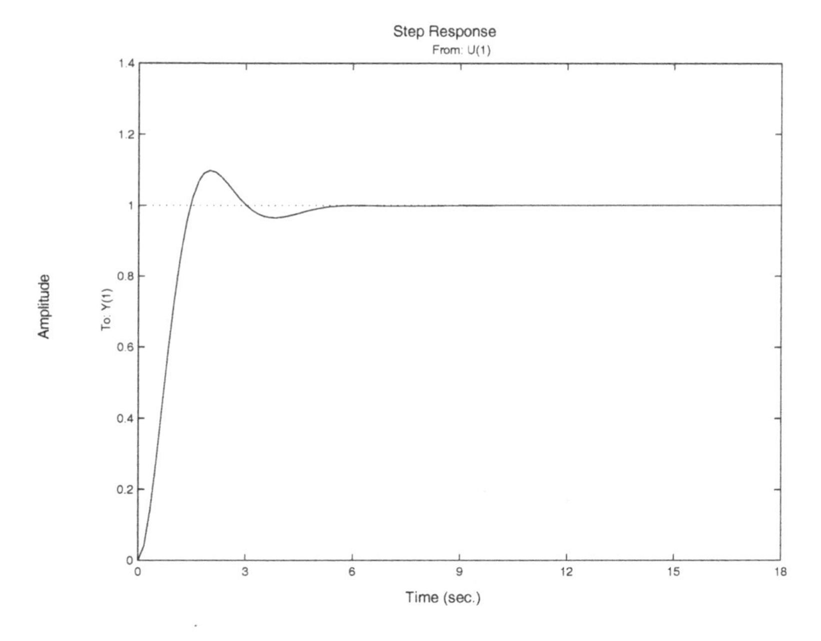

Fig. 7.9 The Step Response of the Phase-Lead Compensated System

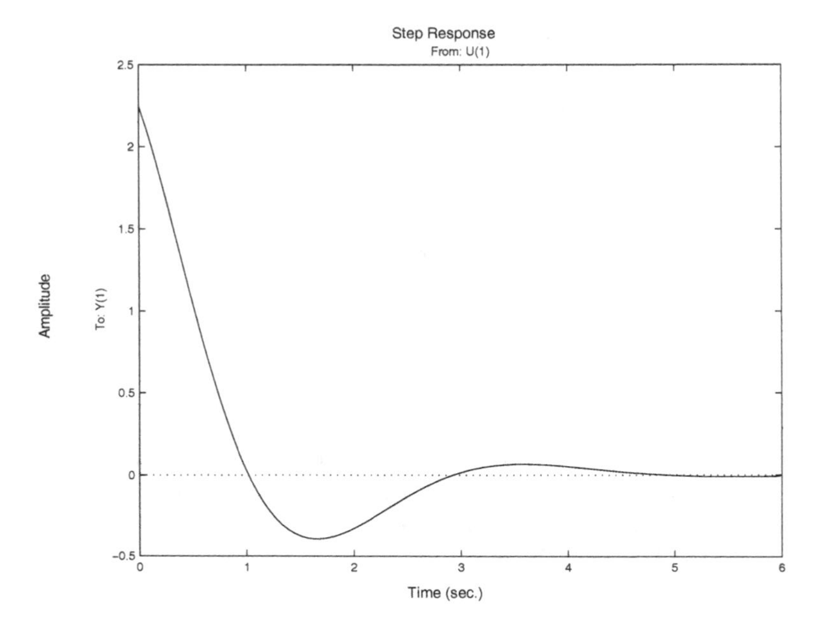

Fig. 7.10 The Response of the Phase-Lead Compensator to a Step at the System's Input

7.5 Lag-Lead Compensation

One way to get some of the benefits of both phase-lag and phase-lead compensation is to use both types of compensation at once. There are analytical methods for designing such compensators, but we will not examine them. We consider an example to see how one can benefit from using lag-lead compensation. Once again consider a system for which:

$$G_p(s)H(s) = \frac{\sqrt{2}}{s(s+1)}.$$

Let us use the same phase-lead compensator that we did in the previous section:

$$G_{lead}(s) = \frac{1.5s+1}{s/1.5+1}$$

However, this time we add a phase-lag compensator to provide an attenuation of $\sqrt{\omega_o/\omega_p} = 1/1.5$ at $\omega_{max} = 1$. From our general rules for phase-lag compensators, we see that $\omega_o = 0.1$ and that $\omega_p = 0.1/1.5 = 0.067$. Thus:

$$G_{lag}(s) = \frac{10s+1}{15s+1},$$

and our compensator is:

$$G_c(s) = \frac{10s+1}{15s+1}\frac{1.5s+1}{s/1.5+1}.$$

In Figure 7.11 we show the Bode plots of $G_p(s)H(s)G_c(s)$, in Figure 7.12 we show the step response, and in Figure 7.13 we show the compensator's response to a unit step input. The settling time of the system is about the same as that of the system with a phase-lead compensator (to understand why, see exercise 17), but the phase margin has improved quite a bit and now appears just where we expected—$\omega = 1.00$, and the compensator's response to a step input is not quite as large as it was previously. The lag-lead compensator is somewhat more complicated than the previous compensators, but it is also a somewhat better compensator.

7.6 The PID Controller

The PID (proportional-integral-derivative) controller[3] is a close relative of the lag-lead compensator and is one of the most commonly used compensators. Its transfer function is:

$$G_c(s) = K_D s + K_P + \frac{K_I}{s}.$$

Looking at the controller's transfer function, we see that the PID controller is composed of three basic elements. The first is a differentiator, the second is a pure gain, and the third element is an integrator. The output of the gain section is proportional to its input; this explains how the PID controller got its name. We note that a proportional-integral (or PI) controller—a PID controller with $K_D = 0$—is just a phase-lag compensator whose pole is located at zero and a proportional-derivative (or PD) controller—a PID controller with $K_I = 0$—is just a phase-lead compensator whose pole is located "at infinity." It is fairly simple to design a PID controller for a given plant, $G_p(s)$.

[3]The three-term controller (now generally referred to as the PID controller) was introduced in 1922 by Nichlas Minorsky for use in the steering of ships [Lew92].

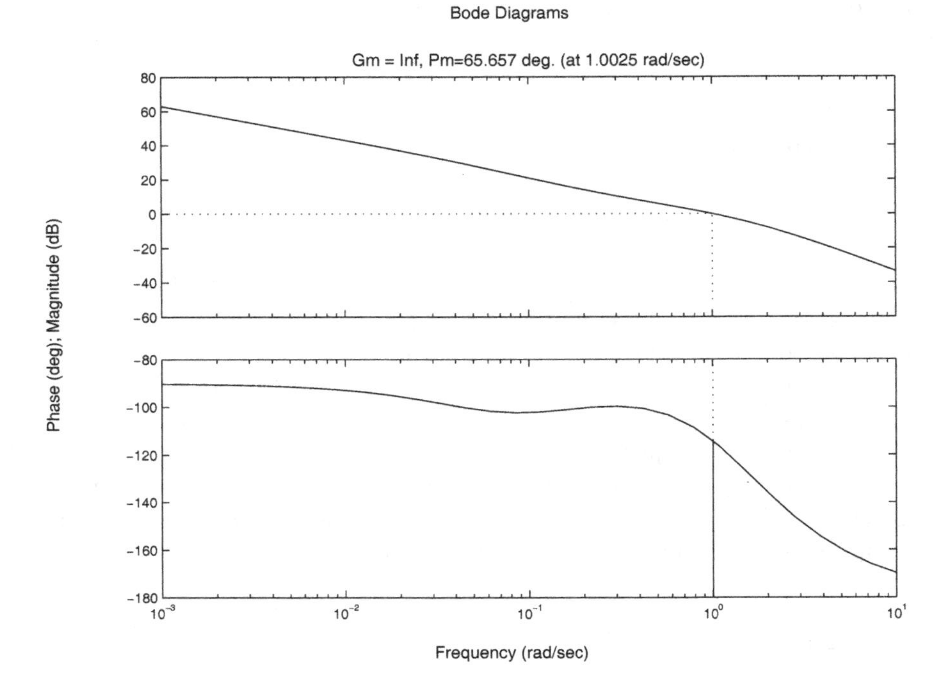

Fig. 7.11 The Bode Plots of $G_p(s)H(s)G_{lag}(s)G_{lead}(s)$

How do each of the components of the PID controller influence the behavior of the closed-loop system? The integral component contributes an additional pole at zero—this increases the order of the system by one and increases the gain at very low frequencies. The proportional component gives a frequency independent gain—it is an ideal amplifier. Finally the derivative component contributes a large gain at high frequencies. This generally speeds up the system.

As we have already seen, any of these three components can force a system into instability. In particular, if one already has one pole at zero, one may well drive one's system into instability by adding an integrator. Similarly adding a large gain to a system can drive the system into instability. Therefore, after adding a PID controller, it is particularly important to check that the resulting system is stable.

Let us consider an analytic technique for designing a PID controller.

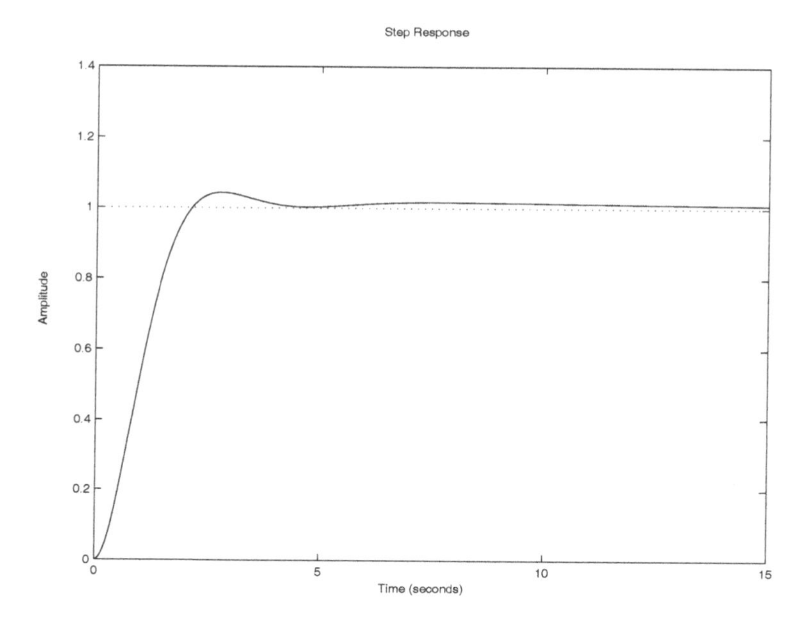

Fig. 7.12 The Response of the Lag-Lead Compensated System to a Unit Step

Suppose that one would like to force the phase margin of one's system to be φ degrees at frequency ω_1. Let $G_{PID}(s)$ be the transfer function of the PID controller. Then one would like to find values of $K_P, K_I,$ andK_D such that:

$$G_{PID}(j\omega_1)G_p(j\omega_1)H(j\omega_1) = 1\angle(\varphi - 180^o).$$

This means that:

$$|G_{PID}(j\omega_1)| = 1/|G_p(j\omega_1)H(j\omega_1)|$$
$$\Theta = \angle G_{PID}(j\omega_1) = \varphi - 180^o - \angle G_p(j\omega_1)H(j\omega_1)$$

We find that:

$$\begin{aligned} G_{PID}(j\omega_1) &= K_P + j(K_D\omega_1 - K_I/\omega_1) \\ &= |G_{PID}(j\omega_1)|(\cos\Theta + j\sin\Theta) \\ &= \frac{1}{|G_p(j\omega_1)H(j\omega_1)|}(\cos\Theta + j\sin\Theta) \end{aligned}$$

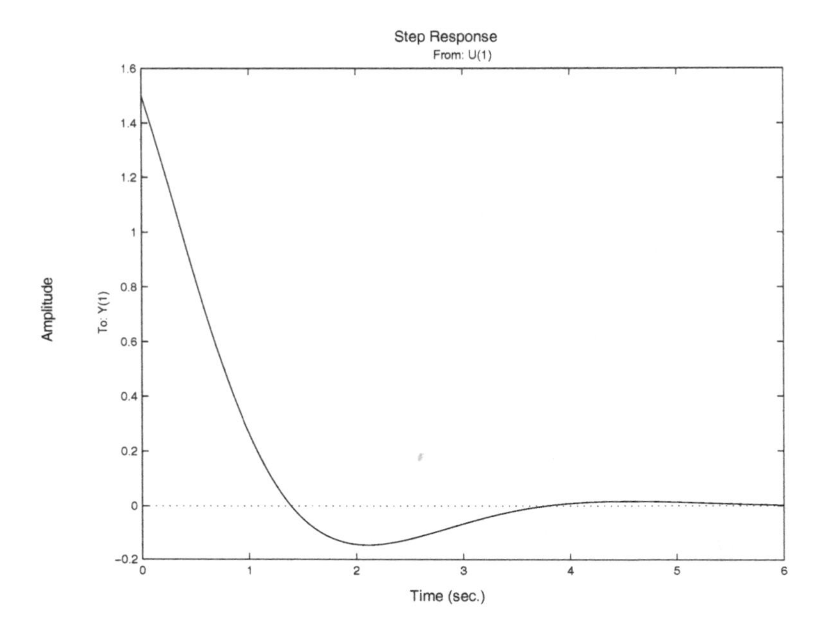

Fig. 7.13 The Response of the Lag-Lead Compensator to a Step at the System's Input

From this we see that:

$$K_P = \frac{\cos\Theta}{|G_p(j\omega_1)H(j\omega_1)|} \tag{7.4}$$

$$K_D\omega_1 - K_I/\omega_1 = \frac{\sin\Theta}{|G_p(j\omega_1)H(j\omega_1)|}. \tag{7.5}$$

Here we have a set of design equations for a PID controller.

Let us design a controller for a system for which

$$G_p(s)H(s) = \frac{\sqrt{2}}{s(s+1)}.$$

Suppose that we want a system with a phase margin of 60° at $\omega_1 = 1$. We find that:

$$|G_{PID}(j\cdot 1)| = 1$$

$$\Theta = 60^\circ - 180^\circ - (-135^\circ) = 15^\circ$$

This leads us to the equations:

$$K_P = \cos(15^o)$$
$$K_D - K_I = \sin(15^o)$$

Assuming that we do not want to add an integrator to the system (as it already has a pole at the origin), we find that $K_P = 0.9659$ and $K_D = 0.2588$. We show the bode plots of the compensated system in Figure 7.14 and the plot of the unit step response in Figure 7.15. We see that the phase margin is indeed 60° at $\omega_1 = 1$, and we see that the unit step response is about as good as the response with a phase lead or a lag-lead compensator. This is reasonable as all three of the systems amplify the high-frequencies and it is precisely the high-frequencies that allow the system to react quickly.

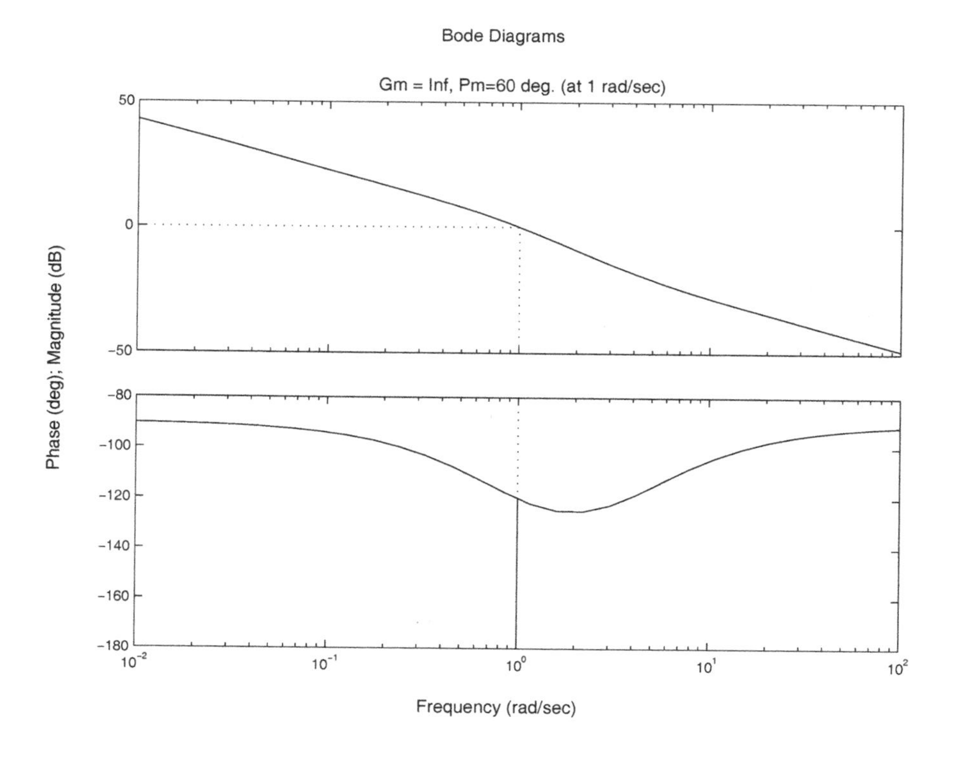

Fig. 7.14 The Bode Plots of $G_p(s)H(s)G_{PID}(s)$

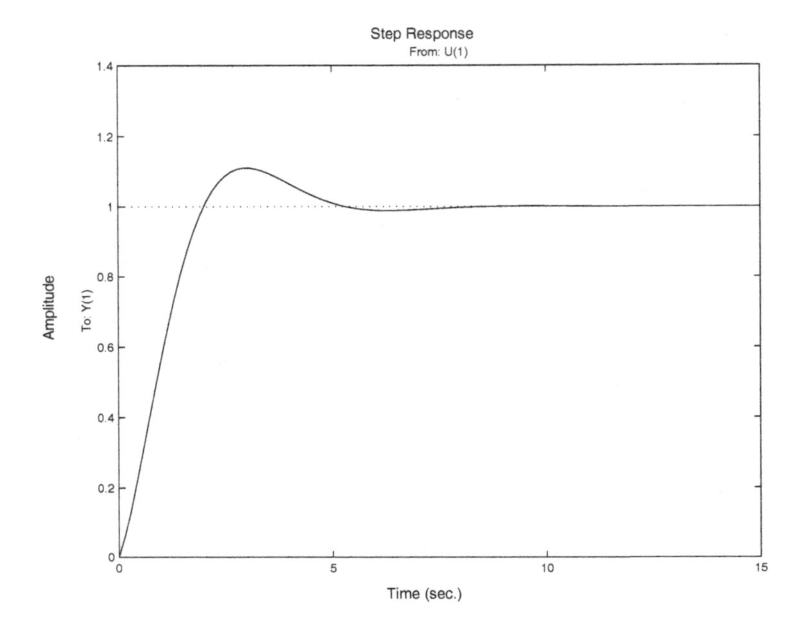

Fig. 7.15 The Response of the PID Compensated System to a Unit Step

If we take the phase margin, φ, as a given, but consider the constants ω_1, and K_P, K_I, and K_D as parameters, we find that ω_1 is a free parameter, K_P is determined by the value of ω_1 and φ, and the combination $K_D\omega_1 - K_I/\omega_1$ is fixed by the values of ω_1 and φ. How do we go about fixing ω_1 and the relation between K_D and K_I?

Often we fix ω_1 by making use of (5.1) which relates the system's settling time, the phase margin, and ω_1. Equation (5.1) states :

$$T_s \approx \frac{8}{\tan(\varphi)\omega_1}.$$

Thus, if we know the settling time and the phase margin we can also choose ω_1 in a sensible fashion. Though (5.1) really only applies to second order systems in standard form, as we have seen (p. 136) it is not unreasonable to use it as a rule of thumb for more general systems.

The choice of K_I and K_D is made on other grounds. As we saw in the previous example, sometimes K_I is set to zero because an additional integrator would be likely to harm the system. One might set K_D to zero if one did not want to add too much gain at high frequencies.

Let us try to make use of what we know to design a PID controller that gives us a phase margin of 60° and a settling time of 1s. To achieve the desired settling time we find that:

$$1 = \frac{8}{\tan(60^o)\omega_1} \Leftrightarrow \omega_1 = 4.6188.$$

We find that $G_p(j\omega_1)H(j\omega_1) = 0.0648\angle - 167.8^o$. Thus, $\Theta = 47.8^o$. We find that:

$$K_P = 10.4$$
$$K_D = 2.5$$

The Bode plots corresponding to this system are shown in Figure 7.16 and the unit step response is given in Figure 7.17. We see that the phase margin is as desired and that the system is substantially faster than the system we designed previously though it does not actually settle in one second.

Suppose that $G_p(s)H(s) = \frac{1}{s+1}$. In this case though the phase margin is very large—it is 135°—the system will not properly track a step input. Suppose that we want the system with a 60° phase margin, a settling time of $T_s = 1.5s$, and we want the system to track a step input. We must add a pole at zero in order for our system to track a step input—if we are to use a PID controller, then $K_I \neq 0$. The larger K_I is the more nearly our system will track a ramp input. Additionally, in order to speed up the system it is reasonable to choose $K_D \neq 0$.

We find that:

$$\omega_1 = \frac{8}{\tan(60^o)1.5} = 3.08$$
$$\Theta = 60^o - 180^o - (-72.0^o) = -48.0^o$$
$$|G_p(\omega_1)H(\omega_1)| = 0.44.$$

With these numbers, we see that:

$$K_P = \frac{\cos(-48^o)}{0.44} = 1.53$$
$$3.08K_D - K_I/3.08 = \frac{\sin(-48^o)}{0.44} = -1.70.$$

If we choose $K_I = 6$, then we find that $K_D = 0.08$. That is, our compensator is $G_{PID}(s) = 1.53 + 0.08s + 6.0/s$. We show the Bode plots of the compensated system and the unit step response in Figures 7.18 and 7.19 respectively. Note that once again the system is not as fast as we hoped

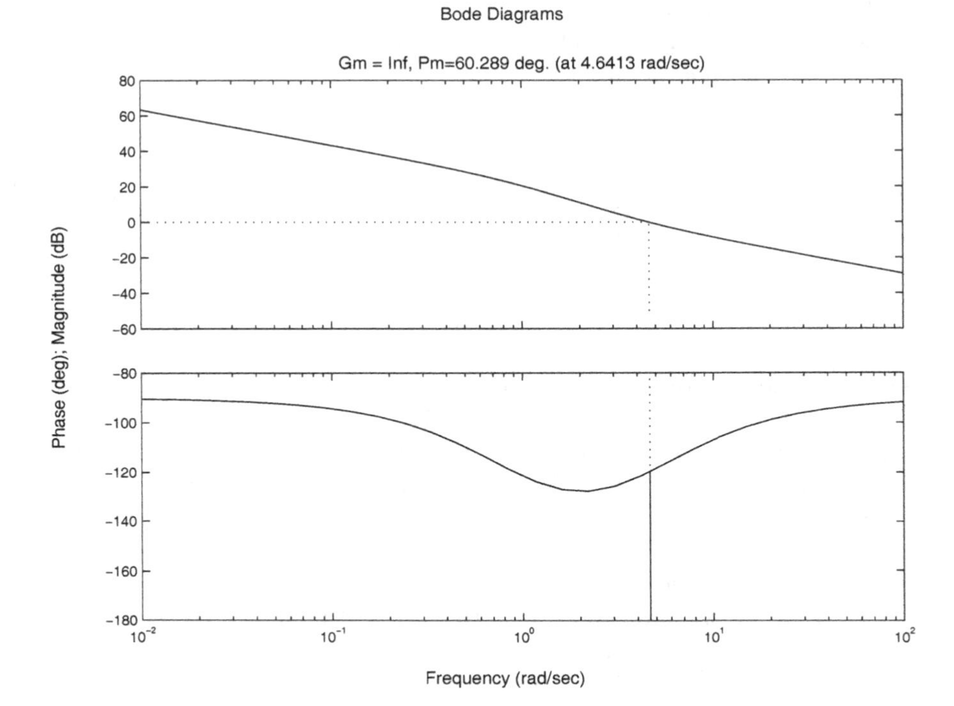

Fig. 7.16 The Bode Plots of $G_p(s)H(s)G_{PID}(s)$

for. It is important to remember that for systems other than the system for which it was derived (5.1) is only a rule of thumb. (For more information about choosing the coefficients of a PID controller, about *tuning* a PID controller, see §7.8.)

7.7 An Extended Example

Let us consider the system of Figure 7.1 but with $\sqrt{2}/(s(s+1))$ replaced by:

$$G_p(s) = \frac{s+30}{s(s+2)(s+3)(s+4)}.$$

The Bode plots for this system with $G_c(s) = 1$ are given in Figure 7.20.

Suppose that we would like to improve the phase margin to 60° and that

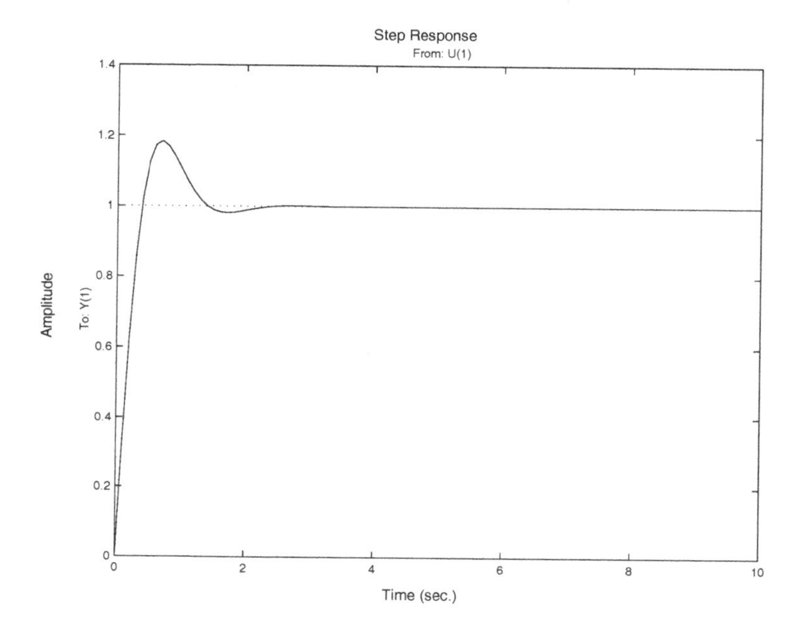

Fig. 7.17 The Response of the PID Compensated System to a Unit Step

we do not care what the crossover frequency is. Then we have several options. Let us start by designing an attenuator and a phase-lag compensator to achieve this phase margin.

7.7.1 *The Attenuator*

In order to use a pure attenuation, we must find the point on the Bode plots at which the phase of our plant is -120°. We must then force the gain at that point to be exactly one. A careful examination of the Bode plots shows that at $\omega \approx 0.5$ the phase is -120° and the magnitude is 7.5 dB. Let $G_c(s) = -7.5$ db $= 1/2.4$. The Bode plots of the compensated system are given in Figure 7.21.

7.7.2 *The Phase-Lag Compensator*

Following our technique for designing phase-lag compensators, we must find the frequency at which $\angle G_p(j\omega_1) = -115^{\circ}$. (We must not forget about the 5° security margin that this technique requires.) A careful examination of

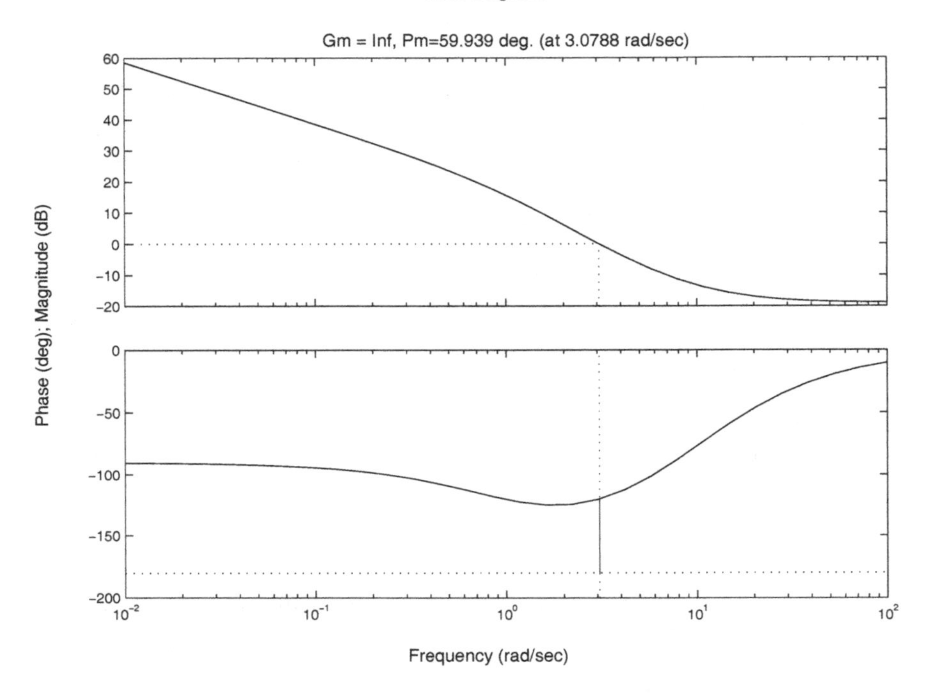

Fig. 7.18 The Bode Plots of $G_p(s)H(s)G_{PID}(s)$

Figure 7.20 shows that $\omega_1 = 0.42$; at this point $|G_p(j\omega_1)| = 9.2$ dB $= 2.9$.

As $\omega_1 = 0.42$, we find that $\omega_o = 0.1 \cdot \omega_1 = 0.042$. As the "gain at infinity", $G_c(\infty) = \omega_p/\omega_o$, must make the gain at ω_1 (which is "at infinity" for our purposes) equal to one in magnitude, we find that:

$$\frac{\omega_p}{\omega_o} = \frac{1}{2.9} \Leftrightarrow \omega_p = \omega_o/2.9 = 0.0145.$$

Thus our phase-lag compensator is:

$$G_c(s) = \frac{s/\omega_o + 1}{s/\omega_p + 1} = \frac{23.8s + 1}{69.0s + 1}.$$

The Bode plots that correspond to the system with the compensator added are given in Figure 7.22

We now add a constraint—we would like to make our system somewhat faster, so we require that ω_1 be equal to 2 rad/sec. To achieve this

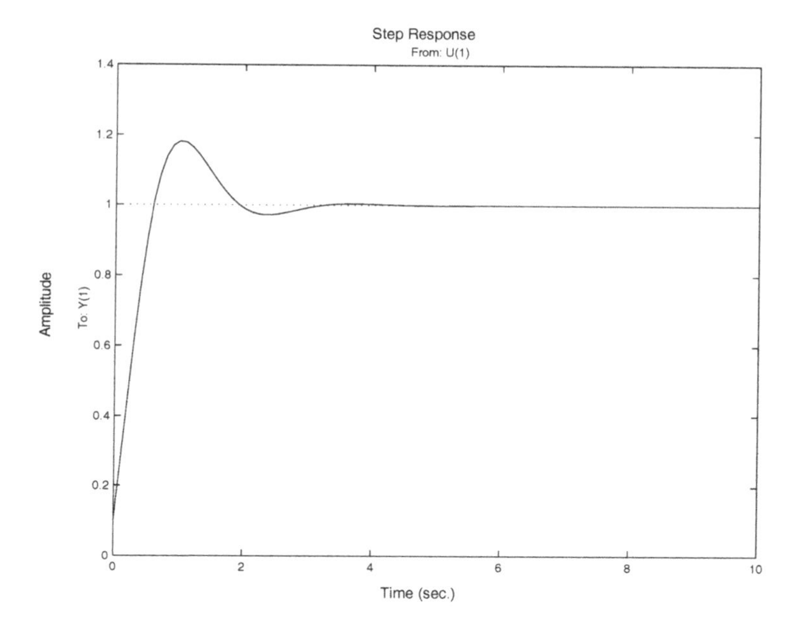

Fig. 7.19 The Response of the PID Compensated System to a Unit Step

goal, we design a phase-lead compensator. After designing the phase-lead compensator, we design a lag-lead compensator to perform the same task.

7.7.3 *The Phase-Lead Compensator*

From Figure 7.20, we find that at $\omega = 2$, we have:

$$|G_p(2j)| = -9.65 \text{ dB} = 1/3.04, \qquad \angle G_p(2j) = -191^\circ.$$

We find that we must add 71° to the system while at the same time increasing the gain by 3.04. In order to choose the ratio of ω_p to ω_o we make use of Figure 7.23—in which the value of the maximum phase and of the gain at the point of maximum phase are plotted as functions of ω_p/ω_o. A careful examination of this plot shows that in order to get the 71° that we need we must let $\omega_p/\omega_o = 36$. At this ratio, the gain at our frequency will be 6. In order to force the point of maximum gain to occur at $\omega_1 = 2$, we must have:

$$\sqrt{\omega_p \omega_o} = 2 \Leftrightarrow \omega_o = 4/\omega_p.$$

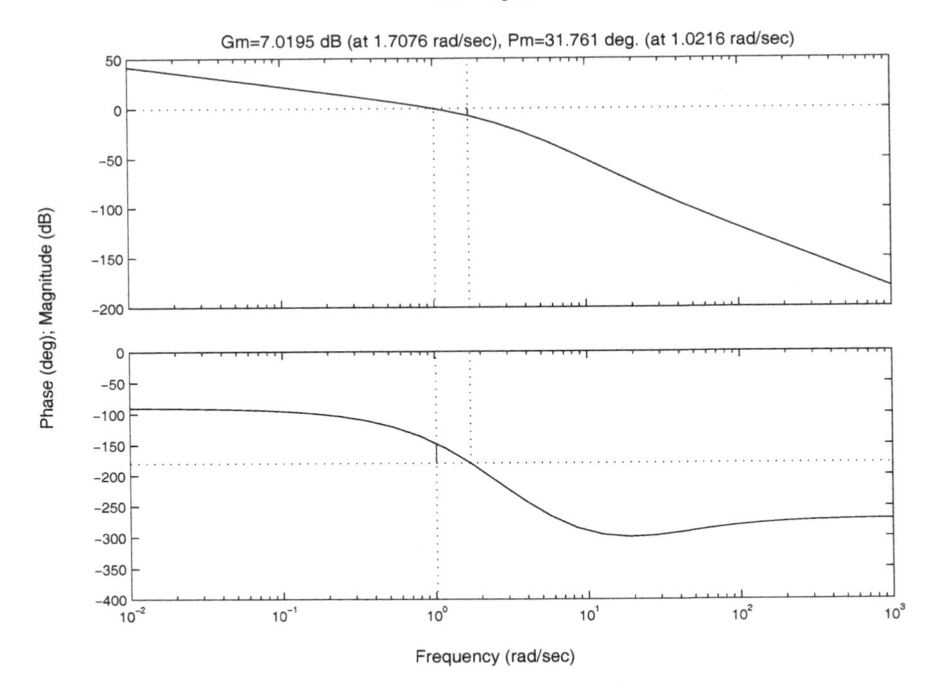

Fig. 7.20 The Bode Plots Corresponding to $G_p(s)$

As $\omega_p/\omega_o = 36$, we find that:

$$\omega_p/\omega_o = \omega_p^2/4 = 36 \Rightarrow \omega_p = \sqrt{144} = 12.$$

Clearly then, $\omega_o = 4/12 = 0.33$.

To make this design work, we will also need to add an attenuator so that the attenuation times $6/3.04 \approx 2.0$ is reduced to one. Thus the attenuation of the attenuator must be 0.5. The Bode plots of the system with the compensator:

$$G_c(s) = 0.5\frac{3.0s + 1}{0.083s + 1}$$

are given in Figure 7.24. We compare the step response of the uncompensated system with the step response of the phase-lead compensated system in Figure 7.25. Note that the phase-lead compensated system is somewhat

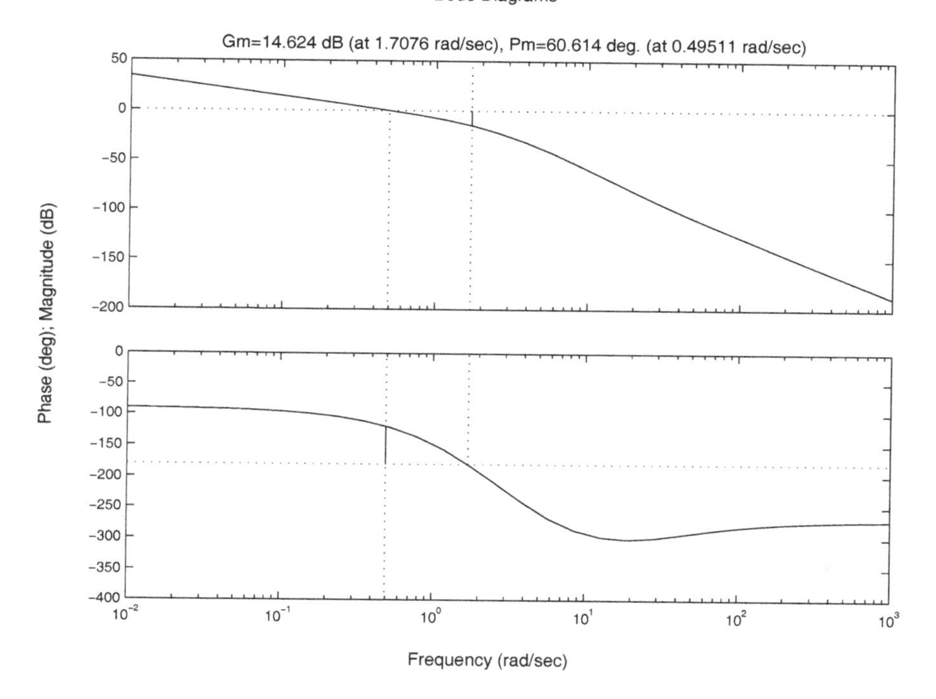

Fig. 7.21 The Bode Plots Corresponding to $G_p(s)/2.4$

faster that the uncompensated system.

7.7.4 *The Lag-Lead Compensator*

In order to design a lag-lead compensator to give us 71° at $\omega = 2$, we design a phase-lead compensator to add $71^o + 5^o = 76^o$ at $\omega = 2$, and then we use the phase-lag compensator to lower the gain at that point to 0 dB = 1. Considering Figure 7.23, we find that in order to achieve a maximum phase of 76° we must make $\omega_p/\omega_o = 66$. Thus, the gain of the compensator at the point of it maximum phase—which we must choose to be $\omega = 2$—is $\sqrt{66} = 8.1$. Thus the attenuation of the phase-lag filter "at infinity" must be $3.04/8.1 = 1/2.7$.

For the phase-lead compensator, we know that $\sqrt{\omega_p \omega_o} = 2$. Thus, $\omega_o = 4/\omega_p$. Additionally, $\omega_p/\omega_o = 66$. Thus, $\omega_p = \sqrt{4 \cdot 66} = 16.2$. From

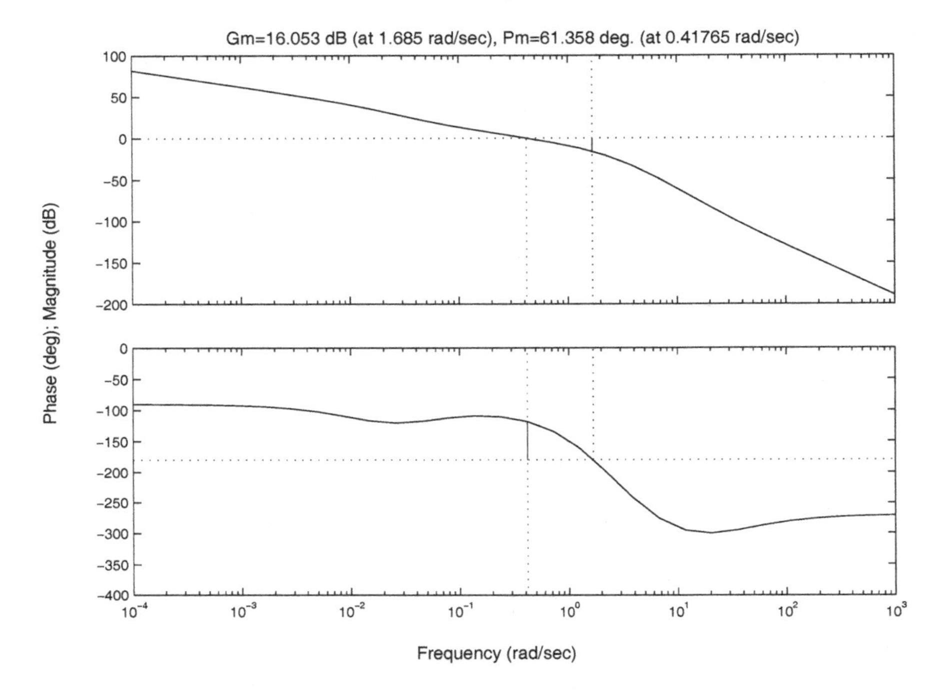

Fig. 7.22 The Bode Plots Corresponding to the System with a Phase-Lag Compensator

this, we find that $\omega_o = 4/\omega_p = 0.246$. Thus the phase-lead compensator is:

$$G_{lead}(s) = \frac{4.06s + 1}{0.062s + 1}.$$

As $\omega_1 = 2$, we find that for the phase-lag compensator, $\omega_o = 0.1 \cdot \omega_1 = 0.2$. As the attenuation "at infinity" must be $1/2.7$, we find that $\omega_p = \omega_o/2.7 = 0.074$. Thus, the phase-lag compensator is:

$$G_{lag}(s) = \frac{5s + 1}{13.5s + 1}.$$

The Bode plots that correspond to the lag-lead compensated system are shown in Figure 7.26 and the step response of the system is shown in Figure 7.27.

Finally let us try to compensate the system using a PID controller. As our system has already got one pole at the origin, it is probably best to

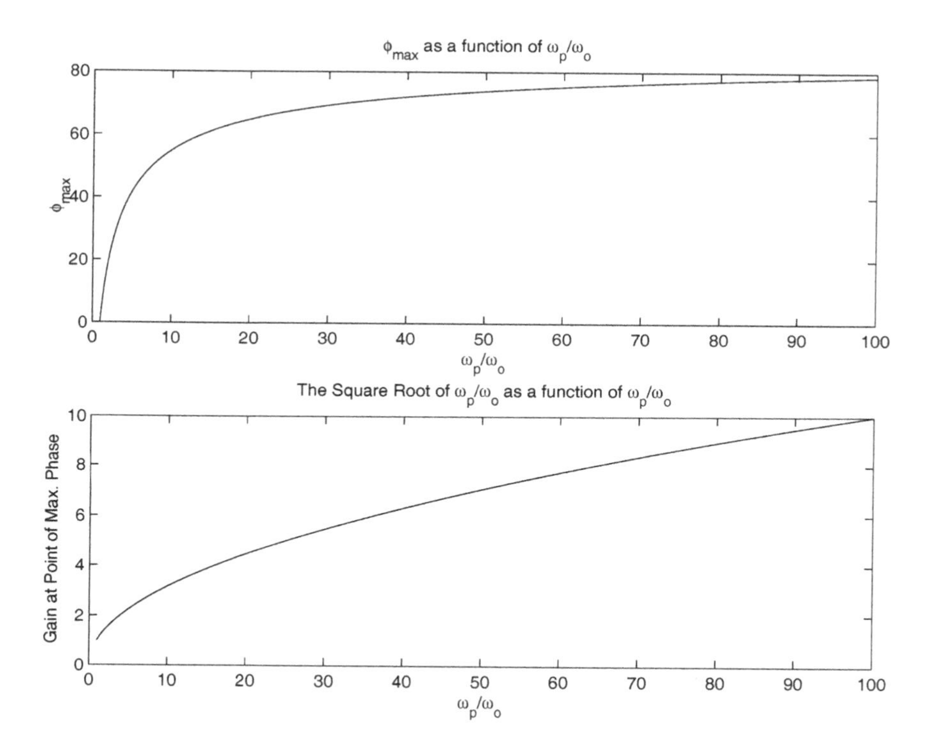

Fig. 7.23 Design Curves for a Phase-Lead Compensator

use a PD controller so that we do not increase the number of poles at the origin.

7.7.5 *The PD Controller*

Making use of the notation and the design equation for the PID controller, we find that:

$$\Theta = 60^o - 180^o - (-191^o) = 71^o.$$

With $K_I = 0$, we find that:

$$K_P = \frac{\cos(\Theta)}{|G_p(2j)|} = \frac{0.326}{1/3.04} = 0.990$$

$$K_D = \frac{1}{\omega_1}\frac{\sin(\Theta)}{|G_p(2j)|} = \frac{1}{2}\frac{0.9455}{1/3.04} = 1.44.$$

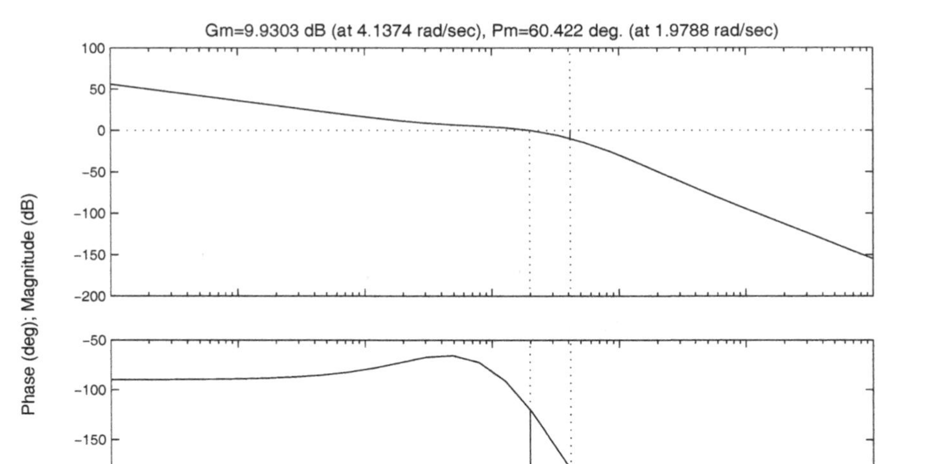

Fig. 7.24 The Bode Plots Corresponding to the System with a Phase-Lead Compensator and an Attenuator

Thus, the PD compensator is:

$$G_c(s) = 1.437s + 0.99.$$

The Bode plots that correspond to the PD compensated system are shown in Figure 7.28 and the step response of the system is shown in Figure 7.29.

7.8 The Ziegler-Nichols Rules

7.8.1 *Introduction*

It is often necessary to choose the coefficients of a controller without having a complete understanding of the system for which the controller is being

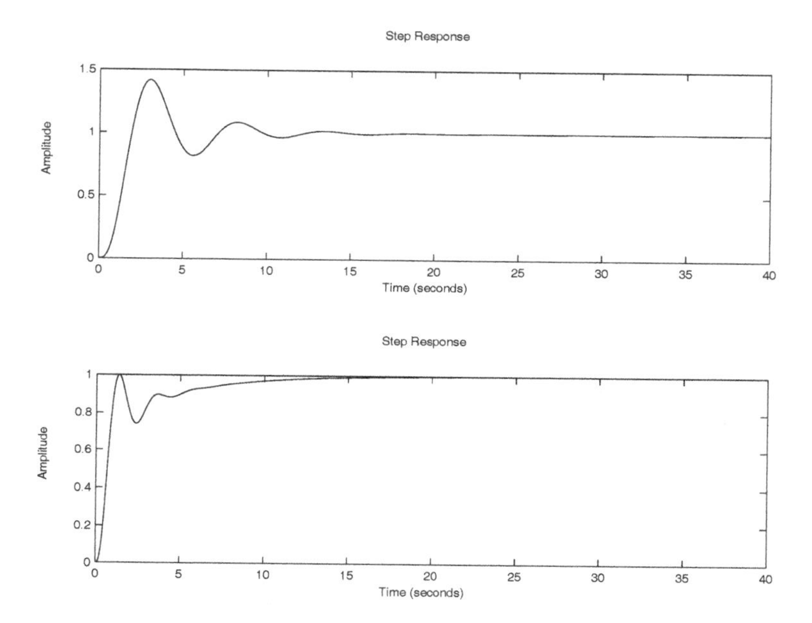

Fig. 7.25 A Comparison of the Step Responses of the Uncompensated System (Top) and the Phase-Lead Compensated System (Bottom)

designed. In 1942, J. G. Ziegler and N. B. Nichols[4] published a set of design rules for use in the design of PID controllers when $G_p(s)$ is not fully characterized [ZN42]. Though these rules are not perfect, they can serve as an intelligent starting point in the design of a PID controller. The rest of this section is a brief introduction to one of the cases in which a version of the design rules can be used.

7.8.2 *The Assumption*

In developing the Ziegler-Nichols rules, we assume that the transfer function of the plant being considered is well approximated by:

$$G_p(s) = \frac{Ce^{-Ls}}{Ts+1}, \qquad T, L > 0. \tag{7.6}$$

[4]This work was performed while Ziegler and Nichols (1914 – 1997) were at Taylor Instruments [Kah97].

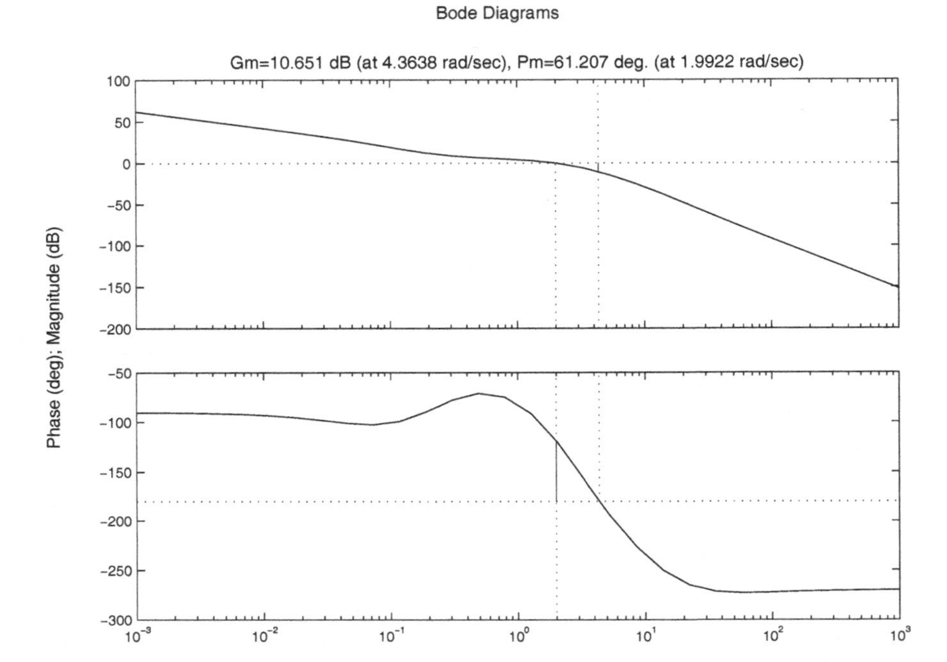

Fig. 7.26 The Bode Plots of the Lag-Lead Compensated System

We assume that C, L, and T are *not* known initially, and we describe how one can make a rather simple measurement on the closed-loop system that serves to partially characterize the system and then use the measurement to choose the coefficients of, to *tune*, the PID controller.

7.8.3 *Characterizing the Plant*

We make use of measurements made on the closed-loop system to characterize the plant. Construct a unity feedback closed-loop system whose plant is the plant being considered and whose compensator is a variable gain—whose compensator's transfer function is $G_c(s) = K_P$. (This can be done by installing a PID controller—which will be done anyhow—and initally setting the coefficients of the integrator and the differentiator to zero.) Initially, take K_P to be a very low gain, and increase the gain until

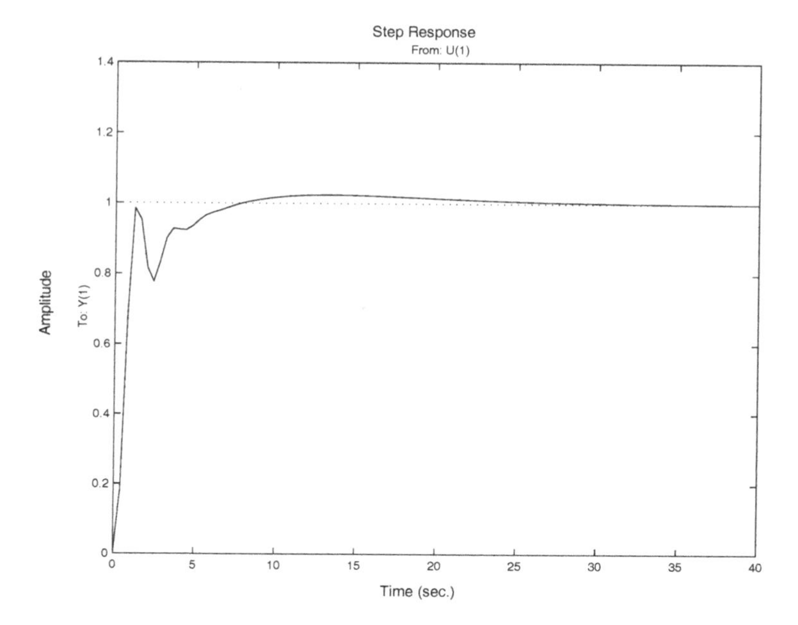

Fig. 7.27 The Step Response of the Lag-Lead Compensated System

the closed-loop system displays persistent oscillations at some fixed frequency. That is, use a proportional controller, start with the gain, K_P, set to be very low, and then increase the gain until the system becomes marginally stable[5]. We refer to the critical gain as K_{cr} and to the period of the oscillations as P_{cr}, and we use these parameters to characterize the system.

The Ziegler-Nichols rules for this case are:

(1) Let $K_P = 0.6K_{cr}$.
(2) Let $K_I = 1.2K_{cr}/P_{cr}$.
(3) Let $K_D = 0.075K_{cr}P_{cr}$.

After making the measurements with $K_I = K_D = 0$, the parameters of the PID controller are set to the values given above.

[5]In [ZN42], it is assumed that the behavior of the physical system agrees with that of the model at least to the extent that the closed-loop system is stable for sufficiently low gains.

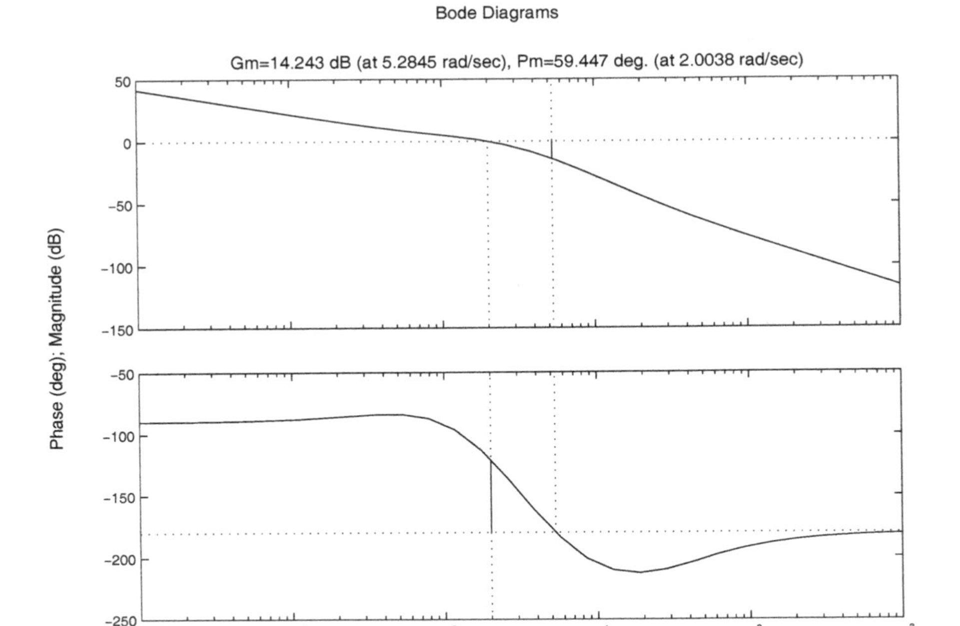

Fig. 7.28 The Bode Plots of the PD Compensated System

7.8.4 *A Partial Verification*

Let us see we what kind of system one ends up with if one uses these rules when $L \ll T$ in (7.6). Considering the nature of the Nyquist plot associated with $K_P G_p(s)$, it is clear that the gain that makes the system critically stable, K_{cr}, satisfies:

$$\frac{K_{cr}Ce^{-Lj\omega}}{Tj\omega+1} = 1\angle(-\pi).$$

As $L \ll T$ and as the angle must be $-\pi$, we find that ωL must be of order one and, therefore, $\omega T \gg 1$. Thus, we find that:

$$\frac{K_{cr}Ce^{-Lj\omega}}{Tj\omega+1} \approx \frac{K_{cr}Ce^{-Lj\omega}}{Tj\omega} \approx 1\angle(-\pi).$$

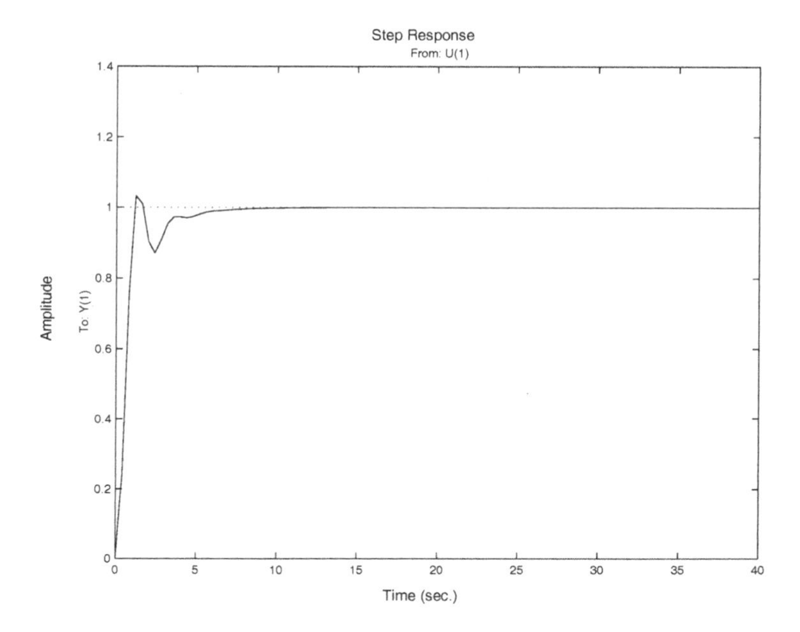

Fig. 7.29 The Step Response of the PD Compensated System

Thus, $-L\omega \approx -\pi/2$, the angular frequency of the oscillations exhibited by the closed-loop system is $\omega \approx \pi/(2L)$, and their period is $P_{cr} \approx 4L$. Additionally, in order for the magnitude of $G_c(s)G_p(s)$ to be approximately one, we must have:

$$K_{cr} \approx \frac{\pi T}{2CL}.$$

Making use of our approximations and the Ziegler-Nichols rules, we find that:

$$\begin{aligned}
G_c(s) &= 0.6K_{cr} + \frac{1.2K_{cr}}{P_{cr}}\frac{1}{s} + 0.075K_{cr}P_{cr}s \\
&= 0.075K_{cr}P_{cr}\left(\frac{8}{P_{cr}} + \frac{16}{P_{cr}^2 s} + s\right) \\
&= \frac{0.075K_{cr}P_{cr}}{s}\left(s + \frac{4}{P_{cr}}\right)^2 \\
&\approx \frac{2\pi T}{C}\frac{0.075}{s}\left(s + \frac{1}{L}\right)^2.
\end{aligned}$$

That is,

$$G_c(s)G_p(s) \approx \frac{0.075}{s} 2\pi T \left(s + \frac{1}{L}\right)^2 \frac{e^{-Ls}}{Ts+1}.$$

7.8.5 *Estimating the System's Phase Margin*

To find the compensated system's phase margin, we must find the frequency for which $|G_c(j\omega)G_p(j\omega)| = 1$ and calculate the system's phase at that frequency. We must find the value of ω for which:

$$\begin{aligned}|G_c(j\omega)G_p(j\omega)| &\approx 0.075 \cdot 2\pi \left|1 + \frac{1}{j\omega L}\right|^2 \left|\frac{Tj\omega}{Tj\omega + 1}\right| \\ &= 1.\end{aligned}$$

Assuming that $T\omega \gg 1$ and noting that $0.075 \cdot 2\pi \approx 0.47$, we find that $\omega L \approx 1$. (As $T \gg L$, the assumption that $T\omega \gg 1$ is now seen to have been justified.) The phase when $\omega = 1/L$ is:

$$\angle G_p(j\omega)G_c(j\omega) \approx \angle(1 + 1/j)^2 e^{-j} = -\pi/2 - 1 \approx -147^{\circ}.$$

Thus, the system's phase margin is about 33°; this is a bit low but not totally unreasonable.

7.8.6 *Estimating the System's Gain Margin*

To find the system's gain margin, we must find the largest value of the system's gain for which the phase is $-\pi + 2n\pi$. From the form of $G_c(s)G_p(s)$, it is clear that we need the smallest value of $\omega > 0$ for which this condition holds. That is, we need to find ω such that:

$$\angle e^{-j\omega L}\left(1 + \frac{1}{j\omega L}\right)^2 = -\pi.$$

Let $\theta \equiv \omega L$. We find that:

$$-\theta - 2\tan^{-1}(1/\theta) = -\pi.$$

As the derivative of $-\theta - 2\tan^{-1}(1/\theta)$ is $-1 + 2/(\theta^2 + 1)$ and this function is positive until $\theta = 1$ and negative thereafter, we find that $-\theta - 2\tan^{-1}(1/\theta)$ increases from a value of $-\pi$ when $\theta \to 0$ to a value of $-1 - \pi/2$ at $\theta = 1$ and then decreases without bound as $\theta \to \infty$. Thus, there is a unique value of $\theta > 0$ for which $-\theta - 2\tan^{-1}(\theta) = -\pi$, and that value is greater than one.

Assuming that the value of θ is reasonably large, we can make use of the approximation

$$-\pi = -\theta - 2\tan^{-1}(1/\theta) \approx -\theta - 2/\theta$$

to solve for θ. We find that we must solve the quadratic equation:

$$\theta^2 - \pi\theta + 2 = 0.$$

The solutions of this quadratic equation are:

$$\theta = \frac{\pi \pm \sqrt{\pi^2 - 8}}{2} \approx \frac{\pi \pm \sqrt{2}}{2} \approx 2.28, 0.86.$$

The quadratic has one solution that is greater than one: $\theta \approx 2.28$. Thus, the angular frequency whose gain determines the gain margin is $\omega \approx 2.28/L$. Substituting this value in the expression for the gain and making use of the assumption that $\omega T \gg 1$, we find that the gain is approximately:

$$|G_c(j2.28/L)G_p(j2.28/L)| \approx \frac{0.075}{\left(\frac{2.28}{L}\right)^2} 2\pi \left(\left(\frac{2.28}{L} \right)^2 + \frac{1}{L^2} \right) \approx \frac{1}{2} \frac{2.28^2 + 1}{2.28^2}.$$

This leads to a gain margin of approximately 4.5 dB which is, once again, somewhat smaller than we might have preferred.

7.8.7 *Estimating the System's Settling Time*

It is simple enough to make use of our previous calculations to estimate the system's settling time. We find that:

$$\begin{aligned} T_{\text{settling}} &\approx \frac{8}{(1/L)\tan(PM)} \\ &\approx \frac{8}{(1/L)\tan(30^o)} \\ &= \sqrt{3} \cdot 8L \\ &\approx 3.4 P_{cr}. \end{aligned}$$

Thus, the settling time of the closed-loop system with a PID controller designed using the Ziegler-Nichols rules presented above is about 3.4 times the period of the oscillations in the marginally stable closed-loop system used to characterize the system.

7.9 Bode's Sensitivity Integrals

7.9.1 *Introduction*

One can achieve much using compensation, but there *are* limits on what is achievable. One of the fundamental limitations comes from Bode's sensitivity integrals [GGS00; Ste03a]. In this section, we make essential use of the methods of complex variables to prove Bode's sensitivity integrals, and we show how these integrals limit what can be achieved using compensation. (If necessary, this section can be skipped without loss of continuity.)

7.9.2 *The Sensitivity Function*

As is clear from §3.3 (p. 70), in a unity-feedback system, the sensitivity of the closed-loop transfer function, $T(s)$, to the plant's transfer function, $G_p(s)$, is

$$S_{G_p(s)}^{T(s)} = \frac{1}{1 + G_c(s)G_p(s)}$$

where $G_c(s)$ is the compensator's transfer function. As we have seen previously (p. 77), this is also the transfer function from the system's input to its error—from $v_{in}(t)$ to to $e(t) = v_{in}(t) - v_o(t)$. Thus, for a unity feedback system, the "transfer function" from the input to the error signal satisfies:

$$T_{V_{in}(s)\to E(s)}(s) = \frac{1}{1 + G_c(s)G_p(s)} = S_{G_p(s)}^{T(s)}(s).$$

In a perfect world, a good engineer would manage to make this function zero, and then the system's response would not depend on the plant, and the error would always be zero. As we will see, this is *impossible*.

7.9.3 *Bode's Sensitivity Integrals*

Consider the following integral:

$$\int_0^\infty \ln \left| S_{G_p(s)}^{T(s)}(j\omega) \right| \, d\omega \tag{7.7}$$

which will be referred to as the *sensitivity integral*. Assuming that $G_c(s)G_p(s)$ is low-pass, for large values of ω the sensitivity is near one, and its logarithm, the integrand, is approximately zero. We now make use of some of the theory of functions of a complex variable to evaluate (7.7) under several different assumptions.

7.9.4 *Case I*

Let the *relative degree* of a transfer function be the degree of its denominator less that of its numerator. First we consider $G_c(s)G_p(s)$ that satisfies:

(1) $G_c(s)G_p(s)$ has relative degree two or more.
(2) The closed-loop system is stable; $1 + G_c(s)G_p(s)$ has no zeros in the right half-plane.
(3) The system is open-loop stable; $G_c(s)G_p(s)$ has no poles in the right half-plane.

We show that under the above conditions, (7.7) equals zero.

Before proving this result, consider what it tells us about the system. The result says that the average value of the logarithm of the absolute value of the sensitivity function is zero. This is *not* what we wanted to hear. This means that in some sense, the "average" value of the absolute value of the sensitivity function is one—and not the zero we want.

Another way of looking at this result is to say that if we make the sensitivity low for some range of frequencies, then we *have no choice* but to make the sensitivity high for some other range of frequencies. We can "move around" the frequency ranges in which such a system is sensitive, but we cannot get rid of them. The fact that "pushing down" the sensitivity in one range causes it to "pop up" in another range is sometimes referred to as *the waterbed effect* [AM08].

To prove that the sensitivity integral is zero, we consider the (complex) logarithm[6] of the sensitivity, $\ln\left(S_{G_p(s)}^{T(s)}(s)\right)$. The last two conditions placed on $G_c(s)G_p(s)$ guarantee that the sensitivity function has neither poles nor zeros in the right half-plane, and this is sufficient to guarantee that the logarithm of the sensitivity function is a single-valued analytic function in the right-half plane [CB90, §37].

Consider:

$$\oint_C \ln\left(S_{G_p(s)}^{T(s)}(s)\right)\,ds$$

where the contour, C, goes from jR to $-jR$ along the imaginary axis and then returns to jR along a semi-circle of radius R that traverses the right half-plane. Because the function is analytic inside the contour, this integral is equal to zero.

[6] We always take $\ln(z)$ to refer to the principle value of the natural logarithm.

Letting $s = j\omega$ on the imaginary axis and letting it equal $Re^{j\theta}$ on the semicircle, we find that the contour integral is equal to:

$$\oint_C \ln\left(S_{G_p(s)}^{T(s)}(s)\right)\,ds = j\int_R^{-R} \ln\left(S_{G_p(s)}^{T(s)}(j\omega)\right)\,d\omega$$
$$+jR\int_{-\pi/2}^{\pi/2} \ln\left(S_{G_p(s)}^{T(s)}(Re^{j\theta})\right)e^{j\theta}\,d\theta. \qquad (7.8)$$

Consider the first integral:

$$j\int_R^{-R} \ln\left(S_{G_p(s)}^{T(s)}(j\omega)\right)\,d\omega.$$

On the real axis, we know that $S_{G_p(s)}^{T(s)}(s)$ is real. As we assume that $G_c(s)G_p(s)$ is low-pass in nature, as $s \to \infty$, we know that $S_{G_p(s)}^{T(s)}(s) \to 1$. As by assumption $S_{G_p(s)}^{T(s)}(s)$ has no zeros in the right half-plane, it must be positive for all $s \geq 0$. Thus, by our choice of logarithm, $\ln\left(S_{G_p(s)}^{T(s)}(s)\right)$ is real on the real axis. This suffices to prove [CB90, §105] that:

$$\overline{\ln\left(S_{G_p(s)}^{T(s)}(s)\right)} = \ln\left(S_{G_p(s)}^{T(s)}(\overline{s})\right).$$

This, in turn, guarantees that the real part of $\ln\left(S_{G_p(s)}^{T(s)}(j\omega)\right)$ is an even function of ω and that the imaginary part is an odd function of ω.

The logarithm of a complex number is the logarithm of the absolute value of the number plus j times the phase of the number. Because the imaginary part of the logarithm has already been shown to be an odd function of ω and the real part to be an even function of ω, we find that:

$$j\int_R^{-R} \ln\left(S_{G_p(s)}^{T(s)}(j\omega)\right)\,d\omega = -2j\int_0^R \ln\left|S_{G_p(s)}^{T(s)}(j\omega)\right|\,d\omega.$$

Now we turn our attention to the second part of (7.8). Recalling that for small values of x:

$$\ln(1+x) = x + O(x^2) \text{ and } \frac{1}{1+x} = 1 - x + O(x^2)$$

we find that for large value of R (for which $G_c(Re^{j\theta})G_p(Re^{j\theta})$ is very small):

$$\begin{aligned}
jR \int_{-\pi/2}^{\pi/2} \ln\left(S_{G_p(s)}^{T(s)}(Re^{j\theta})\right) e^{j\theta}\, d\theta
&= jR \int_{-\pi/2}^{\pi/2} \ln\left(\frac{1}{1+G_c(Re^{j\theta})G_p(Re^{j\theta})}\right) e^{j\theta}\, d\theta \\
&= jR \int_{-\pi/2}^{\pi/2} (-G_c(Re^{j\theta})G_p(Re^{j\theta}) \\
&\quad + O((G_c(Re^{j\theta})G_p(Re^{j\theta}))^2))\, e^{j\theta}\, d\theta.
\end{aligned}$$

Because we assume that for large s the function $G_p(s)G_c(s)$ decays as $1/s^2$, is is clear that as $R \to \infty$ the integral tends to zero.

Putting everything together and letting $R \to \infty$, we find that:

$$\int_0^{\infty} \ln\left|S_{G_p(s)}^{T(s)}(j\omega)\right| d\omega = 0. \tag{7.9}$$

This is the first form of Bode's sensitivity integral.

7.9.5 *Case II*

Now consider $G_c(s)G_p(s)$ that satisfies:

(1) The relative degree of $G_c(s)G_p(s)$ is *one* and

$$A = \lim_{s\to\infty} sG_c(s)G_p(s).$$

(2) The closed-loop system is stable; $1 + G_c(s)G_p(s)$ has no zeros in the right half-plane.
(3) The system is open-loop stable; $G_c(s)G_p(s)$ has no poles in the right half-plane.

The only change that we need to make to the preceding development is connected to the second integral in (7.8)—the integral over the semicircle. We now perform that calculation again making use of our current assumptions.

$$
\begin{aligned}
jR\int_{-\pi/2}^{\pi/2} & \ln\left(S_{G_p(s)}^{T(s)}(Re^{j\theta})\right)e^{j\theta}\,d\theta \\
&= jR\int_{-\pi/2}^{\pi/2}\ln\left(\frac{1}{1+G_c(Re^{j\theta})G_p(Re^{j\theta})}\right)e^{j\theta}\,d\theta \\
&= jR\int_{-\pi/2}^{\pi/2}(-G_c(Re^{j\theta})G_p(Re^{j\theta}) \\
&\quad + O(G_c(Re^{j\theta})G_p(Re^{j\theta}))^2))\,e^{j\theta}\,d\theta \\
&\overset{R\to\infty}{=} j\int_{-\pi/2}^{\pi/2}\frac{-A}{Re^{j\theta}}Re^{j\theta}\,d\theta \\
&= -j\pi A.
\end{aligned}
$$

Putting everything together and considering what happens as $R \to \infty$, we find that:

$$-2j\int_0^\infty \ln\left|S_{G_p(s)}^{T(s)}(j\omega)\right|\,d\omega - j\pi A = 0$$

or that:

$$\int_0^\infty \ln\left|S_{G_p(s)}^{T(s)}(j\omega)\right|\,d\omega = -A\pi/2. \tag{7.10}$$

This is the second form of Bode's sensitivity integral we consider. As A gets more positive, the average of the logarithm of the absolute value of the sensitivity function gets more and more negative. In principle, this means that the sensitivity function can be small over larger and larger regions. A word of caution is in order: In general, the relative degree of $G_c(s)G_p(s)$ in real, practical systems is greater than one.

7.9.6 *Case III*

Next consider the case that:

(1) The relative degree of $G_c(s)G_p(s)$ is two or more.
(2) The closed-loop system is stable; $1 + G_c(s)G_p(s)$ has no zeros in the right half-plane.
(3) The system is open-loop unstable; $G_c(s)G_p(s)$ has N poles in the right half-plane at the locations $\zeta_k, \quad k = 1, \ldots, N$.

Because $G_c(s)G_p(s)$ has poles in the right half-plane, the sensitivity function has zeros at the same points in the right half-plane. In order to analyze this case, we consider the related function:

$$H(s) \equiv S_{G_p(s)}^{T(s)} \prod_{k=1}^{N} \frac{s+\zeta_k}{s-\zeta_k}.$$

The new function has neither poles nor zeros in the right half-plane.

Making use of the results about the first case, it is easy to see that:

$$\oint_C \ln\left(H(s)\right)\, ds = 0$$

where the contour, C, is unchanged. From the properties of the logarithm and our previous results, it is clear that:

$$-2j\int_0^{\infty} \ln\left|S_{G_p(s)}^{T(s)}(j\omega)\right|\, d\omega = -\oint_C \ln\left(\prod_{k=1}^{N}\frac{s+\zeta_k}{s-\zeta_k}\right) ds.$$

In order to calculate the contour integral we consider some of the properties of:

$$\prod_{k=1}^{N}\frac{s+\zeta_k}{s-\zeta_k}$$

First we consider the terms $(s+\zeta_k)/(s-\zeta_k)$ for which $\zeta_k \in \mathcal{R}$. For such terms, it is obvious that when $s = j\omega$ the term's absolute value is one, and, as the polynomials all have real coefficient, it is obvious that the term's phase is an odd function of ω.

As the sensitivity function is composed of polynomials in s with real coefficients, a complex zero is always accompanied by its complex conjugate. For complex ζ_k, we consider the second order function:

$$\frac{s+\zeta_k}{s-\zeta_k}\frac{s+\overline{\zeta_k}}{s-\overline{\zeta_k}} = \frac{s^2+2\mathcal{R}(\zeta_k)s+|\zeta_k|^2}{s^2-2\mathcal{R}(\zeta_k)s+|\zeta_k|^2}.$$

This is also a rational function with real coefficients. Thus, when considered as a function of $s = j\omega$, its phase is an odd function of ω. Additionally, it is clear that when $s = j\omega$, its absolute value is always one.

Making use of these facts, it is not hard to see that the contribution to the second integral from the piece of the contour on the imaginary axis is zero. We have already seen that the imaginary part of the logarithm is the phase of the function—which is odd, so its integral over a symmetric region is zero. Additionally, when $s = j\omega$, the real part of the integrand is the logarithm of one—which is always zero. Thus, we need only calculate:

$$\oint_C \ln\left(\prod_{k=1}^{N} \frac{s+\zeta_k}{s-\zeta_k}\right) ds = \sum_{k=1}^{N} \oint_{\text{large semicircle}} \ln\left(\frac{s+\zeta_k}{s-\zeta_k}\right) ds$$
$$= \sum_{k=1}^{N} \oint_{\text{large semicircle}} \ln\left(1 + 2\frac{\zeta_k}{s-\zeta_k}\right) ds.$$

On the large semicircle, each of these integrals tends to:

$$\oint_{\text{large semicircle}} 2\frac{\zeta_k}{s-\zeta_k}\, ds.$$

Taking $s = Re^{j\theta}$, this becomes:

$$j\int_{-\pi/2}^{\pi/2} 2\frac{\zeta_k}{Re^{j\theta}-\zeta_k} Re^{j\theta} d\theta \overset{R\to\infty}{=} 2\pi j \zeta_k.$$

Putting everything together, we find that:

$$\int_0^{\infty} \ln\left|S_{G_p(s)}^{T(s)}(j\omega)\right| d\omega = \pi \sum_{k=1}^{N} \zeta_k. \tag{7.11}$$

This is the third form of Bode's sensitivity integral we consider.

7.9.7 *Case IV*

The last case we consider is the case for which:

(1) The relative degree of $G_c(s)G_p(s)$ is one, and

$$A = \lim_{s\to\infty} sG_c(s)G_p(s).$$

(2) The closed-loop system is stable; $1 + G_c(s)G_p(s)$ has no zeros in the right half-plane.

(3) The system is open-loop unstable; $G_c(s)G_p(s)$ has N poles in the right half-plane at the locations $\zeta_k, \quad k = 1, \ldots, N$.

All of the real work has already been done. Combining our previous results, we find that:

$$\int_0^{\infty} \ln\left|S_{G_p(s)}^{T(s)}(j\omega)\right| d\omega = -A\pi/2 + \pi \sum_{k=1}^{N} \zeta_k. \tag{7.12}$$

This is the fourth and most general form of Bode's sensitivity integral. Noting that when the conditions of cases I and III obtain, A is equal to zero and that when the conditions of case II obtain, the sum of the ζ_k is zero, one can say that *(7.12) gives the value of (7.7), the sensitivity integral, for all four cases we have examined.*

$G_c(s)G_p(s) = 1/(s+1)$—An Example

To fix ideas, we calculate the value of the sensitivity integral when $G_c(s)G_p(s) = 1/(s+1)$ and compare the value to that given by (7.10). The sensitivity function is:

$$S_{G_p(s)}^{T(s)} = \frac{s+1}{s+2}.$$

Thus, we need to calculate:

$$\int_0^\infty \ln\left(\frac{\sqrt{\omega^2+1^2}}{\sqrt{\omega^2+2^2}}\right) d\omega = \frac{1}{2}\lim_{R\to\infty}\left(\int_0^R \ln(\omega^2+1^2)d\omega - \int_0^R \ln(\omega^2+2^2)d\omega\right).$$

It behooves us to calculate:

$$\int_0^R \ln(\omega^2+a^2)\,d\omega.$$

Integrating by parts and letting $u = \ln(\omega^2+a^2)$ and $dv = d\omega$, we find that:

$$\begin{aligned}
\int \ln(\omega^2+a^2)\,d\omega &= \omega\ln(\omega^2+a^2) - 2\int \frac{\omega^2}{\omega^2+a^2}\,d\omega \\
&= \omega\ln(\omega^2+a^2) - 2\int \left(1 - \frac{a^2}{\omega^2+a^2}\right) d\omega \\
&= \omega\ln(\omega^2+a^2) - 2\omega + 2\int \frac{a^2}{\omega^2+a^2}\,d\omega \\
&= \omega\ln(\omega^2+a^2) - 2\omega + 2a\tan^{-1}(\omega/a).
\end{aligned}$$

Returning to the integral we actually need to calculate, we find that:

$$\begin{aligned}\frac{1}{2}\lim_{R\to\infty}\left(\int_0^R \ln(\omega^2+1^2)\,d\omega - \int_0^R \ln(\omega^2+2^2)\,d\omega\right)\\ &= (1/2)\lim_{R\to\infty}\,(R\ln(R^2+1^2)\\ &\quad -2R+2\cdot 1\tan^{-1}(R/1) - (R\ln(R^2+2^2)\\ &\quad -2R+2\cdot 2\tan^{-1}(R/2)))\\ &= -\pi/2.\end{aligned}$$

As the value of A is 1, this is just what (7.10) gives.

Using MATLAB® to Verify What We Have Seen—An Example

Consider a unity feedback system for which:

$$G_c(s)G_p(s) = \frac{K}{(s+2)(s-1)} = \frac{K}{s^2+s-2}.$$

We find that:

$$S_{G_p(s)}^{T(s)} = \frac{s^2+s-2}{s^2+s+K-2}.$$

Clearly the system is stable as long as $K > 2$. For all such K, we are in Case III, and the value of the integral of the logarithm of the sensitivity function must be π.

It is clear that as $K \to \infty$, the sensitivity function is near zero for an ever wider range of low frequencies. To compensate for this nice behavior, we must find that at somewhat higher frequencies the sensitivity function is larger than one. (Because $G_c(s)G_p(s)$ is low pass in nature, at very high frequencies, the sensitivity tends to one.) Considering Figures 7.30 and 7.31 we see that this is indeed what happens. Note that the value of

$$\int_0^\infty \ln\left|S_{G_p(s)}^{T(s)}(j\omega)\right|\,d\omega$$

as numerically evaluated by MATLAB is printed above each plot of the logarithm of the sensitivity function. The estimate of the integral is always close to π—as it should be.

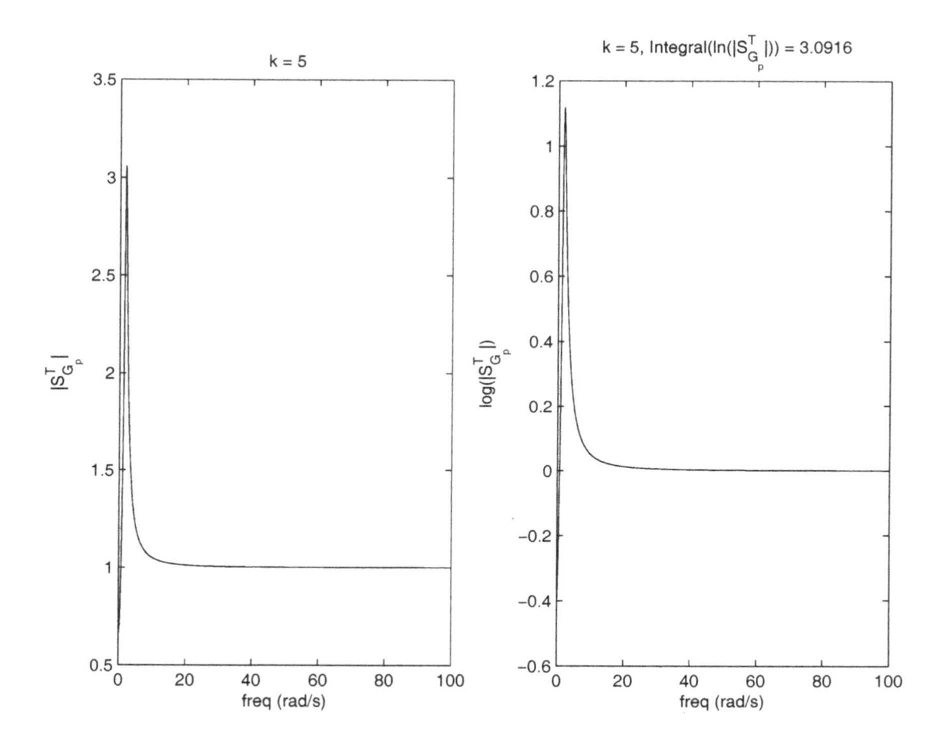

Fig. 7.30 Plots of the Sensitivity and Log Sensitivity Functions when $K = 5$

7.9.8 *Available Bandwidth*

As $s \to \infty$, we know that $G_p(s)G_c(s) \to 0$. In general, we do not know precisely *how* it tends to zero, so we cannot say more than "for all $\omega > \Omega$, $G_c(j\omega)G_p(j\omega)$ is very small." Thus, we find that:

$$\int_0^\Omega \ln\left|S^{T(s)}_{G_p(s)}\right| d\omega \approx -A\pi/2 + \pi\sum_{k=1}^{N}\zeta_k, \qquad \ln\left|S^{T(s)}_{G_p(s)}\right| \approx 0, \omega > \Omega.$$

When designing a system, both the improvements and the intentional degradations (made to "balance" the improvements) must happen within the range of frequencies $[0, \Omega]$—within the *available bandwidth* [Ste03a]. Because we cannot control precisely what happens above $\omega = \Omega$, we cannot "send the degradation of to infinity"—where it could be smeared out over an infinitely long region so it would not bother us much.

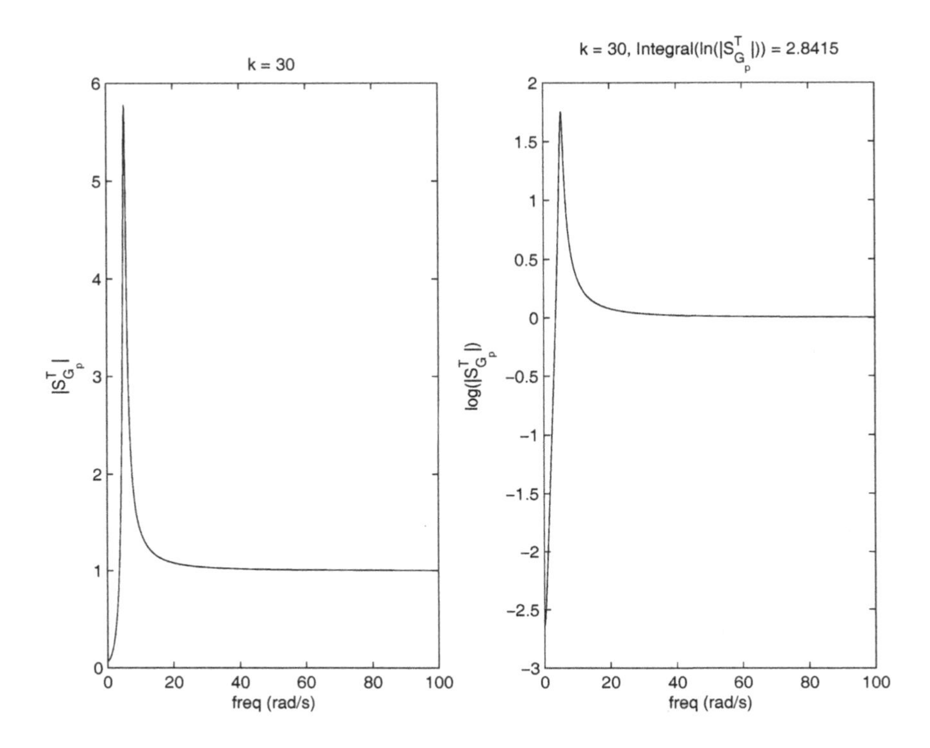

Fig. 7.31 Plots of the Sensitivity and Log Sensitivity Functions when $K = 30$

7.9.9 *Open-Loop Unstable Systems with Limited Available Bandwidth*

Assume that we are in Case III and that the available bandwidth is Ω. Then we know that:

$$\int_0^{\Omega} \ln \left| S_{G_p(s)}^{T(s)}(j\omega) \right| \, d\omega \approx \pi \sum_{k=1}^{N} \zeta_k.$$

In order for this to hold, there must be some ω_0 within $[0, \Omega]$ for which

$$\ln \left| S_{G_p(s)}^{T(s)}(j\omega_0) \right| \geq \frac{\pi}{\Omega} \sum_{k=1}^{N} \zeta_k \Rightarrow \left| S_{G_p(s)}^{T(s)}(j\omega_0) \right| \geq \exp\left(\frac{\pi}{\Omega} \sum_{k=1}^{N} \zeta_k \right).$$

In particular, as Ω gets smaller, the maximum value of the absolute value of the sensitivity function gets larger. Clearly, it is quite difficult to control

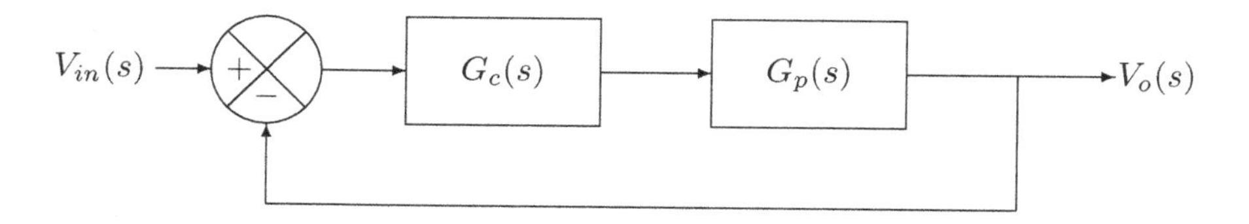

Fig. 7.32 A Generic Unity-Gain Feedback System with Compensation

low-available-bandwidth open-loop unstable systems.

7.10 Exercises

(1) We are interested in controlling the angle of a simple satellite whose transfer function is $G_p(s) = K/s^2, \quad K > 0$. Let $G_p(s) = K/s^2$ in Figure 7.32. Use the root locus diagram to show that using a phase-lag compensator, $G_c(s)$, will make the system *unstable*.

(2) In the system of Figure 7.32, let

$$G_p(s) = \frac{40}{s^3 + 6s^2 + 11s + 6} = \frac{40}{(s+1)(s+2)(s+3)}.$$

(a) Using root locus techniques, show that (provided that K_I is small enough) one can add integral compensation ($G_c(s) = K_I/s$) to this system without driving it into instability.

(b) Now find the largest value of K_I for which the step response of the system will not be oscillatory. (Use MATLAB to solve for the zeros of any polynomials whose zeros must be found.)

(c) Calculate the time constant of the system found in the previous section. (The time constant of a system is the reciprocal of the of the absolute value of the pole nearest the imaginary axis. One expects the settling time of a system to be approximately three times the system's time constant. Why?) Have MATLAB calculate the step response of the closed-loop system, and see if the result is in reasonable agreement with the time constant found.

(3) Let the plant of Figure 7.32 be a motor whose transfer function is:

$$G_p(s) = \frac{50}{s(2s+1)}.$$

Show that using a PD controller you can make the system as fast as desired while maintaining the accuracy of the system. (I.e. show that the steady state output of the system to a unit step is always 1.)

(4) Let the plant of Figure 7.32 be a motor whose transfer function is:

$$G_p(s) = \frac{50}{s(2s+1)}.$$

(a) Design a simple attenuator to cause the systems phase margin to be 45°.

(b) Use the `step` command to plot the step response of the compensated system.

(c) What is the (approximate value of the) system's time constant as seen from the step response?

(d) What is the exact value of the time constant?

(5) Once again, let the plant of Figure 7.32 be a motor whose transfer function is:

$$G_p(s) = \frac{50}{s(2s+1)}.$$

(a) Use a phase-lag compensator to achieve a phase margin of 45°.

(b) Using MATLAB, find the unit step response of the compensated system.

(c) Explain the nature of the step response by considering both $G_p(s)$ and $G_c(s)$.

(6) Compare and contrast the results of the previous two problems. Explain the differences in the performance by making use of your knowledge of the systems.

(7) Let the plant of Figure 7.32 be:

$$G_p(s) = \frac{2}{s(s+1)^2}.$$

(a) Find the phase margin of the system when $G_c(s) = 1$.

(b) Design a lag-lead compensator that will give the system a phase margin of 55° at the same location as the phase margin of the uncompensated system.

(c) Find the gain and phase margins of the system with the lag-lead compensator.

(8) Repeat problem 7 using a PID controller rather than a lag-lead compensator.

(9) Show that if $G_p(s) = 1/s$ in the system of Figure 7.32, then adding a PI compensator leaves the system stable and leaves one with a system that tracks a ramp input in the steady state.

(10) Let:

$$G_p(s) = \frac{s+1}{s^2(s+10)}$$

in the system of Figure 7.32.

(a) Find the gain and phase margins of the system.

(b) Show that a PD compensator leaves the system stable for all $K_P, K_D > 0$.

(c) Find a condition on K_P and K_I that guarantees that a PI compensator will make the system *un*stable.

(11) Consider a unity-feedback system for which $G_c(s) = K$ and $G_p(s) = 1/s$.

(a) Calculate the system's transfer function, $T(s)$, and the sensitivity of the transfer function to changes in $G_p(s)$, $S_{G_p}^T(s)$.

(b) Calculate

$$\int_0^\infty \ln \left| S_{G_p}^T(j\omega) \right| \, d\omega,$$

and check your answer using the Bode sensitivity integral.

(c) What does this integral tell us about our ability to make the system less sensitive to changes in the plant by increasing K?

(d) What property of $G_c(s)G_p(s)$ is responsible for this result?

(12) Consider a unity-feedback system for which:

$$G_p(s) = \frac{1}{(s+1)(s+10)(s+20)}.$$

(a) Calculate K_{cr} and P_{cr} for this system.

(b) Use the Ziegler-Nichols rules to design a PID controller, $G_c(s)$, for this system.

(c) Use MATLAB to find the phase margin, gain margin, and settling time for the system with the controller. (You may wish to use the `step` command to determine the settling time.)

(d) To what extent do your results agree with the theory developed in this chapter?

(e) Speculate intelligently on the reason or reasons that your results do not agree entirely with the theory developed in this chapter.

(13) Consider a unity feedback system for which $G_c(s)$ is (initially) 1—for which there is (initially) no compensator. The Bode plots of $G_p(s)$, a marginally stable system, are given in Figure 7.33, and the Bode plots of $G_p(s)$ at low frequencies are given in Figure 7.34.

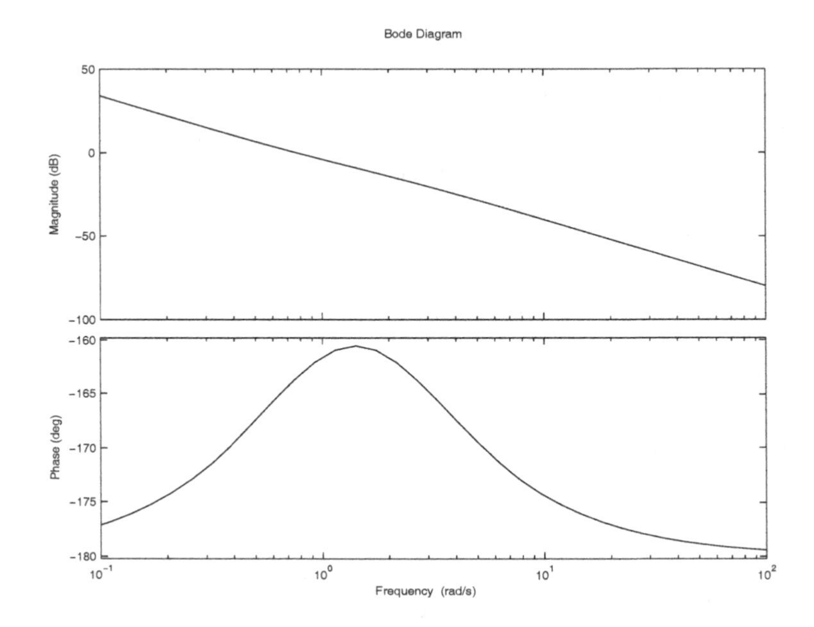

Fig. 7.33 The Bode Plots of $G_p(s)$

(a) Use Figure 7.34 to determine the behavior of $G_p(s)$ near the origin.
(b) Use the results of the previous section and Figure 7.33 to determine the Nyquist plot of the (closed-loop) system.
(c) Use the results of the previous section and the fact that $G_p(s)$ is marginally stable to determine the stability of the closed-loop system.
(d) Design a PD compensator, $G_c(s)$, that causes the closed-loop system to have a phase margin of (approximately) 60° and a settling time of (approximately) 0.2 s.

(14) Consider a unity feedback system for which:

$$G_p(s) = \frac{1}{s(s+1)(s+2)}.$$

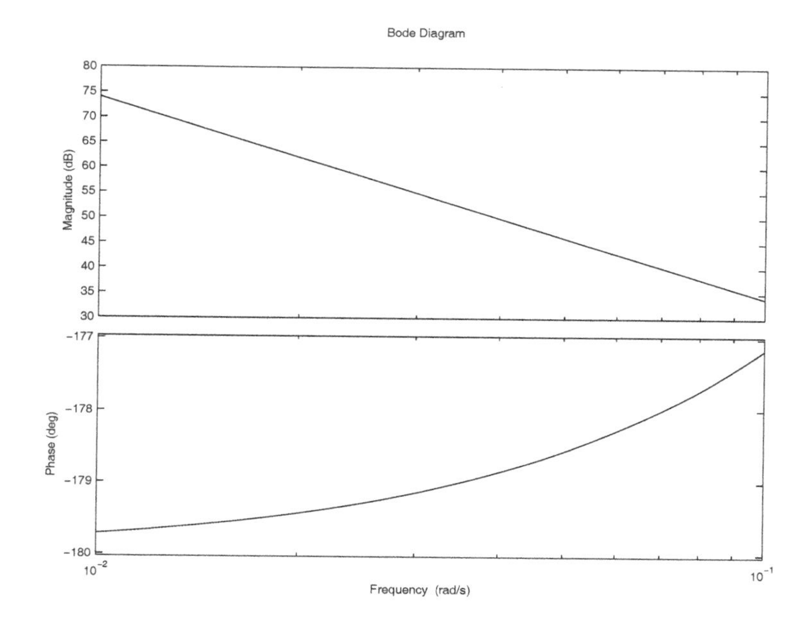

Fig. 7.34 The Bode Plots of $G_p(s)$ at Low Frequencies

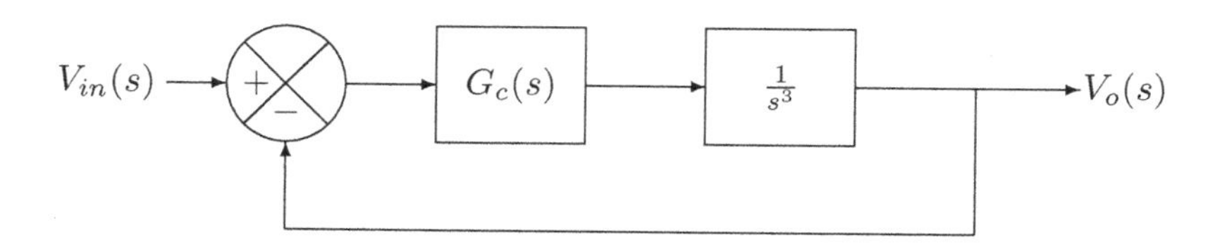

Fig. 7.35 A System with a Triple Integrator, $1/s^3$

(a) Design a PD compensator that causes the closed-loop system to have a phase margin of 60° and that causes the system to settle in about 2 s.

(b) Check that the system with the compensator is stable.

(c) Why shouldn't one use a D-type compensator (a compensator for which $G_c(s) = K_D s$)? (Consider the results of §3.7.)

(15) Consider the system of Figure 7.35.

(a) Prove that a compensator of the form $G_c(s) = K_P + K_D s$, a PD compensator, cannot stabilize the system.

(b) Design a compensator of the form $G_c(s) = K_2 s^2 + K_1 s + K_0$ that stabilizes the system, causes the closed-loop system to have a phase margin of 60°, to have a settling time of 5 s, and to track quadratic inputs—inputs of the form $(at^2 + bt + c)u(t)$.

(c) If we were to take the compensator designed in the preceding section and multiply it by an additional gain, K, for which values of K would the system be stable? (I.e., if we were to keep K_0, K_1, and K_2 fixed but we were to add an additional gain K so that the transfer function of the new compensator would be $G_c(s) = K(K_2 s^2 + K_1 s + K_0)$, for which values of K would the closed-loop system be stable?)

(d) Is it reasonable to speak about the gain margin of this system? Explain.

(16) Consider a unity feedback system for which:

$$G_p(s) = \frac{1}{0.1s + 1}.$$

(a) Design a PI compensator, $G_c(s)$, that causes the closed-loop system to have a phase margin of 60°, to track DC, and to have settling time of 100 ms.

(b) Use the Routh array to examine the stability of the closed-loop system designed in the previous section.

(17) Comparing the step responses of Figures 7.9 and 7.12, we find that the two systems have very similar settling times. Make use of the systems' Bode plots, given in Figures 7.8 and 7.11, and of (5.1) to explain why the settling times are so similar.

Chapter 8

Some Nonlinear Control Theory

8.1 Introduction

A circuit is linear if it obeys the principal of superposition—if when the output corresponding to v_{in1} is v_{out1} and the output corresponding to v_{in2} is v_{out2} then the output corresponding to $av_{in1} + bv_{in2}$ is $av_{out1} + bv_{out2}$. Systems of the type that we have discussed until now—those that have transfer functions—obey the principal of superposition (under the standard condition that the systems' initial conditions are all zero). Many practical circuits are not composed solely of linear elements; circuits often contain one or more elements that behave in a nonlinear fashion. A typical nonlinear circuit element, η, is a comparator. If x is the input to a comparator, then the output of the comparator, $f_{comp}(x)$, is:

$$f_{comp}(x) = \text{signum}(x) = \begin{cases} 1 & x > 0 \\ 0 & x = 0 \\ -1 & x < 0 \end{cases}.$$

Let us check that the comparator *does not* obey the principle of superposition. If the input to the comparator is $x = a > 0$, then its output is $f_{comp}(x) = 1$. For the comparator to obey the principle of superposition, the output when $x = a + a = 2a$ would have to be $1 + 1 = 2$. In fact, the output is 1; the comparator is not a linear element.

When dealing with linear elements, our main concern has been with stability. We have checked that the output of our system cannot become unbounded when the input is bounded. When dealing with nonlinear elements, the boundedness of the output is often not our main concern. Most of the nonlinear elements that we encounter in practice are like the comparator—their output is inherently bounded. We do not worry that the system may become wildly unstable; we worry that the system may start oscillating.

In this chapter, our goal is to determine whether or not a feedback system that incorporates a nonlinear element, η (see Figure 8.1), supports periodic oscillations—in engineering terminology "limit cycles." We are interested in seeing whether or not the system can itself sustain the oscillations—therefore we consider systems with no input. We rely on noise or the initial conditions to start the oscillations. Since we are dealing with periodic functions, we make use of Fourier Series.

In this chapter we introduce two techniques for analyzing circuits that contain nonlinear elements—the describing function technique and the method of Tsypkin. The first method is an approximate method that is widely applicable; the second method is an exact method that applies only to a relatively small class of systems.

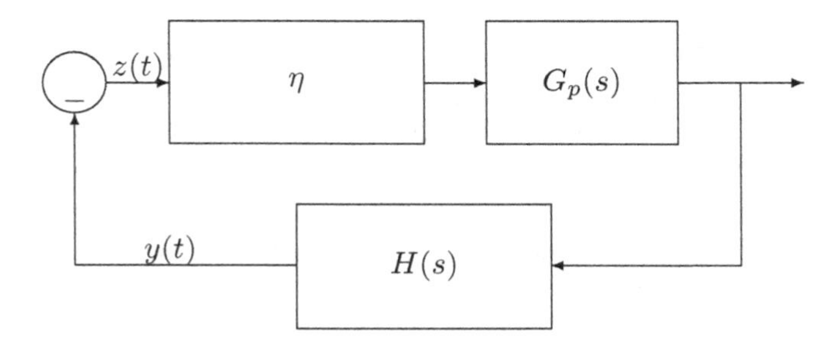

Fig. 8.1 Block Diagram of a Nonlinear Feedback Circuit

8.2 The Describing Function Technique

8.2.1 *The Describing Function Concept*

Assume that our feedback system does support limit cycles. Then the input to the "inverter" will be a periodic function $y(t)$. To simplify our analysis, we approximate $y(t)$ by a sine wave—we assume that $y(t) \approx -M\sin(\omega t)$. If $G_p(j\omega)H(j\omega)$ is sufficiently low-pass in nature, then this assumption is not too unreasonable. The output of the nonlinear element, η, which may have some energy at higher frequencies, will be filtered by $G_p(j\omega)H(j\omega)$ into a reasonable facsimile of a pure sine wave. The output of the nonlinear element will be $f(-y(t)) \approx f(M\sin(\omega t))$. Because we assume that the linear elements remove the harmonics from the output of the nonlinear element, it follows that the output of the nonlinear element can be usefully approximated by the portion of the output located at the fundamental frequency. It follows that $f(M\sin(\omega t)) \approx A\sin(\omega t) + B\cos(\omega t)$ where A and B are the first Fourier sine and cosine coefficients respectively. That is:

$$A = \frac{2}{T}\int_0^T \sin(\omega t) f(M\sin(\omega t))\,dt$$
$$B = \frac{2}{T}\int_0^T \cos(\omega t) f(M\sin(\omega t))\,dt$$

where $T = \frac{2\pi}{\omega}$.

We define the describing function[1] of the nonlinearity as:

$$D(M,\omega) \equiv \frac{A + jB}{M}.$$

The function $D(M,\omega)$, the describing function, can be treated as an *amplitude dependent gain.* The phase angle of the describing function gives the phase change of the output of the nonlinear device (relative to the phase of the input) just as the phase angle of the transfer function gives the phase change of its output.

The Describing Function of the Comparator—An Example

To calculate the describing function of the comparator, we find the first Fourier coefficients of the function:

$$g(x) = f_{comp}(M\sin(\omega t)).$$

The output of the the comparator is a square wave with period T. (See Figure 8.2.) We must calculate the integrals:

$$A = \frac{2}{T}\left(\int_0^{T/2} 1\cdot\sin(\omega t)\,dt + \int_{T/2}^{T} -1\cdot\sin(\omega t)\,dt\right)$$
$$B = \frac{2}{T}\left(\int_0^{T/2} 1\cdot\cos(\omega t)\,dt + \int_{T/2}^{T} -1\cdot\cos(\omega t)\,dt\right).$$

[1]The describing function technique was first used by the J. Groszkowski in radio transmitter design before the Second World War and was formalized in 1964 by J. Kudrewicz [Lew92].

From the symmetries of the integrands, we see that:

$$A = \frac{8}{T}\int_0^{T/4} 1 \cdot \sin(\omega t)\,dt$$
$$B = 0.$$

As $\int \sin(\omega t)\,dt$ is $-\cos(\omega t)/\omega + C$, we find that the value of A is:

$$A = \frac{8}{T}\frac{1}{\omega} = \frac{8}{T\omega} = \frac{8}{2\pi} = \frac{4}{\pi}.$$

Thus, the describing function of the comparator is $\frac{A+jB}{M} = \frac{4}{\pi M}$.

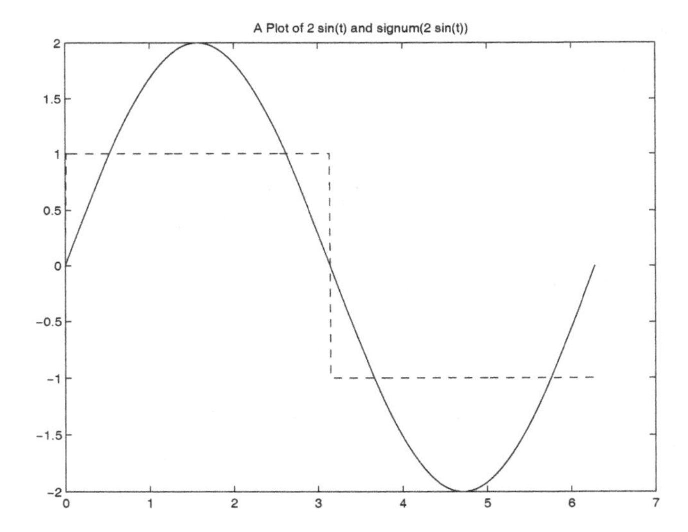

Fig. 8.2 A Plot of $M\sin(\omega t)$ and signum$(M\sin(\omega t))$ When $M = 2, \omega = 1$

What does this function tell us about the behavior of the comparator? First of all, since the function is real, it tells us that there is no phase shift. Indeed the apparent phase of the sine wave and the square wave are the same. Additionally, we find that as M goes up the "amplification" of the comparator goes down. This too is to be expected; the level of the comparator's output is always the same. The comparator amplifies small signals and attenuates large ones.

8.2.2 *Predicting Limit Cycles*

One standard technique for predicting limit cycles makes use of the describing function. One argues that if a limit cycle with angular frequency ω and amplitude M exists, then the total gain the limit cycle sees as it makes "one full circuit" around the circuit of Figure 8.1 must be precisely 1. The gain that the linear elements provide is clearly $-G_p(j\omega)H(j\omega)$ (where the minus sign comes from the inverter). What is the gain introduced by the nonlinear element? It is just $D(M,\omega)$. Thus the total gain that the sine wave sees is $-G_p(j\omega)H(j\omega)D(M,\omega)$. In order for this product to equal one, we find that:

$$G_p(j\omega)H(j\omega) = -\frac{1}{D(M,\omega)}.$$

In general solutions to this equation are pairs of numbers (M_k, ω_k)—frequencies and magnitudes. There may be (and often are) more than one solution for any given circuit. Each pair specifies the frequency of a solution and its magnitude. Here we find a fundamentally nonlinear phenomenon—our possible solutions have fixed magnitudes.

Oscillations in a Circuit Using a Comparator—An Example

Suppose that the nonlinear element in Figure 8.1 is a comparator, that $G_p(s) = \frac{1}{s^2}$, and that $H(s) = 1$. We would like to determine whether or not the circuit supports limit cycles.

Note that were our nonlinear comparator replaced by a linear gain, A, then the transfer function of the closed-loop system would be:

$$T(s) = \frac{G_p(s)}{1 + AG_p(s)} = \frac{1/s^2}{1 + A/s^2} = \frac{1}{s^2 + A}.$$

From the linear theory, we find that the circuit oscillates with angular frequency $\omega = \sqrt{A}$. For any given gain, A, the circuit will tend to oscillate at a specific frequency. The amplitude of the oscillation, however, is not determined by the gain.

When the comparator is present, we must look for solutions of the equation:

$$-\frac{1}{\omega^2} = -\frac{\pi M}{4}.$$

Clearly the solutions are $\omega = \sqrt{\frac{4}{\pi M}}$. Equivalently, $T = \pi\sqrt{\pi}\sqrt{M}$.

We find that there are oscillatory solutions possible at any given frequency—just as in the linear case. What distinguishes the nonlinear case from the linear case is that in the nonlinear case *there is only one magnitude for each frequency.*

Because the circuit that we are analyzing is not too complicated, we can find the true oscillations of the system. Suppose that the system starts with $y(0) = 0, \dot{y}(0) = D > 0$. Then for $t > 0$, and until some point that we will determine shortly, $y(t) > 0$. Thus, the output of the comparator will be -1 until the solution hits zero again. That is $y(t)$ solves the equation:

$$\int_0^t \int_0^r \text{signum}(-y(s) = -1)\, ds\, dr + Dt = y(t). \tag{8.1}$$

The term Dt appears to take care of the initial condition $\dot{y}(0) = D$.

Integrating (8.1), we find that $y(t) = -t^2/2 + Dt$. Clearly this will hit zero when $t = 2D$. At this point, the derivative is $\dot{y}(2D) = -D$. Following the procedure outlined above, we find that until $t = 4D$, the solution is $(t-2D)^2/2 - D(t-2D)$. At this point we start the process again. Since at $t = 4D$ the function and its first derivative are equal to what they were initially, we know that from this point on the pattern repeats itself every $4D$ seconds. That is, we have found that the function is periodic with period $T = 4D$. We graph the output of the circuit in Figure 8.3. Note how very sinusoidal the function looks. The maximum amplitude of the solution, M, is $D^2/2$. Thus, the relationship between the period and the amplitude is $T = 4\sqrt{2}\sqrt{M}$. This is the same type of relationship that our approximate technique gave. In fact, the constants in the two equations are almost the same—$\pi\sqrt{\pi} \approx 5.57$ and $4\sqrt{2} \approx 5.66$.

8.2.3 *The Stability of Limit Cycles*

In the previous section, we discussed a condition for the existence of limit cycles. Mere existence of oscillatory solutions is not all the knowledge that we need—we need to know if such solutions are stable. We want to determine whether or not a limit cycle, once it is entered into, is likely to be self-maintaining. We want to make sure that if one perturbs the limit

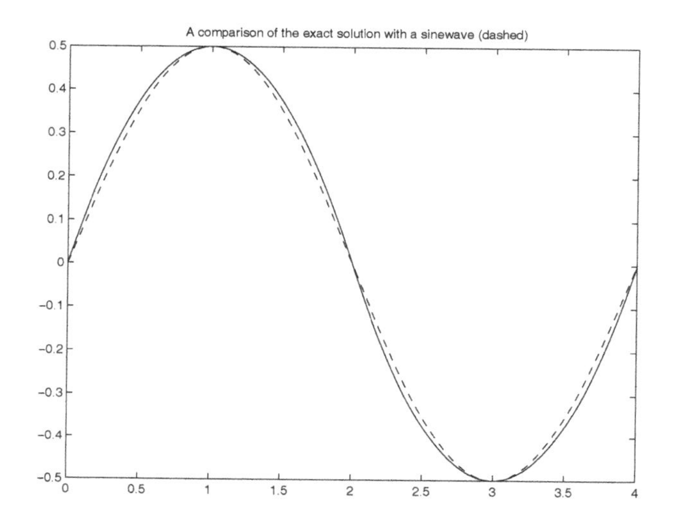

Fig. 8.3 A Comparison of the Output of Our Circuit with a Sine Wave

cycle a little bit that the limit cycle will not simply go away[2].

We make use of the describing function to test for stability. Suppose that some system supports a limit cycle with angular frequency ω and magnitude M. That is, the system output can be $M\sin(\omega t)$. We test for stability by assuming that something changed the magnitude of the oscillation from M to $M + \Delta M$. We want to know the total gain the solution "sees" as it traverses the circuit. From our previous discussions it is clear that that gain is $-G_p(j\omega)H(j\omega)D(\omega, M + \Delta M)$.

What is a reasonable condition for stability? What condition should guarantee that if the the magnitude goes up for some reason that it will go back down to its equilibrium position? The condition is that the total gain around the loop be less than one if the magnitude of the oscillation is increased. And what condition guarantees that if something decreases the magnitude of the oscillation that the oscillation will rise back to it equilibrium level? Clearly in this case the gain around the loop needs to

[2]This type of stability seems to be quite different from the type of stability we considered for linear systems. For linear systems we wanted to know whether the system could behave "badly" when its inputs were reasonable; for limit cycles we want to know whether a small perturbation will tend to stop the predicted oscillations. On the other hand, as we saw in §1.6.3, in an unstable linear system small perturbations can also change the nature of the system's behavior. Thus the two types of stability are related.

be greater than one. Putting the conditions for the existence and stability into a set of formulas, we find that the conditions for a stable limit cycle are that there exist an M, an ω, and an $\epsilon > 0$ such that:

$$\begin{array}{ll} |G_p(j\omega)H(j\omega)||D(\omega, M + \Delta M)| > 1 & -\epsilon < \Delta M < 0 \\ -G_p(j\omega)H(j\omega)D(\omega, M + \Delta M) = 1 & \Delta M = 0 \\ |G_p(j\omega)H(j\omega)||D(\omega, M + \Delta M)| < 1 & 0 < \Delta M < \epsilon. \end{array}$$

As $G_p(j\omega)H(j\omega)$ do not depend on the value of M, this is a condition on the describing function. The condition is that the magnitude of the describing function be a decreasing function of M in the vicinity of the limit cycle. Putting this condition another way, if $|D(\omega, M)|$ is a differentiable function of M, then $\frac{\partial}{\partial M}|D(\omega, M)| < 0$ at the values of ω and M that correspond to the limit cycle.

Stability of Oscillations—An Example

We have shown that when the nonlinear element in Figure 8.1 is a comparator and the linear elements are $G_p(s) = \frac{1}{s^2}, H(s) = 1$, then the system supports limit cycles at all possible frequencies. We would now like to determine the stability of the limit cycles that we have predicted.

To see that the limit cycles are stable it is enough to look at the describing function of the comparator. Since it is $D(\omega, M) = \frac{4}{\pi M}$ and since for all positive M this function is decreasing, we see that the limit cycles should indeed be stable.

Following the logic of our "stability proof," what should stability mean? It should mean that "a long time" after increasing the amplitude a little one finds that the system's amplitude has returned to its original amplitude. Thus, our prediction is that a long time after making this small change to the system's output the system's output will return to its original amplitude.

This is not what happens in our system! We have already seen that the frequency of the output is connected to the maximum of the output. Suppose that at the point at which the output was at it maximum—the point at which $M = D^2/2$—one made a small change in the amplitude—one changed it to $M + \Delta M = (D + \Delta D)^2/2$. Then what happens is that one *forever* changes the period to $4(D + \Delta D)$. Admittedly the output will be very similar to the previous output. However, the output will never actually return to its previous state.

> Here we see one of the fundamental limitations of this approach to the stability of limit cycles. Even though this approach implies that the limit cycles are stable in a very strong sense, we find that they are only stable in a weaker sense.

The describing function technique for the determination of the existence of limit-cycles is an *approximate technique.* It *generally* works well if one is careful to use it only when the system under consideration is low-pass. The rule of thumb is that for best results the degree of the numerator of $G_p(s)H(s)$ should be at least two less than the degree of the denominator of $G_p(s)H(s)$. As we will see, if this is not the case the technique need not work well. The test for the stability is even more *ad hoc*—and it should not be trusted too far.

8.2.4 *More Examples*

8.2.4.1 *A Nonlinear Oscillator*

We now design an oscillator using a comparator. Suppose that in Figure 8.1 the nonlinear element is a comparator and the linear elements are $G_p(s) = -\frac{1}{2}\frac{s}{s^2+s+1}, H(s) = 1$. The condition for the existence of limit cycles is that:

$$G_p(j\omega) = -1/D(\omega, M) \Leftrightarrow \frac{1}{2}\frac{j\omega}{-\omega^2+1+j\omega} = \frac{\pi M}{4}.$$

Since the right hand side of this equation is real, the left hand side must be as well. In order for this to happen, the denominator of the expression on the right hand side must be a pure imaginary number. This happens when $\omega = 1$. At this point, the left hand side is equal to $1/2$, so the right hand side must be one half as well. We find that $M = 2/\pi$. We have found that there is precisely one limit cycle with angular frequency $\omega = 1$ and amplitude $M = 2/\pi$. As we have already seen that the describing function is a decreasing function of M, we find that the limit cycle is stable.

Why does this system oscillate? The linear portion of the system has the transfer function $-\frac{1}{2}\frac{s}{s^2+s+1}$. When the input to this system is a unit step, its output is $-\frac{1}{2}\frac{1}{s^2+s+1}$. The inverse Laplace transform of this function is $-\frac{1}{2}e^{-t/2}\sin(\sqrt{3/4}t)$. That is, its output is a damped exponential—it naturally oscillates. The problem with the linear system is that its output will decay to zero. The function of the comparator is to "kick" the linear element every half period and in that way to add some "energy" to the system.

We note something else of interest. The natural frequency at which the system oscillates seems to be $\sqrt{3/4} \approx .866$ and not one. A rigorous treatment of the system shows that the frequency of oscillation is in fact $\omega = \sqrt{3/4}$ and not $\omega = 1$. This lack of accuracy in the results of the describing function technique is not surprising—the linear part of this system is not very low-pass. (The degree of the denominator is only one greater than the degree of the numerator.)

8.2.4.2 *A Comparator with a Dead Zone*

Consider a comparator with a dead zone—a circuit whose output profile is shown in Figure 8.4. The output of the circuit is always zero as long as the absolute value of the input is less than one. Let us consider the case that the input to the circuit is $V_{in}(t) = M\sin(\omega t)$ where $M > 1$—the case one must consider in order to calculate the interesting part of the describing function of the nonlinear circuit element. In this case, the output will be a function with the same period as the sine wave, but the output will equal zero until the points at which $M\sin(\omega t) = 1$. A typical example of the output of a comparator with a dead zone is given in Figure 8.5. The smallest positive value of t for which $M\sin(\omega t) = 1$ is:

$$\tau = \frac{\sin^{-1}(1/M)}{\omega}.$$

From the symmetry of Figure 8.5, it is clear that:

$$\begin{aligned} A &= \frac{8}{T}\int_{\tau}^{\frac{\pi}{2\omega}} \sin(\omega t)\,dt = \frac{4}{\pi}\cos(\omega\tau) \\ B &= 0. \end{aligned}$$

From the definition of τ we find that $A = \frac{4}{\pi}\cos(\sin^{-1}(1/M))$. As the cosine term is positive and as we know that $\sin^2(x) + \cos^2(x) = 1$, it is clear that $\cos(x) = \sqrt{1 - \sin^2(x)}$. Thus, $\cos(\sin^{-1}(1/M)) = \sqrt{1 - 1/M^2}$. We find that:

$$D(\omega, M) = \begin{cases} 0 & M < 1 \\ \frac{4}{\pi M}\sqrt{1 - 1/M^2} & M \geq 1 \end{cases}.$$

We plot this function if Figure 8.6.

Suppose that we now use this nonlinear element in the circuit of Figure 8.1 with $G_p(s)H(s) = -\frac{2s}{s^2+s+1}$. We would like to know whether or not there are limit cycles, and if there are limit cycles, we would like to know

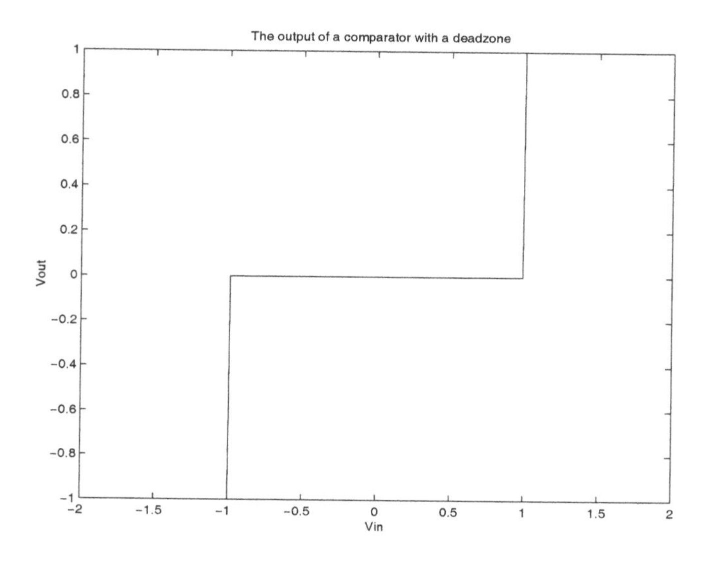

Fig. 8.4 The Output Profile of a Comparator with a Dead Zone

if they are stable. When checking for limit cycles, we once again find that $\omega = 1$. We also find that M must satisfy the equation $\frac{4}{\pi M}\sqrt{1 - 1/M^2} = 1/2$. Though it is simple enough to solve this equation by hand, it is still simpler to look at Figure 8.6. We see that the describing function equals 1/2 twice. Thus, there are two limit cycles. Are the limit cycles stable? Since the describing function is increasing at the first intersection, the first limit cycle is not stable. As the describing function is decreasing at the second intersection, the second limit cycle it is stable.

8.2.4.3 *A Simple Quantizer*

Next we consider a simple quantizer—a circuit element whose input-output relationship is as shown in Figure 8.7. Though this looks like something where it would be quite unpleasant to calculate the describing function, it is actually quite easy. If one looks at the output of this quantizer, one finds that it is precisely the output of our comparator added to the output of the comparator with a dead zone. It is not difficult to see that the describing function is just the "sum" of the describing functions of the two items just

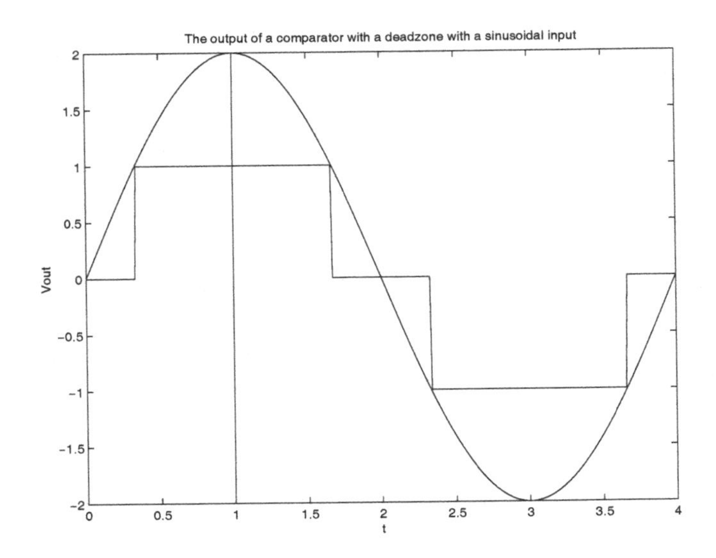

Fig. 8.5 The Output of a Comparator with a Dead Zone when the Input Is a Sine Wave

mentioned. Thus, the describing function of the element is:

$$D(\omega, M) = \begin{cases} \frac{4}{\pi M} & M < 1 \\ \frac{4}{\pi M} + \frac{4}{\pi M}\sqrt{1 - 1/M^2} & M \geq 1 \end{cases}.$$

We plot this function in Figure 8.8.

Suppose that in the circuit of Figure 8.1 we let the nonlinear element be this simple quantizer, and we let $G_p(s) = \frac{2}{3}\frac{-s}{s^2+s+1}$, $H(s) = 1$. Once again we find that the only place at which there might be a limit cycle is $\omega = 1$. At this point, $G_p(j) = -2/3$. Thus, we find that $1/D(\omega, M) = 2/3$. That is, we must find all places where $D(\omega, M) = 3/2$. Looking at Figure 8.8, we find that there are three such points. At two of those points, the describing function is decreasing. Thus, there are *two* stable limit cycles. The other limit cycle is not stable.

8.2.5 *Graphical Method*

As we saw in §8.2.2, the describing function technique predicts the existence of a limit cycle at those points for which:

$$G_p(j\omega)H(j\omega) = -\frac{1}{D(\omega, M)}.$$

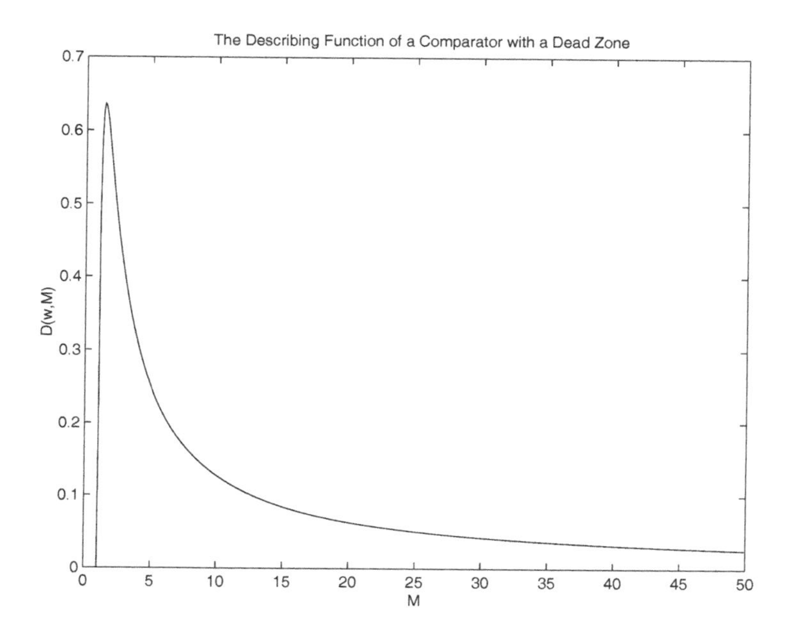

Fig. 8.6 The Describing Function of a Comparator with a Dead Zone

Such a condition lends itself to graphical analysis when $D(\omega, M)$ is only a function of M. Let us try a simple example.

Consider the system of Figure 8.1 where the nonlinear element is a comparator with a dead zone, $G_p(s) = -\frac{2s}{s^2+s+10}$, and $H(s) = 1$. We graph the functions:

$$G_p(j\omega)H(j\omega) = -\frac{2j\omega}{10-\omega^2+j\omega} \qquad \omega \geq 0$$

$$-\frac{1}{D(\omega,M)} = \begin{cases} -\infty & M < 1 \\ -\frac{\pi M}{4}\frac{1}{\sqrt{1-1/M^2}} & M \geq 1 \end{cases} \quad M > 0$$

as functions of ω and M respectively. The graph is shown in Figure 8.9. We note that there is an intersection of the two curves. Thus, there is a limit cycle.

Using the graphical method, one can often see whether or not one can easily get rid of the limit cycle. In our case, a careful examination of Figure 8.9 (or a simple calculation) shows that if we introduce an attenuation of more than $\pi/4$, then we will no longer have a limit cycle—the ellipse and the dashed line will no longer intersect.

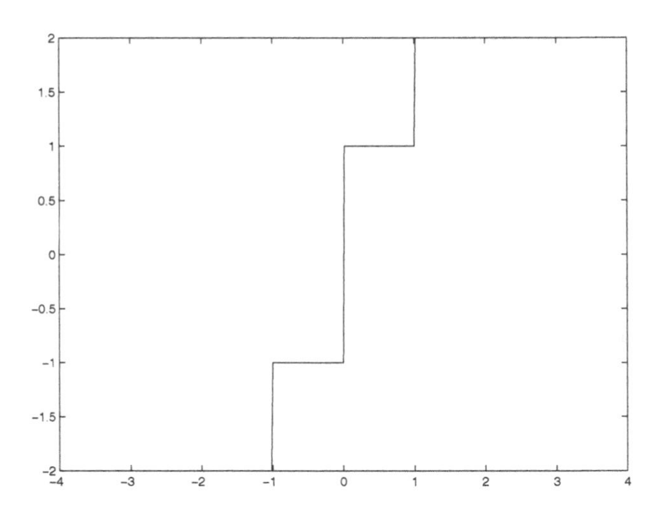

Fig. 8.7 The Input-Output Relationship of the Simple Quantizer

8.3 Tsypkin's Method

When dealing with a comparator-type nonlinearity, there is an exact method for determining the existence of limit cycles—Tsypkin's method[3]. This method, introduced by Tsypkin [Tsy84] and Hamel [Ham49] separately in the late forties and early fifties, allows one to determine the frequency of a limit cycle to any desired accuracy. It also allows one to calculate the form of the limit cycle to any desired accuracy.

This method deals with continuous limit cycles that are non-negative for half of their period and non-positive in the next half-period. The idea of the method is as follows. Suppose that in Figure 8.1 the function $z(t)$—the output of the entire linear portion of the circuit—is periodic of period T, is positive in $(0, T/2)$ and negative in $(T/2, T)$. Then the output of the nonlinear component—a comparator—is a square wave. Assume that the comparator used is the comparator of Figure 8.2; then the amplitude of the square wave is 1. The Fourier series representation of the output of the

[3]Named after Yakov Zalmanovich Tsypkin (1919-1997)—one of its inventors [Sto].

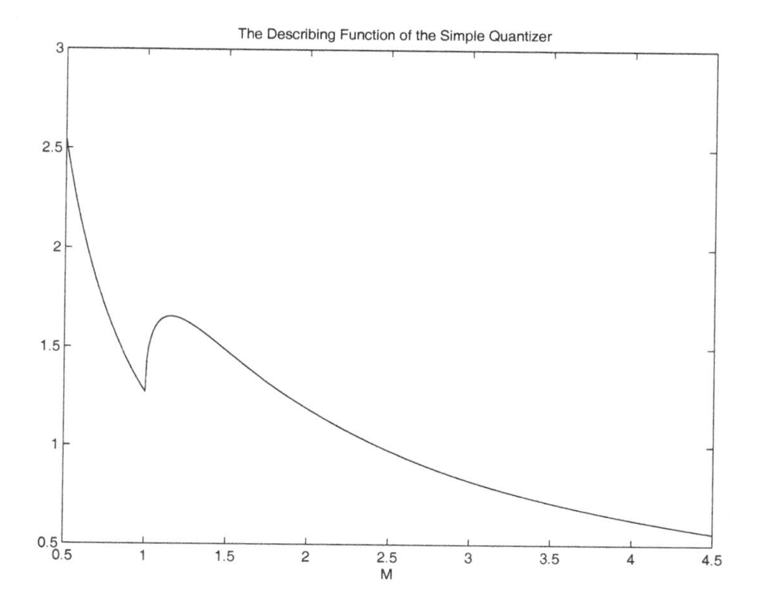

Fig. 8.8 The Describing Function of the Simple Quantizer

comparator is:

$$\text{square}_T(t) = \frac{4}{\pi}\sum_{n=1}^{\infty}\frac{\sin((2n-1)\omega t)}{2n-1}, \quad \omega = \frac{2\pi}{T}.$$

The frequency response of the linear part is just $-G_p(j\omega)H(j\omega)$. When the square wave passes through the linear part, each component is affected by the linear part separately (according to the principle of superposition). In the steady state, we find that (with $K(j\omega) \equiv -G_p(j\omega)H(j\omega)$) the square wave is transformed into:

$$z(t) = \frac{4}{\pi}\sum_{n=1}^{\infty}\frac{|K(j(2n-1)\omega)|\sin((2n-1)\omega t + \angle(K(j(2n-1)\omega)))}{2n-1}. \quad (8.2)$$

Evaluating this at $t = 0$ gives us:

$$z(0) = \frac{4}{\pi}\sum_{n=1}^{\infty}\frac{|K(j(2n-1)\omega)|\sin(\angle(K(j(2n-1)\omega)))}{2n-1}.$$

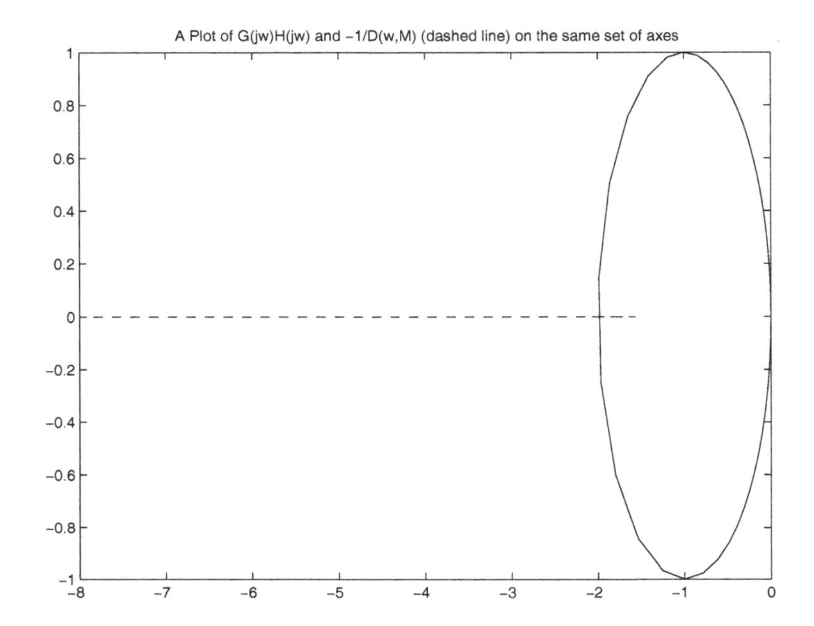

Fig. 8.9 The Graphical Method for Checking for the Existence of Limit Cycles

Clearly, this is equal to:

$$z(0) = -\frac{4}{\pi}\sum_{n=1}^{\infty}\frac{\operatorname{Im}(G_p(j(2n-1)\omega)H(j(2n-1)\omega))}{2n-1}. \tag{8.3}$$

Assuming that $G_p(s)H(s)$ is the quotient of two polynomials, as long as the degree of the denominator is greater than the degree of the numerator:

$$|G_p(j(2n-1)\omega)H(j(2n-1)\omega)| \leq \frac{\mathrm{C}}{(2n-1)\omega}. \tag{8.4}$$

As the terms in (8.2) and (8.3) are of order $1/n^2$ and are continuous functions, the sums converge absolutely and uniformly and $z(t)$ is a continuous function[4]. As by assumption $z(t)$ is positive in its first half-period and negative in its second half-period, $z(0) = z(T/2) = 0$. We find that a necessary condition for limit cycles that are positive through one half-period

[4]Let:

$$y(s) = \sum_{n=0}^{\infty} y_n(s)$$

where the $y_n(s)$ are continuous functions. As the sum is assumed to converge absolutely

and negative in the other half-period is that:

$$\lambda(\omega) \equiv -\frac{4}{\pi}\sum_{n=1}^{\infty}\frac{\mathrm{Im}(G_p(j(2n-1)\omega)H(j(2n-1)\omega))}{2n-1} = z(0) = 0. \qquad (8.5)$$

This is a condition that is easily checked using a computer and letting n range from 1 to a "sufficiently large" number. Also note that once a value of ω for which $\lambda(\omega) = 0$ has been found it is easy to plug this value back into Equation (8.2) in order to find $z(t)$—the form of the limit cycle. If $z(t)$ is positive until $T/2$ and is negative from $T/2$ to T then there is indeed a limit cycle with angular frequency ω.

$G_p(s) = -s/(s^2+s+1), H(s) = 1$—An Example

Once we are dealing with a specific problem we need to find the constant, C, (that appears in (8.4)) and we need to work out

and uniformly, for any $\delta > 0$ one can pick a number N such that:

$$\sum_{n=N+1}^{\infty} |y_n(s)| < \delta/4.$$

Furthermore, as all of the $y_n(t)$ are assumed to be continuous:

$$S_N(s) = \sum_{n=0}^{N} y_n(s)$$

is continuous too. Thus, for any $\delta/2$, one can pick an $\epsilon > 0$ so that if $|s - s'| < \epsilon$, then $|S_n(s) - S_N(s')| < \delta/2$. Combining these results, we find that for any $\delta > 0$ there exists an $\epsilon > 0$ such that if $|s - s'| < \epsilon$, then:

$$\begin{aligned}|y(s) - y(s')| &\leq \left|\sum_{n=0}^{\infty} y_n(s) - \sum_{n=0}^{\infty} y_n(s')\right| \\ &\leq \left|\sum_{n=0}^{N} y_n(s) - \sum_{n=0}^{N} y_n(s')\right| + \left|\sum_{n=N+1}^{\infty} y_n(s) - \sum_{n=N+1}^{\infty} y_n(s')\right| \\ &\leq \delta/2 + \left|\sum_{n=N+1}^{\infty} y_n(s)\right| + \left|\sum_{n=N+1}^{\infty} y_n(s')\right| \\ &\leq \delta.\end{aligned}$$

That is, we find that $y(s)$ is continuous.

what a "sufficiently large" number is. It is easy to see that:

$$\begin{aligned}|G_p(j\omega)| &= \left|\frac{-j\omega}{1-\omega^2+j\omega}\right|\\ &= \frac{\omega}{\sqrt{(1-\omega^2)^2+\omega^2}}\\ &\leq \frac{\omega}{|1-\omega^2|}\\ &\leq \frac{\omega}{\frac{24}{25}\omega^2} = \frac{25}{24}\frac{1}{\omega}, \qquad \omega \geq 5.\end{aligned}$$

We find that[5] as long as $(2N+1)\omega > 5$:

$$\begin{aligned}\left|-\frac{4}{\pi}\sum_{n=N+1}^{\infty}\operatorname{Im}\left(\frac{G_p(j(2n-1)\omega)}{(2n-1)}\right)\right| &\leq \frac{4}{\pi}\frac{25}{24}\sum_{n=N+1}^{\infty}\frac{1}{\omega(2n-1)^2}\\ &\leq \frac{4}{\pi\omega}\frac{25}{24}\int_N^\infty \frac{1}{(2x-1)^2}\,dx\\ &= \frac{25}{12\pi\omega}\frac{1}{2N-1}.\end{aligned}$$

We find that if $N = 1000$, the value of the tail of the sum cannot exceed $\frac{25}{12\pi\omega}\frac{1}{1999}$. If $\omega > 1/2$, then this does not exceed $\frac{50}{12\pi}\frac{1}{1999} \approx 0.00066$. It would seem that $N = 1000$ ought to be sufficiently large.

Using a simple MATLAB program that calculates the first thousand terms of $\lambda(\omega)$, we find that $\lambda(\sqrt{3/4}-0.001) \approx 0.0018$ and $\lambda(\sqrt{3/4}+0.001) \approx -0.0012$. From our previous error estimate we know that this is sufficient to prove that $\lambda(\sqrt{3/4}-0.001) > 0$ and $\lambda(\sqrt{3/4}+0.001) < 0$. Thus, there is a zero of $\lambda(\omega)$ in the interval $\left[\sqrt{3/4}-0.001, \sqrt{3/4}+0.001\right]$. Using $\omega = \sqrt{3/4}$ and (8.2) we find that the form of the limit cycle is as shown in Figure 8.10. We evaluated the sum using a MATLAB program that calculates (8.2) at 1001 points using the first hundred odd harmonics of ω. We note that the picture does not look terribly sinusoidal. It is to be expected (as noted on p. 256) that in this case the describing function technique will not work very well.

[5] Recall that:

$$\sum_{i=N+1}^{\infty} f(i) \leq \int_N^\infty f(x)\,dx$$

if $f(x)$ is a decreasing function on $[N,\infty]$.

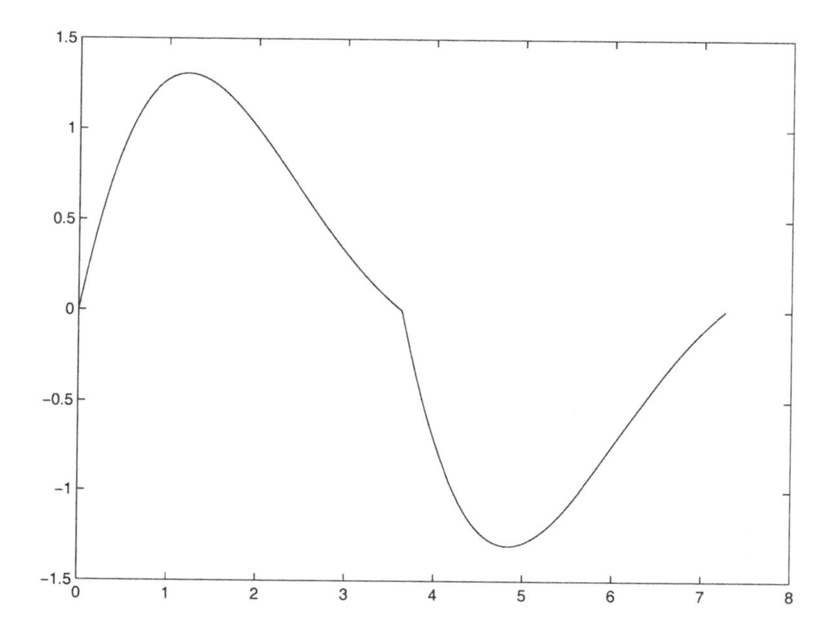

Fig. 8.10 The Limit Cycle for the System with $G_p(s) = -s/(s^2 + s + 1), H(s) = 1$

8.4 The Tsypkin Locus and the Describing Function Technique

The describing function technique is an approximate method for finding limit cycles; the Tsypkin locus is an exact method for finding limit cycles. By using the Tsypkin locus one can determine some of the limits of the describing function technique[6]

From the definition of $\lambda(\omega)$ in (8.5) we see that $\lambda(\omega)$ is a weighted average of a sequence of values of the imaginary part of $G_p(j\omega)H(j\omega)$. The only way that $\lambda(\omega)$ can equal zero–as it must for a limit cycle to exist–is for the imaginary part to be positive for some values of ω and negative for others (or for the imaginary part to be identically zero—which does not generally happen). In order for a limit cycle to exist in a system whose nonlinear element is a comparator, the describing function analysis requires

[6]This section is adapted from [Eng02]. Parts of this section are ©2002 IEEE. Those portions are reprinted, with permission, from "Limitations of the Describing Function for Limit Cycle Prediction," *IEEE Transactions on Automatic Control*, Vol. 47, No. 11 (2002).

that there exist a value ω_o such that $G_p(j\omega_o)H(j\omega_o) < 0$—such that the frequency response is negative at that frequency. It is perfectly possible to produce systems for which the latter condition is fulfilled but for which the former condition is not. For such systems the describing function technique erroneously predicts the existence of a limit cycle.

As an example of a system for which the describing function technique erroneously predicts a limit cycle consider:

$$G_p(s)H(s) = \frac{1}{s}\left(\frac{s/\sqrt{3}+1}{\sqrt{3}s+1}\right)^3.$$

We find that $1/(j\omega)$ contributes -90° to the phase of $G_p(j\omega)H(j\omega)$ for all values of ω. The second term in the transfer function is just a thrice repeated phase-lag compensator. It is easy to see (using (7.2) and (7.3)) that the phase-lag compensator achieves its minimum phase at $\omega = 1$ and at this point the phase of the compensator is -30°. Thus, the phase of $G_p(j\omega)H(j\omega)$ will always be between 0° and -180° and it reaches -180° at only one point—$\omega = 1$. At that point the frequency response is negative. According to the describing function analysis the fact that $G_p(j)H(j)$ is negative implies the existence of a limit cycle. The Tsypkin locus analysis guarantees that in this case there will be no limit cycle because the weighted sum of the imaginary part of $G_p(j\omega)H(j\omega)$ cannot be zero.

This example is a bit unfair—it is well known that the describing function technique works well with low-pass systems, and our system is not very low-pass. However, it is easy to show that there are other, more low-pass system for which the describing function fail to work properly. As a simple example consider the system for which:

$$G_p(s)H(s) = \frac{1}{s(s+1)}\left(\frac{10s+1}{(28+12\sqrt{5})s+1}\right)^2.$$

The linear part of the system has two zeros at infinity—it is sufficiently low-pass that we expect the describing function technique to work properly. The Bode plots corresponding to the system are given in Figure 8.11. We see that the phase hits -180° once and is never less that this. Though the describing function analysis indicates that there will be limit cycles, from the exact method we know that there will not be any limit cycles.

Plotting the graphs of $G_p(j\omega)H(j\omega)$ and $-1/D(M,\omega)$ (see figure 8.12), we find that rather than the two graphs crossing one another's path, the two graphs have a single point of tangency. In an approximate method such as the describing function method, it is questionable whether such intersections are meaningful. By using Tsypkin's method, we have shown that at least at times such intersections are not meaningful—such intersections need not imply that the system supports a limit cycle.

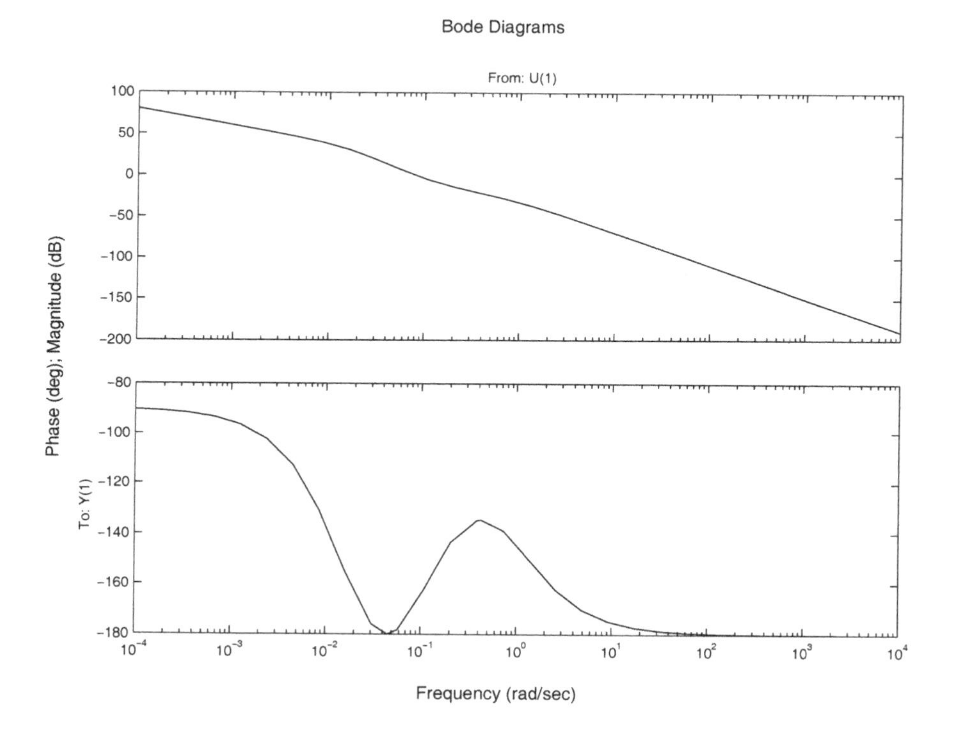

Fig. 8.11 The Bode Plots

8.5 Exercises

(1) Suppose that in Figure 8.13 the nonlinearity is a limiter—a non-ideal amplifier. Consider, for example:

$$f(x) = \begin{cases} C & x \geq C \\ x & |x| < C \\ -C & x \leq -C \end{cases}$$

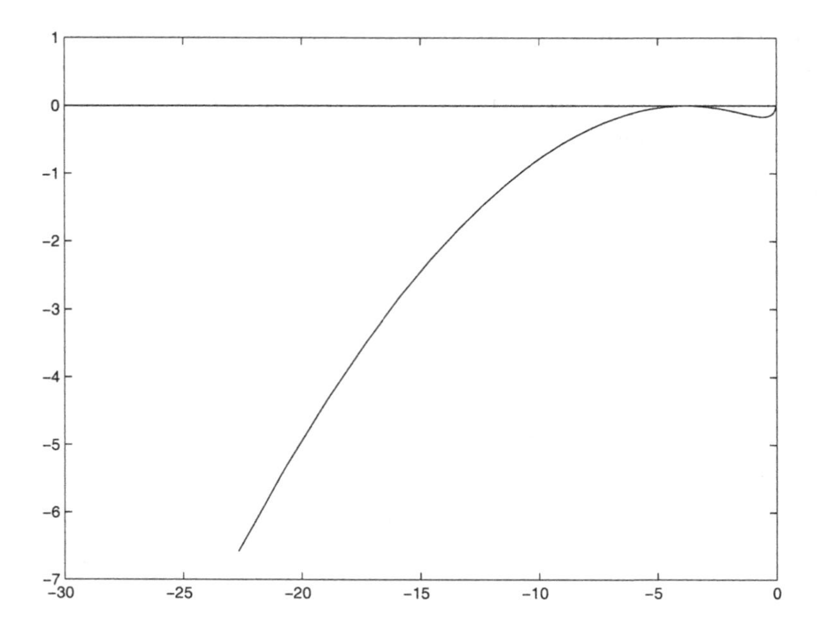

Fig. 8.12 The Intersection of $G_p(j\omega)H(j\omega)$ and $-1/D(M,\omega)$

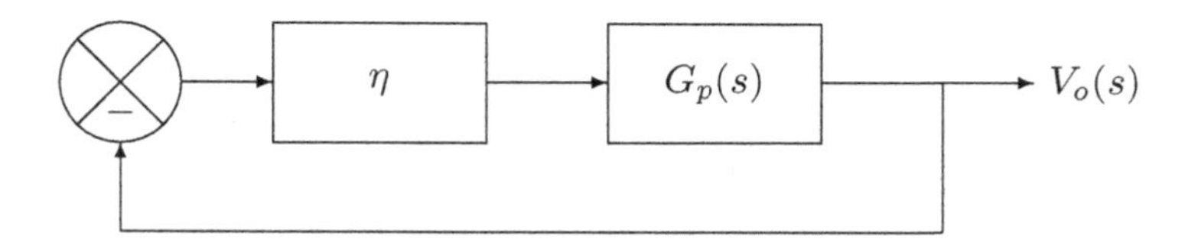

Fig. 8.13 A Generic Nonlinear Unity-Gain Feedback System

where $f(x)$ is the output of the nonlinear element when its input is x.

(a) Find the describing function of the nonlinear element.

(b) To what function does the describing function tend as $C \to 0$? Is this reasonable? Explain.

(2) Suppose that in Figure 8.13:

$$G_p(s) = \frac{6}{s^3 + s^2 + 4s + 1}$$

and the nonlinearity is the limiter of Problem 1 with $C = 1/2$. Does the describing function technique predict that the system supports a limit cycle? If it does, does it predict a stable limit cycle or an unstable limit cycle?

(3) Suppose that in the system of Figure 8.13 we have:

$$G_p(s) = \frac{\mathrm{K}}{2}\frac{-s}{s^2 + s + 1}$$

and the nonlinear element is the limiter of Problem 1.

(a) What is the smallest value of K for which the describing function technique predicts a limit cycle?

(b) Should one expect the limit cycle predicted in the previous part to be stable? Explain.

(4) (a) Calculate the describing function of the nonlinear element which is described by the function:

$$f(x) = \begin{cases} 3.5 & x \geq 3 \\ 2.5 & 2 \leq x < 3 \\ 1.5 & 1 \leq x < 2 \\ 0 & -1 \leq x < 1 \\ -1.5 & -2 \leq x < -1 \\ -2.5 & -3 \leq x < -2 \\ -3.5 & x < -3 \end{cases}.$$

(b) What limit cycles does one expect to see in the output of the circuit of Figure 8.13 with this nonlinear element and with a linear element whose transfer function is:

$$G_p(s) = \frac{1}{s(s+1)^2}?$$

(5) We generally assume that the transfer function of a DC motor is:

$$G_p(s) = \frac{K}{s(\tau s + 1)}. \tag{8.6}$$

This is a reasonably good approximation, but there is actually a small third-order term in the denominator. A better approximation to the transfer function is:

$$G_p(s) = \frac{K}{\epsilon s^3 + s(\tau s + 1)}. \tag{8.7}$$

Suppose that the nonlinearity in Figure 8.13 is a comparator and the plant is a DC motor.

(a) Making use of the inexact transfer function (8.6) does one predict stable limit cycles?
(b) Using the more exact transfer function (8.7) does one predict stable limit cycles?
(c) What is the crucial difference between the two approximate transfer functions?

(6) Suppose that in Figure 8.13 the nonlinearity is a comparator and the transfer function of the linear system is:

$$G_p(s) = \frac{1}{s^5}\left(\frac{s/\sqrt{3}+1}{\sqrt{3}s+1}\right)^3$$

Note that:

$$\frac{s/\sqrt{3}+1}{\sqrt{3}s+1}$$

is the transfer function of a phase-lag filter.

(a) Show that the describing function technique predicts a limit cycle in this system.
(b) Describe the limit cycle. In particular discuss the stability of the limit cycle.
(c) Plot the function $G_p(j\omega)H(j\omega)$ and $-1/D(M,\omega)$. What does the manner in which they intersect imply about the accuracy of the predictions in first two parts of this question?

(7) Use Tyspkin's method to show that the limit cycle predicted in Problem 6 does not really exist. Note that this is so despite the fact that Problem 6 concerns a very low-pass system.

(8) Use the describing function technique to show that if the linear part of the system of Problem 1 is replaced by:

$$G_p(s) = \frac{K}{s(\tau s + 1)}, \quad K, \tau > 0$$

and the nonlinear part is as given in Problem 1, then the system does not support limit cycles.

Chapter 9

An Introduction to Modern Control

9.1 Introduction

When controlling a system using the techniques of classical control, one generally uses the system's output in the feedback. In this way one can improve the gain and phase margins, make the system faster, and generally improve the system's performance.

When controlling a system using the techniques of modern control, one can (generally) do much more than just improve the margins; one can place the system's poles as one pleases. This improvement, however, comes at a price. In order to place the poles of the system, one must generally know the *complete internal state* of the system. This is often not a trivial task.

9.2 The State Variables Formalism

In classical control, one generally describes a system using high-order differential equations or, equivalently, a transfer function. In modern control we use systems of first order equations. We consider systems that have a single input, $u(t)$, a single output, $y(t)$, but many states. To describe systems, we make use of one matrix, $\mathbf{A}$ and three vectors, $\vec{x}(t)$—which is the vector that is made up of the system's states, and the vectors $\vec{b}$ and $\vec{c}$. The equations that describe the evolution of the system's states and the system's output must be put in the form:

$$\vec{x}'(t) = \mathbf{A}\vec{x}(t) + \vec{b}u(t), \quad \vec{x}(0) = \vec{x}_0 \tag{9.1}$$

$$y(t) = \vec{c}^T\vec{x}(t). \tag{9.2}$$

The first equation describes the evolution of the state of the system and is called the *state equation.* The second equation gives the output as a linear combination of the states.

Let us consider a simple example. If our system is a motor whose transfer function is:

$$G(s) = \frac{1}{s(s+1)},$$

and if we let the input to the motor be $u(t)$ and the output of the motor be $y(t)$—the motor shaft's angle, then we find that:

$$\frac{Y(s)}{U(s)} = \frac{1}{s(s+1)}.$$

Taking inverse Laplace transforms (and adding the appropriate initial conditions), we find that:

$$y''(t) + y'(t) = u(t), \quad y(0) = \alpha, y'(0) = \beta.$$

If we let $x(t) = y(t)$ and we let the state of the system be:

$$\vec{x}(t) = \begin{bmatrix} x(t) \\ x'(t) \end{bmatrix},$$

then we find that we can write our equations as:

$$\overbrace{\begin{bmatrix} x(t) \\ x'(t) \end{bmatrix}'}^{\vec{x}'(t)} = \overbrace{\begin{pmatrix} 0 & 1 \\ 0 & -1 \end{pmatrix}}^{\mathbf{A}} \overbrace{\begin{bmatrix} x(t) \\ x'(t) \end{bmatrix}}^{\vec{x}(t)} + \overbrace{\begin{bmatrix} 0 \\ 1 \end{bmatrix}}^{\vec{b}} u(t), \quad \begin{bmatrix} x(0) \\ x'(0) \end{bmatrix} = \begin{bmatrix} \alpha \\ \beta \end{bmatrix}$$

$$y(t) = \overbrace{(1\ 0)}^{\vec{c}^T} \overbrace{\begin{bmatrix} x(t) \\ x'(t) \end{bmatrix}}^{\vec{x}(t)}.$$

The first equation is the state equation. It describes the evolution of the state of the system. (In this case the state is the motor shaft's angular position and its angular velocity.) The second equation states that the output of the system is the motor shaft's angular position.

9.3 Solving Matrix Differential Equations

Given an equation of the form:

$$\vec{x}'(t) = \mathbf{A}\vec{x}(t), \quad \vec{x}(0) = \vec{x_0}, \tag{9.3}$$

what can we say about $\vec{x}(t)$? Recalling that the solution of the scalar equation:

$$x'(t) = ax(t), \quad x(0) = x_0$$

is $x(t) = e^{at}x_0$, a reasonable guess at the solution of (9.3) is:

$$\vec{x}(t) = e^{\mathbf{A}t}\vec{x_0}.$$

It remains to *define* the matrix exponential in such a way that this equation is well defined and in such a way that (9.3) holds.

Let us *define* the matrix exponential by using the Taylor series expansion of the exponential. (See §9.15 for some technical results about series of matrices.) That is, let:

$$e^{\mathbf{A}} \equiv \mathbf{I} + \mathbf{A} + \frac{\mathbf{A}^2}{2!} + \cdots + \frac{\mathbf{A}^n}{n!} + \cdots. \tag{9.4}$$

Considering the derivative of $e^{\mathbf{A}t}$, we find that:

$$\begin{aligned}\frac{d}{dt}e^{\mathbf{A}t} &= \frac{d}{dt}\left(\mathbf{I} + \mathbf{A}t + \frac{\mathbf{A}^2t^2}{2!} + \cdots + \frac{\mathbf{A}^nt^n}{n!} + \cdots\right)\\ &= \left(\mathbf{0} + \mathbf{A} + \mathbf{A}^2t + \cdots + \frac{\mathbf{A}^nt^{n-1}}{(n-1)!} + \cdots\right)\\ &= \mathbf{A}e^{\mathbf{A}t}.\end{aligned}$$

If we consider the derivative of $e^{\mathbf{A}t}\vec{x}_0$, we find that:

$$\frac{d}{dt}\left(e^{\mathbf{A}t}\vec{x}_0\right) = \mathbf{A}e^{\mathbf{A}t}\vec{x}_0.$$

Thus, $e^{\mathbf{A}t}\vec{x}_0$ is a solution of the differential equation. We also find that at time zero this function is equal to $\vec{x}_0$. Thus:

$$x(t) = e^{\mathbf{A}t}\vec{x}_0$$

solves the differential equation subject to the given initial conditions.

Recall that for a scalar a, $e^{-at}e^{at} = 1$. That is, e^{-at} is the multiplicative inverse of e^{at}. We now prove that the equivalent statement is true for matrices: $e^{-\mathbf{A}t}e^{\mathbf{A}t} = \mathbf{I}$. Considering the derivative of the product of the two matrix exponentials, we find that:

$$\frac{d}{dt}\left(e^{-\mathbf{A}t}e^{\mathbf{A}t}\right) = -\mathbf{A}e^{-\mathbf{A}t}e^{\mathbf{A}t} + e^{-\mathbf{A}t}\mathbf{A}e^{\mathbf{A}t}.$$

From the definition of the matrix exponential, it is clear that $e^{-\mathbf{A}t}\mathbf{A} = \mathbf{A}e^{-\mathbf{A}t}$. Combining this with the previous result, we find that

$$\frac{d}{dt}\left(e^{-\mathbf{A}t}e^{\mathbf{A}t}\right) = 0.$$

As when $t = 0$ we find that $e^{-\mathbf{A}0}e^{\mathbf{A}0} = \mathbf{I}$ and as the derivative of $e^{-\mathbf{A}t}e^{\mathbf{A}t}$ is zero, we find that

$$e^{-\mathbf{A}t}e^{\mathbf{A}t} = \mathbf{I};$$

we find that the matrices $e^{\mathbf{A}t}$ and $e^{-\mathbf{A}t}$ are inverses of one another.

We make use of this fact to show that (9.3) has a unique solution; the proof is by contradiction. Suppose that there were a second solution of (9.3), $\vec{z}(t)$. Considering $e^{-\mathbf{A}t}\vec{z}(t)$, we find that this function's derivative is:

$$\frac{d}{dt}\left(e^{-\mathbf{A}t}\vec{z}(t)\right) = -\mathbf{A}e^{-\mathbf{A}t}\vec{z}(t) + e^{-\mathbf{A}t}\mathbf{A}\vec{z}(t) = \vec{0}.$$

As the value of the function itself when $t = 0$ is $\mathbf{I}\vec{z}(0) = \vec{x}_0$ and as the function's derivative is zero, we find that:

$$e^{-\mathbf{A}t}\vec{z}(t) = \vec{x}_0 \Rightarrow \vec{z}(t) = e^{\mathbf{A}t}\vec{x}_0$$

which is the solution we found previously. That is, the homogeneous ordinary differential equation has a *unique* solution.

9.4 The Significance of the Eigenvalues of the Matrix

We find that to understand the solutions of (9.3) we must understand the behavior of the matrix exponential $e^{\mathbf{A}}$. Let us assume that $\mathbf{A}$ has a full set of eigenvectors and eigenvalues. That is, if $\mathbf{A}$ is an $n \times n$ matrix let us assume that there are n linearly independent vectors, $\vec{v}_i$, and n scalars λ_i that satisfy:

$$\mathbf{A}\vec{v}_i = \lambda_i\vec{v}_i.$$

Writing these vectors as a matrix:

$$\mathbf{M} \equiv (\vec{v}_1 \cdots \vec{v}_n),$$

and defining the matrix $\mathbf{\Lambda}$ as:

$$\mathbf{\Lambda} \equiv \begin{pmatrix} \lambda_1 & 0 & \cdots & \cdots & 0 \\ 0 & \lambda_2 & 0 & \cdots & 0 \\ & & \vdots & & \\ 0 & \cdots & 0 & \lambda_{n-1} & 0 \\ 0 & \cdots & \cdots & 0 & \lambda_n \end{pmatrix}.,$$

we find that:

$$\mathbf{AM} = \mathbf{M\Lambda}.$$

As the eigenvectors are linearly independent, $\mathbf{M}$ is invertible. We find that:

$$\mathbf{A} = \mathbf{M\Lambda M}^{-1}$$

It follows that:

$$\begin{aligned} \mathbf{I} &= \mathbf{MM}^{-1} \\ \mathbf{A} &= \mathbf{M\Lambda M}^{-1} \\ \mathbf{A}^2 &= \mathbf{M\Lambda M}^{-1}\mathbf{M\Lambda M}^{-1} \\ &= \mathbf{M\Lambda}^2\mathbf{M}^{-1} \\ \vdots &= \vdots \\ \mathbf{A}^n &= \mathbf{M\Lambda}^n\mathbf{M}^{-1}. \end{aligned}$$

Let us consider the matrix exponential again. Assuming that $\mathbf{A}$ has n

independent eigenvectors, we find that:

$$\begin{aligned}
e^{\mathbf{A}} &= \mathbf{I} + \mathbf{A} + \frac{\mathbf{A}^2}{2!} + \cdots + \frac{\mathbf{A}^n}{n!} + \cdots \\
&= \mathbf{M}\mathbf{M}^{-1} + \mathbf{M}\mathbf{\Lambda}\mathbf{M}^{-1} + \frac{\mathbf{M}\mathbf{\Lambda}^2\mathbf{M}^{-1}}{2!} + \cdots \\
&= \mathbf{M}\begin{pmatrix} e^{\lambda_1} & 0 & \cdots & \cdots & 0 \\ 0 & e^{\lambda_2} & 0 & \cdots & 0 \\ & & \vdots & & \\ 0 & \cdots & 0 & e^{\lambda_{n-1}} & 0 \\ 0 & \cdots & \cdots & 0 & e^{\lambda_n} \end{pmatrix}\mathbf{M}^{-1} \\
&= \mathbf{M}e^{\mathbf{\Lambda}}\mathbf{M}^{-1}.
\end{aligned}$$

We see three things:

(1) The exponential of a diagonal matrix is just the original matrix with the diagonal elements exponentiated.

(2) Calculating a matrix exponential can be done in two steps:

   (a) Calculate the eigenvectors and eigenvalues corresponding to the matrix.

   (b) Use the results of the previous part to find $\mathbf{M}$, $e^{\mathbf{\Lambda}}$, and then $e^{\mathbf{A}}$.

(3) The exponents in the exponentials are just the eigenvalues of the original matrix.

9.5 Poles and Eigenvalues

We now return to the solution of (9.3). We have seen that the solution is:

$$\vec{x}(t) = e^{\mathbf{A}t}\vec{x}_0.$$

We now know that as long as $\mathbf{A}$ is diagonalizable—as long as it can be written as:

$$\mathbf{A} = \mathbf{M}\mathbf{\Lambda}\mathbf{M}^{-1},$$

then $\vec{x}(t) = e^{\mathbf{A}t}\vec{x}_0$ can be written:

$$\begin{aligned}\vec{x}(t) &= \mathbf{M}e^{\mathbf{\Lambda t}}\mathbf{M}^{-1}\vec{x}_0 \\ &= \mathbf{M}\begin{pmatrix} e^{\lambda_1 t} & 0 & \cdots & \cdots & 0 \\ 0 & e^{\lambda_2 t} & 0 & \cdots & 0 \\ & & \vdots & & \\ 0 & \cdots & 0 & e^{\lambda_{n-1} t} & 0 \\ 0 & \cdots & \cdots & 0 & e^{\lambda_n t} \end{pmatrix}\mathbf{M}^{-1}\vec{x}_0.\end{aligned}$$

We see that there are generally n exponential functions here and their exponents are $\lambda_1 t, ..., \lambda_n t$. When solving systems using the transfer function approach, one generally finds that the exponents are the poles times t. We see that *the eigenvalues of* $\mathbf{A}$ *are just the poles of the transfer function.*

9.6 Inhomogeneous Matrix Differential Equations

We would like to solve an equation of the form:

$$\vec{x}'(t) = \mathbf{A}\vec{x}(t) + \vec{b}u(t), \quad \vec{x}(0) = \vec{x}_0. \tag{9.5}$$

We solve (9.5) using the *method of variation of parameters.* I.e. we assume that the solution is of the form:

$$\vec{x}(t) = e^{\mathbf{A}t}\vec{z}(t)$$

and then we solve for $\vec{z}(t)$. Note that:

$$\vec{x}(0) = \vec{x}_0 = e^{\mathbf{A}0}\vec{z}(0) = \vec{z}(0).$$

Thus, $\vec{z}(0) = \vec{x}_0$.

Assume that the solution of (9.5) is of the given form. Then we find that:

$$\begin{aligned}(e^{\mathbf{A}t}\vec{z}(t))' &= \mathbf{A}e^{\mathbf{A}t}\vec{z}(t) + \vec{b}u(t) \quad \Rightarrow \\ \mathbf{A}e^{\mathbf{A}t}\vec{z}(t) + e^{\mathbf{A}t}\vec{z}\,'(t) &= \mathbf{A}e^{\mathbf{A}t}\vec{z}(t) + \vec{b}u(t) \quad \Rightarrow \\ e^{\mathbf{A}t}\vec{z}\,'(t) &= \vec{b}u(t) \quad \Rightarrow \\ \vec{z}\,'(t) &= e^{-\mathbf{A}t}\vec{b}u(t).\end{aligned}$$

We solve this equation by integrating it from 0 to t and imposing the initial condition. Upon performing the indicated integration and imposing the initial condition, we find that:

$$\vec{z}(t) = \int_0^t e^{-\mathbf{A}\tau}\vec{b}u(\tau)\,d\tau + \vec{x}_0.$$

Multiplying through by $e^{\mathbf{A}t}$, we find that:

$$\vec{x}(t) = e^{\mathbf{A}t}\left(\int_0^t e^{-\mathbf{A}\tau}\vec{b}u(\tau)\,d\tau + \vec{x}_0\right).$$

Making use of this equality, and moving the exponential inside the integral, we find that:

$$\vec{x}(t) = e^{\mathbf{A}t}\left(\int_0^t e^{-\mathbf{A}\tau}\vec{b}u(\tau)\,d\tau + \vec{x}_0\right) \tag{9.6}$$

$$= \int_0^t e^{-\mathbf{A}(\tau-t)}\vec{b}u(\tau)\,d\tau + e^{\mathbf{A}t}\vec{x}_0. \tag{9.7}$$

Using this formulation it is not difficult to show that if all of the eigenvalues of $\mathbf{A}$ are negative (if all of the poles of the transfer function are negative), then if $u(t)$ is bounded so is the state of the system, and so is the output.

It is also easy to show that (9.6) is the *unique* solution of (9.5). Suppose that there were two different solutions $\vec{x}(t)$ and $\vec{z}(t)$. Let $\vec{d}(t) = \vec{x}(t) - \vec{z}(t)$. We find that:

$$\vec{d}'(t) = \mathbf{A}\vec{x}(t) + \vec{b}u(t) - \mathbf{A}\vec{z}(t) - \vec{b}u(t) = \mathbf{A}\vec{d}(t), \qquad \vec{d}(0) = \vec{0}.$$

As the *unique* solution to this *homogeneous* differential equation is $\vec{0}$, we find that $\vec{x}(t)$ must equal $\vec{z}(t)$—that there is only one solution to the inhomogeneous differential equation.

9.7 The Cayley-Hamilton Theorem

A matrix's characteristic equation is the equation:

$$\alpha(\lambda) \equiv \det(\mathbf{A} - \lambda\mathbf{I}) = a_n\lambda^n + \cdots a_1\lambda^1 + a_0\lambda^0 = 0.$$

The Cayley-Hamilton theorem states that any matrix, $\mathbf{A}$, satisfies its own characteristic equation—that:

$$\alpha(\mathbf{A}) = a_n\mathbf{A}^n + \cdots a_1\mathbf{A}^1 + a_0\mathbf{A}^0 = \mathbf{0} \tag{9.8}$$

(where $\mathbf{A}^0 = \mathbf{I}$). We prove the Cayley-Hamilton theorem when $\mathbf{A}$ is diagonalizable.

Consider $\alpha(\mathbf{A})$. Because $\mathbf{A}$ can be written as $\mathbf{A} = \mathbf{M}\mathbf{\Lambda}\mathbf{M}^{-1}$, we find that:

$$\alpha(\mathbf{A}) = \mathbf{M}\alpha(\mathbf{\Lambda})\mathbf{M}^{-1}.$$

But $\alpha(\mathbf{\Lambda})$ acts on each element of a *diagonal* matrix separately. As each element satisfies $\alpha(\lambda) = 0$, we find that $\alpha(\mathbf{\Lambda}) = 0$. From this we conclude that $\alpha(\mathbf{A}) = 0$, which was to be proved.

A simple consequence of the Cayley-Hamilton theorem is that any power of $\mathbf{A}$ can be expressed in term of $\mathbf{I}, \mathbf{A}, ..., \mathbf{A}^{n-1}$. Consider, for example, $\mathbf{A}^n$. From (9.8) we see that:

$$\mathbf{A}^n = -\frac{1}{a_n}\left(a_{n-1}\mathbf{A}^{n-1} + \cdots + a_1\mathbf{A} + a_0\mathbf{I}\right).$$

To find a representation for $\mathbf{A}^{n+1}$ multiply the preceding equation by $\mathbf{A}$ and then replace $\mathbf{A}^n$ by its value as given in the preceding equation. Proceeding inductively, we find that *any power of* $\mathbf{A}$ *can be expressed as a linear combination of* $\mathbf{I}, \mathbf{A}, ..., \mathbf{A}^{\mathbf{n}-\mathbf{1}}$. In particular, this means that:

$$e^{\mathbf{A}t} = f_0(t)\mathbf{I} + f_1(t)\mathbf{A} + \cdots + f_{n-1}(t)\mathbf{A}^{n-1}. \tag{9.9}$$

9.8 Controllability

Given a system whose state is described by (9.1), it is reasonable to ask if one can control the state of the system—which is a vector—by using the scalar input, $u(t)$, alone. A system for which one can force the state to zero in finite time from any initial condition using the input, $u(t)$, alone is said to be *controllable*.

Let us see when a system may be controllable. From (9.6) we see the solution of (9.1) is:

$$\vec{x}(t) = e^{\mathbf{A}t}\left(\int_0^t e^{-\mathbf{A}\tau}\vec{b}u(\tau)\,d\tau + \vec{x}_0\right).$$

As the matrix exponential is always invertible, the only way that $\vec{x}(t)$ can become zero is if:

$$\int_0^t e^{-\mathbf{A}\tau}\vec{b}u(\tau)\,d\tau = -\vec{x}_0$$

From (9.8) we deduced that the matrix exponential can be written as a combination of the powers of $\mathbf{A}$ from the zeroth power to the $(n-1)^{th}$

power. We find that a necessary condition for controllability is that:

$$\int_0^t \left(f_0(t)\mathbf{I} + f_1(t)\mathbf{A} + \cdots + f_{n-1}(t)\mathbf{A}^{n-1}\right)\vec{b}u(\tau)\,d\tau = -\vec{x}_0.$$

Performing the indicated integrations (and denoting the integral $\int_0^t f_j(\tau)u(\tau)\,d\tau$ by β_j) we find that we must have:

$$\beta_0\mathbf{I}\vec{b} + \beta_1\mathbf{A}\vec{b} + \cdots + \beta_{n-1}\mathbf{A}^{n-1}\vec{b} = -\vec{x}_0.$$

That is, we must be able to set a linear combination of the n vectors, $\vec{b}, \mathbf{A}\vec{b}, ..., \mathbf{A}^{n-1}\vec{b}$ equal to an arbitrary $-\vec{x}_0$. That is the n vectors must span $\mathcal{R}^n$. A simple condition for this is that the determinant of the matrix:

$$\mathrm{Con} \equiv \left(\vec{b}\ \mathbf{A}\vec{b}\ \cdots\ \mathbf{A}^{n-1}\vec{b}\right)$$

must be nonzero. This matrix is known as *the controllability matrix.* We have shown that in order for a system to controllable it is necessary that:

$$\det(\mathrm{Con}) \neq 0.$$

It can be shown that this is also a sufficient condition for a system to be controllable.

9.9 Pole Placement

It has been shown (see, for example, [Bro74]) that if a system is controllable, then using state feedback—using feedback that is a *linear* combination of the elements of the elements of the state vectors—the poles of the system with feedback can be placed in any desired position in the complex plane. If one would like the linear combination of the states to be real, then one must make sure that all complex poles appear with their conjugates. We consider some examples of pole placement shortly.

A generic system with *state feedback* is given in Figure 9.1. Note that a linear combination of the states—and not a function of the output—is being fed back to the input. In order to use state feedback one must either have access to the states, or one must be able to estimate the states. In the next section we discuss the conditions under which the state of the system can be estimated from a knowledge of the system's input and output. A nice example of pole placement is presented in §9.14.1.

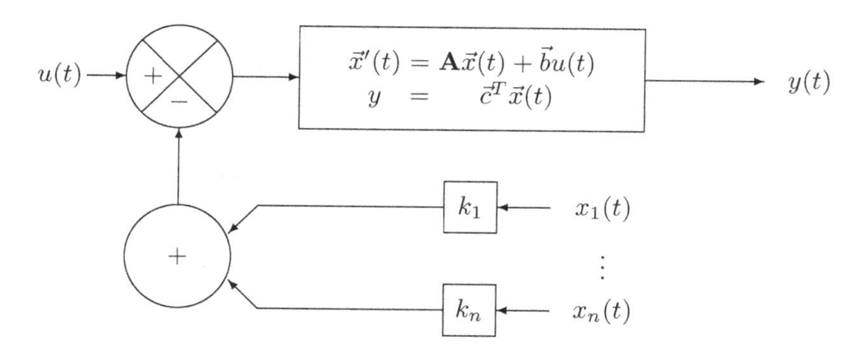

Fig. 9.1 State Feedback

9.10 Observability

In order to use pole placement techniques, one must know the internal state of one's system. This can be very difficult as often the system is a "black box," and one does not have access to its internal state. When this is the case, it is important to know if it is possible to determine the internal state of the system by looking at the signals that enter and leave the system. We consider the following question. Given a system that can be described by (9.1), (9.2), for which $\mathbf{A}, \vec{b}$, and $\vec{c}^T$ are known, but whose initial state is not known, is it possible to determine the system's initial state by observing $y(t)$ and $u(t)$? A system for which such measurements of external signals provide enough information to determine the initial state of the system (and via (9.1) the state of the system for all time) is said to be *observable*.

Once again we make use of (9.6). We know that:

$$\vec{x}(t) = e^{\mathbf{A}t}\left(\int_0^t e^{-\mathbf{A}\tau}\vec{b}u(\tau)\,d\tau + \vec{x}_0\right).$$

As we are now interested in the output of the system—in $y(t)$—we multiply this expression by $\vec{c}^T$. We find that:

$$\vec{y}(t) = \vec{c}^T e^{\mathbf{A}t}\left(\int_0^t e^{-\mathbf{A}\tau}\vec{b}u(\tau)\,d\tau + \vec{x}_0\right).$$

The contribution of the initial condition to $y(t)$ is:

$$y_{init}(t) \equiv \vec{c}^T e^{\mathbf{A}t}\vec{x}_0.$$

If this is zero for any nonzero vector $\vec{x}_0$ then the system is not observable as it is impossible to distinguish between a non-zero initial condition and the initial condition $\vec{0}$.

Making use of (9.9) we find that $y_{init}(t)$ is:

$$\begin{aligned} y_{init}(t) &= \vec{c}^T\left(f_0(t)\mathbf{I} + f_1(t)\mathbf{A} + \cdots + f_{n-1}(t)\mathbf{A}^{n-1}\right)\vec{x}_0 \\ &= \left(f_0(t)\vec{c}^T + f_1(t)\vec{c}^T\mathbf{A} + \cdots + f_{n-1}(t)\vec{c}^T\mathbf{A}^{n-1}\right)\vec{x}_0. \end{aligned}$$

If the vectors $\vec{c}^T, \vec{c}^T\mathbf{A}, \ldots, \vec{c}^T\mathbf{A}^{n-1}$ do not span $\mathcal{R}^n$, then there exists some vector $\vec{x}_0$ that is orthogonal to all of the vectors and for which $y_{init}(t) \equiv 0$. Thus in order for a system to be observable it is necessary that the vectors $\vec{c}^T, \vec{c}^T\mathbf{A}, \ldots, \vec{c}^T\mathbf{A}^{n-1}$ be linearly independent. Another way of saying this is to say that the matrix whose rows these vectors are:

$$\text{Obs} \equiv \begin{pmatrix} \vec{c}^T \\ \vec{c}^T\mathbf{A} \\ \vdots \\ \vec{c}^T\mathbf{A}^{n-1} \end{pmatrix}$$

must have nonzero determinant.

We have shown that a necessary condition for observability is that the matrix Obs, known as the observability matrix, have non-zero determinant—that:

$$\det(\text{Obs}) \neq 0.$$

To show that this condition is also sufficient, consider $y_{init}(t) \equiv \vec{c}^T e^{\mathbf{A}t}\vec{x}_0$ again. We find that:

$$y_{init}(t) = \vec{c}^T e^{\mathbf{A}t}\vec{x}_0 = \sum_{k=0}^{\infty} \frac{\vec{c}^T\mathbf{A}^k\vec{x}_0}{k!}t^k.$$

This is a Maclaurin series, it is not difficult to show that it must converge, and, clearly, in order for it to be identically zero, all of its coefficients must be zero. That is, if $y_{init}(t) \equiv 0$, then for all $k \geq 0$, $\vec{c}^T\mathbf{A}^k$ is orthogonal to the vector $\vec{x}_0^T$. In particular, the n vectors $\vec{c}^T\mathbf{A}^k, \quad k = 0, \ldots, n-1$ are all orthogonal to this non-zero vector. As we are dealing with n vectors in an n dimensional space, these n vectors cannot be linearly independent. Thus, if the contribution of $y_{init}(t)$ is identically zero for some initial condition, the determinant of the observability matrix is necessarily zero, and we have proved that the condition is necessary and sufficient.

9.11 Observer Design

Suppose that the state of a particular system is not available to onlookers. (The system can be thought of as being hermetically sealed so that no information about the system's internal state other than the system's output is available to someone looking at the system.) An *observer* is a system designed to take as it inputs the input to and output from such a system and that is designed to ensure that its state tends to that of the system being observed. (See Figure 9.2.) The state of the observer is available to the observer's user, so that (after a little while) the state of a properly designed observer can be used as a proxy for the state of the system being observed.

Assume that the observer has access to the input to and output from the system being observed and that the observer's designer knows the internal structure of the system being observed. (That is, the designer knows $\mathbf{A}, \vec{b}$, and $\vec{c}^T$.) The one thing the designer does not have access to is the internal state of the system being observed.

Consider a system whose input is $u(t)$, whose output is $y(t)$, whose state is $\vec{x}(t)$, whose state equation is $\vec{x}'(t) = \mathbf{A}\vec{x}(t) + \vec{b}u(t)$, and whose output satisfies $y(t) = \vec{c}^T\vec{x}(t)$. We now describe how to design an observer whose state tends to $\vec{x}(t)$ arbitrarily quickly.

The observer's inputs are $u(t)$ and $y(t)$—the two signals known to everyone, its state is $\vec{s}(t)$, and its output is $r(t)$. The observer's output's dependence on its state is the same as that of the system being observed: It is $r(t) = \vec{c}^T\vec{s}(t)$. The observer's state evolves according to the equation:

$$\vec{s}'(t) = \mathbf{A}\vec{s}(t) + \vec{b}u(t) + \vec{l}(y(t) - r(t)).$$

Consider the difference of the states of the two systems in Figure 9.2: $\vec{d}(t) \equiv \vec{x}(t) - \vec{s}(t)$. The derivative of the difference satisfies:

$$\begin{aligned}
\vec{d}'(t) &= \vec{x}'(t) - \vec{s}'(t) \\
&= \mathbf{A}\vec{x}(t) + \vec{b}u(t) - \left(\mathbf{A}\vec{s}(t) + \vec{b}u(t) + \vec{l}(y(t) - r(t))\right) \\
&= \mathbf{A}\vec{d}(t) - \vec{l}(y(t) - r(t)) \\
&= \mathbf{A}\vec{d}(t) - \vec{l}\vec{c}^T(\vec{x}(t) - \vec{s}(t)) \\
&= \left(\mathbf{A} - \vec{l}\vec{c}^T\right)\vec{d}(t).
\end{aligned}$$

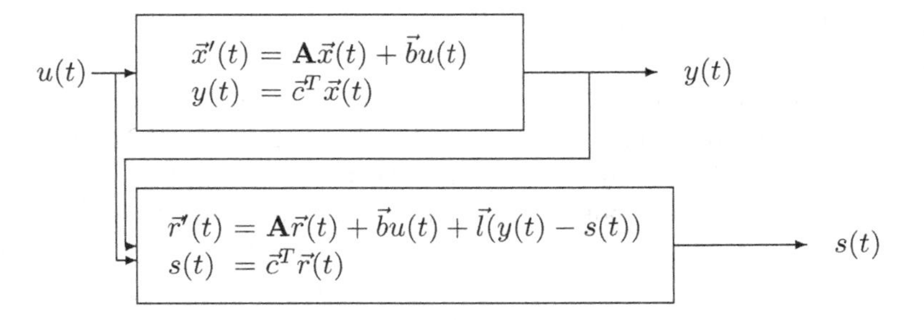

Fig. 9.2 A System with an observer. The upper block is the system being observed; the lower one is the observer. The observer's inputs are the input to and the output of the system being observed. "Designing" such an observer means choosing the values of the components of $\vec{l}$.

Defining $\mathbf{A}_f \equiv \mathbf{A} - \vec{l}\vec{c}^T$, it is clear that:

$$\vec{d}(t) = e^{\mathbf{A}_f t}\vec{d}_0.$$

We would like to force $\vec{d}(t)$ to zero as quickly as possible so that $\vec{s}(t) \to \vec{x}(t)$ as quickly as possible—so that the state of the observer, which we have access to, tends to that of the system being observed as quickly as possible.

In order to fix the rate of convergence at some particular value, we must be able to control the location of the eigenvalues of $\mathbf{A}_f$. Thus, we need to know whether or not we can *place* the eigenvalues of $\mathbf{A} - \vec{l}\vec{c}^T$ at arbitrary locations. This is same issue we considered when dealing with pole placement, but in that case we needed to place the eigenvalues of $\mathbf{A} - \vec{b}\vec{k}^T$. We know that this is possible precisely when the determinant of the controllability matrix is non-zero.

Let us consider *our* question again. When can we place the eigenvalues of $\mathbf{A} - \vec{l}\vec{c}^T$? Recalling that the eigenvalues of a matrix and its transpose are the same, we find that our question is equivalent to asking when can we place the eigenvalues of $\mathbf{A}^T - \vec{c}\vec{l}^T$? In this form, we see that we are asking the exact same question that we ask when we want to know whether or not pole placement is possible. Thus, the answer must be the same: We can place the poles of $\mathbf{A}_f$ as we please when the determinant of the relevant controllability matrix:

$$\left(\vec{c}\;\mathbf{A}^T\vec{c}\cdots\;\left(\mathbf{A}^T\right)^{n-1}\vec{c}\right),$$

is non-zero. As the determinant of a matrix and its transpose are equal, the condition we just found is equivalent to requiring that the determinant of:

$$\begin{pmatrix} \vec{c}^T \\ \vec{c}^T \mathbf{A} \\ \vdots \\ \vec{c}^T \mathbf{A}^{n-1} \end{pmatrix}$$

not be zero. The matrix we have found is precisely the observability matrix, and the condition is the necessary and sufficient condition for observability. Thus, it is possible to design an observer whose state converges to that of the observed system arbitrarily quickly precisely when the system is observable.

9.12 Choosing States When a System is Characterized by an ODE

Consider a system whose transfer function is:

$$T(s) = \frac{a}{b_n s^n + \ldots + b_0}.$$

Let $u(t)$ be the system's input and let $y(t)$ be its output. Then:

$$\frac{Y(s)}{U(s)} = \frac{a}{b_n s^n + \ldots + b_0} \Rightarrow b_n y^{(n)}(t) + \ldots + b_0 y(t) = au(t).$$

That is, the system's output satisfies an ordinary differential equation (ODE)—it is characterized by an ODE. It is clear that this ODE can be written in matrix form as:

$$\overbrace{\begin{bmatrix} y(t) \\ y'(t) \\ y''(t) \\ \vdots \\ y^{(n-2)}(t) \\ y^{(n-1)}(t) \end{bmatrix}'}^{\vec{x}'(t)} = \overbrace{\begin{pmatrix} 0 & 1 & 0 & \cdots & 0 \\ \vdots & \vdots & \ddots & \vdots & \vdots \\ \vdots & \vdots & \vdots & \ddots & \vdots \\ 0 & 0 & \ldots & 0 & 1 \\ -b_0 & -b_1 & -b_2 & \ldots & -b_{n-1} \end{pmatrix}}^{\mathbf{A}} \overbrace{\begin{bmatrix} y(t) \\ y'(t) \\ y''(t) \\ \vdots \\ y^{(n-2)}(t) \\ y^{(n-1)}(t) \end{bmatrix}}^{\vec{x}(t)} + \overbrace{\begin{bmatrix} 0 \\ 0 \\ 0 \\ \vdots \\ 0 \\ a \end{bmatrix}}^{\vec{b}} u(t)$$

with

$$y(t) = \overbrace{[1\ 0\ 0 \ldots 0]}^{\vec{c}^T} \overbrace{\begin{bmatrix} y(t) \\ y'(t) \\ y''(t) \\ \vdots \\ y^{(n-2)}(t) \\ y^{(n-1)}(t) \end{bmatrix}}^{\vec{x}(t)} \text{ and } \vec{x}(0) = \begin{bmatrix} y(0) \\ y'(0) \\ y''(0) \\ \vdots \\ y^{(n-2)}(0) \\ y^{(n-1)}(0) \end{bmatrix}.$$

We have found an appropriate way to describe the system using the state variables formalism, and we see that the "appropriate" states for a system described by an nth order differential equation are the output and its first $n-1$ derivatives..

9.13 Converting Transfer Functions to State Equations

Suppose that one is given a system that is described by a transfer function of the form:

$$\frac{Y(s)}{U(s)} = T(s) = \frac{a_0 + a_1 s + \cdots + a_m s^m}{b_0 + b_1 s + \cdots + s^n},$$

and one would like to define a set of states for this system and to produce the state equations which correspond to the system. It turns out that there is an easy way to do this *as long as one has the freedom to define the states of the system.*

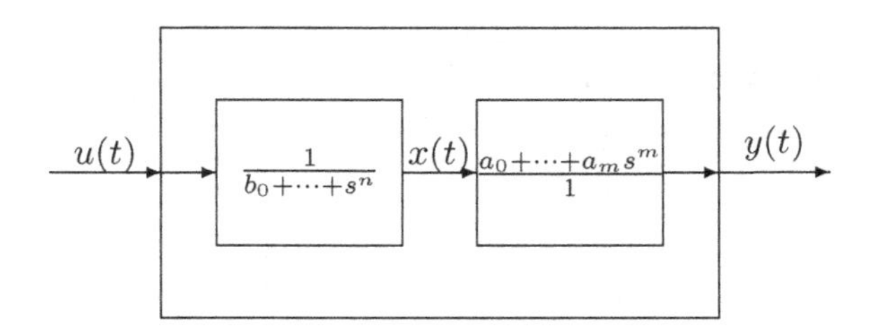

Fig. 9.3 A System Divided into Two Subsystems

Let us consider a system that is made up of two subsystems as shown in Figure 9.3. Clearly $x(t)$ satisfies:

$$b_0x(t) + b_1x'(t) + \cdots + x^{(n)}(t) = u(t).$$

If we define the states of our system as $x(t), ..., x^{(n-1)}(t)$ then we find that the state vector satisfies:

$$\begin{bmatrix} x(t) \\ x'(t) \\ \vdots \\ x^{(n-1)}(t) \end{bmatrix}' = \begin{pmatrix} 0 & 1 & \cdots & \cdots & \cdots & 0 \\ 0 & 0 & 1 & \cdots & \cdots & 0 \\ & & \vdots & \vdots & & \\ 0 & \cdots & \cdots & \cdots & 0 & 1 \\ -b_0 & -b_1 & \cdots & \cdots & \cdots & -b_{n-1} \end{pmatrix} \begin{bmatrix} x(t) \\ x'(t) \\ \vdots \\ x^{(n-1)}(t) \end{bmatrix} + \begin{bmatrix} 0 \\ \vdots \\ 0 \\ 1 \end{bmatrix} u(t).$$

Also, clearly:

$$y(t) = a_0x(t) + a_1x'(t) + \cdots + a_mx^{(m)}(t) = \begin{bmatrix} a_0 & \cdots & a_m & 0 & \cdots & 0 \end{bmatrix} \begin{bmatrix} x(t) \\ x'(t) \\ \vdots \\ x^{(n-1)}(t) \end{bmatrix}.$$

As long as $m < n$, this technique works. Note however, that here the states have been chosen in a mathematically—rather than practically—convenient fashion.

Designing an Observer—An Example

Consider the system whose transfer function is:

$$\frac{Y(s)}{U(s)} = G(s) = \frac{s}{s^2 + 3s + 2}.$$

We would like to:

- Define a reasonable set of states for the system.
- Check to make sure that the system is observable.
- Design an observer that takes 30 ms to "settle."

Using our method of choosing states, we let $X(s)$ satisfy:

$$\frac{X(s)}{U(s)} = \frac{1}{s^2 + 3s + 2},$$

and we take $Y(s) = sX(s)$. We find that:

$$Y(s) = \frac{s}{s^2 + 3s + 2} U(s)$$

as it must be.

We take $x(t)$ and $x'(t)$ to be the system's states, and we find that:

$$\begin{bmatrix} x(t) \\ x'(t) \end{bmatrix}' = \begin{pmatrix} 0 & 1 \\ -2 & -3 \end{pmatrix} \begin{bmatrix} x(t) \\ x'(t) \end{bmatrix} + \begin{bmatrix} 0 \\ 1 \end{bmatrix} u(t), \qquad y(t) = [0\ 1] \begin{bmatrix} x(t) \\ x'(t) \end{bmatrix}.$$

This system's observability matrix is:

$$\text{Obs} = \begin{pmatrix} 0 & 1 \\ -2 & -3 \end{pmatrix}.$$

As $\det(\text{Obs}) = 2 \neq 0$, the system is observable.

In order to cause the observer to settle within 30 ms, we pick the eigenvalues of $\mathbf{A}_f = \mathbf{A} - \vec{l}\vec{c}^{\mathrm{T}}$ in such a way that the associated time constant is 10 ms. To do this, we choose to locate both eigenvalues at -100.

To proceed, we calculate $\mathbf{A}_f$ and then choose the components of $\vec{l}$ to force the eigenvalues to be in the desired locations. Given our values of $\mathbf{A}$ and $\vec{c}^T$, we find that:

$$\mathbf{A}_f = \begin{pmatrix} 0 & 1 \\ -2 & -3 \end{pmatrix} - \begin{bmatrix} l_1 \\ l_2 \end{bmatrix} [0\ 1] = \begin{pmatrix} 0 & 1 - l_1 \\ -2 & -3 - l_2 \end{pmatrix}.$$

To determine the eigenvalues we consider the solutions of $\det(\mathbf{A}_f - \lambda \mathbf{I}) = 0$. That is, we consider the solutions of:

$$\det \begin{pmatrix} -\lambda & 1 - l_1 \\ -2 & -3 - l_2 - \lambda \end{pmatrix} = \lambda^2 + (2 + l_2)\lambda + 2 - 2l_1 = 0.$$

As we would like $\lambda = -100, -100$, we must force this polynomial to be $(\lambda + 100)^2 = \lambda^2 + 200\lambda + 10000$. That is, $l_2 = 198$ and $l_1 = -4,999$. This completes the design of the observer. Note that the values of the components of $\vec{l}$, and of l_1 in particular, are large. Because the observer is generally implemented as a computer program, mildly large values such as these need not worry us too much as long as we have sufficiently accurate knowledge of $\mathbf{A}$, $\vec{b}$, and $\vec{c}^{\mathrm{T}}$.

9.14 Examples

9.14.1 *Pole Placement*

Let us consider a simple motor control problem. Suppose that one would like to control a motor whose input is $u(t)$ and whose output, $y(t)$, is the motor's shaft angle. Assume that the motor's transfer function is:

$$\frac{Y(s)}{U(s)} = G(s) = \frac{3}{s(s+2)}.$$

Picking our states to be $y(t)$ and $y'(t)$, we find that the state equations that correspond to this transfer function are:

$$\vec{x}'(t) = \begin{bmatrix} y(t) \\ y'(t) \end{bmatrix}' = \overbrace{\begin{pmatrix} 0 & 1 \\ 0 & -2 \end{pmatrix}}^{\mathbf{A}} \begin{bmatrix} y(t) \\ y'(t) \end{bmatrix} + \overbrace{\begin{bmatrix} 0 \\ 3 \end{bmatrix}}^{\vec{b}} u(t) \tag{9.10}$$

$$y(t) = \overbrace{[1 \;\; 0]}^{\vec{c}^T} \begin{bmatrix} y(t) \\ y'(t) \end{bmatrix}$$

Let us calculate the observability and the controllability matrices. We find that:

$$\text{Obs} = \begin{pmatrix} \vec{c}^T \\ \vec{c}^T \mathbf{A} \end{pmatrix} = \begin{pmatrix} 1 & 0 \\ 0 & 1 \end{pmatrix}, \quad \text{Con} = \begin{pmatrix} \vec{b} & \mathbf{A}\vec{b} \end{pmatrix} = \begin{pmatrix} 0 & 3 \\ 3 & -6 \end{pmatrix}.$$

We see that $\det(\text{Obs}) = 1$ and $\det(\text{Con}) = -9$. Thus, this system is both observable and controllable. As the system is controllable, we should be able to place the system's poles anywhere we please by the addition of state feedback. Let us add the appropriate feedback and see what happens.

In our case, the states of the system are the output and the output's derivative. The system with state feedback added is shown in Figure 9.4.

Let us determine the equations that govern the system with state feedback added. The effect of the state feedback is to cause $u(t)$ to equal:

$$u(t) = -\begin{bmatrix} k_1 & k_2 \end{bmatrix} \begin{bmatrix} y(t) \\ y'(t) \end{bmatrix} + u_{ext}(t).$$

Combining this with (9.10) we find that the equations that describe the system with feedback are:

$$\begin{bmatrix} y(t) \\ y'(t) \end{bmatrix}' = \begin{pmatrix} 0 & 1 \\ 0 & -2 \end{pmatrix} \begin{bmatrix} y(t) \\ y'(t) \end{bmatrix} - \begin{bmatrix} 0 \\ 3 \end{bmatrix} \left(\begin{bmatrix} k_1 & k_2 \end{bmatrix} \begin{bmatrix} y(t) \\ y'(t) \end{bmatrix} + u_{ext}(t). \right)$$

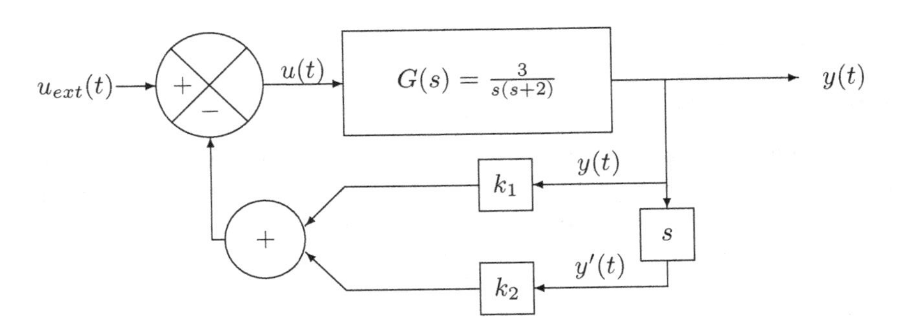

Fig. 9.4 A Simple Feedback Circuit

This can be rewritten as:

$$\begin{bmatrix} y(t) \\ y'(t) \end{bmatrix}' = \overbrace{\begin{pmatrix} 0 & 1 \\ -3k_1 & -2-3k_2 \end{pmatrix}}^{\mathbf{A_f}} \begin{bmatrix} y(t) \\ y'(t) \end{bmatrix} + \begin{bmatrix} 0 \\ 3 \end{bmatrix} u_{ext}(t).$$

Let us find the eigenvalues of the matrix that characterizes the system with state feedback—$\mathbf{A_f}$. The eigenvalues satisfy:

$$\det(\mathbf{A_f} - \lambda\mathbf{I}) = \det\begin{pmatrix} -\lambda & 1 \\ -3k_1 & -2-3k_2-\lambda \end{pmatrix} = \lambda^2 + (2+3k_2)\lambda + 3k_1 = 0.$$

If one would like λ_1 and λ_2 to be the system's eigenvalues, then one would like to be able to write characteristic polynomial, $\lambda^2+(2+3k_2)\lambda+3k_1$ in the form:

$$(\lambda - \lambda_1)(\lambda - \lambda_2) = \lambda^2 - (\lambda_1 + \lambda_2) + \lambda_1\lambda_2.$$

To make this possible, all that we need to do is to pick k_1 and k_2 to satisfy the equations:

$$\begin{aligned} 3k_1 &= \lambda_1\lambda_2 \\ 2 + 3k_2 &= -(\lambda_1 + \lambda_2). \end{aligned}$$

We see that for any λ_1 and λ_2 this is possible, and if we restrict λ_1 and λ_2 to be either real or complex conjugates, then k_1 and k_2 will be real as well.

9.14.2 *Adding an Integrator*

Consider the following scenario. One is given a first order system whose nominal transfer function is $1/(s+3)$. Suppose, however, that it is known

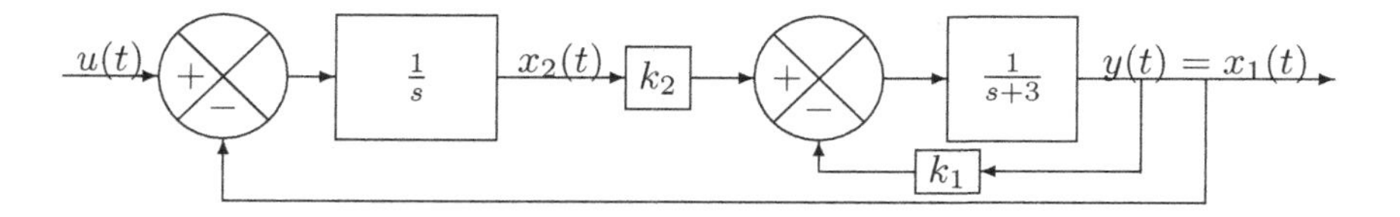

Fig. 9.5 A First Order System with an Integrator Added.

that the location of the pole can vary substantially. (I.e. the pole, nominally located at -3, may vary over a wide range.) One would like to design the system to have a DC gain of 1. One would also like the system's time constant to be less than or equal to $\tau = (1/3)s$.

It is easy to see (see Problem 11) that state feedback for a first order system leads to proportional control. Just using proportional control one cannot make the steady state response totally insensitive to the location of the system's pole. (Why not?) One solution to the DC gain problem is to add a block (and a state) to the system. In Figure 9.5 we have added an integrator before the state-variable feedback that we add to the plant; we have added the output of the integrator to the set of state variables; and we have added a gain block after the integrator. It is clear that the forward gain of our system at DC is infinite and that the overall gain of the system at DC is one—independent of the actual location of the pole that is nominally located at $s = -3$.

The state equations of the system are:

$$\vec{x}(t) = \begin{bmatrix} x_1(t) \\ x_2(t) \end{bmatrix}, \vec{x}'(t) = \underbrace{\begin{pmatrix} -3-k_1 & k_2 \\ -1 & 0 \end{pmatrix}}_{A_f} \vec{x}(t) + \begin{bmatrix} 0 \\ 1 \end{bmatrix} u(t), \quad y(t) = [1\ 0]\vec{x}(t).$$

The eigenvalues of the matrix A_f of the final system are the solutions of:

$$\det(A_f - \lambda I) = \lambda^2 + (3 + k_1) + k_2 = 0.$$

As the second order polynomial with roots at λ_1 and λ_2 and with quadratic term λ^2 is $\lambda^2-(\lambda_1+\lambda_2)\lambda+\lambda_1\lambda_2$, we find that by a suitable choice of the gains k_1 and k_2 we can place the poles wherever we would like to. In particular, to keep $\tau = (1/3)s$, we put one pole at $s = -3$ and one at $s = -4$. With this choice of poles, the gains must be $k_1 = 4$ and $k_2 = 12$. Adding an integrator in this fashion is often possible in more complex systems too.

9.14.3 *Modern Control Using MATLAB®*

The MATLAB Control System Toolbox™ allows one to describe a system using its state equation. To define one's system, one uses the command `ss(A,B,C,D)`. Here, `A` is our matrix $\mathbf{A}$, `B` is our vector $\vec{b}$, `C` is our vector $\vec{c}^T$, and `D` allows one to make the output contain a linear combination of the system's inputs as well as its states. This can be summarized by saying that if a system is defined using `ss(A,B,C,D)`, then the state equation that the system satisfies is:

$$\vec{x}'(t) = \mathtt{A}\vec{x}(t) + \mathtt{B}\vec{u}(t), \quad \vec{y}(t) = \mathtt{C}\vec{x}(t) + \mathtt{D}\vec{u}(t).$$

Note that MATLAB allows both the input and the output to the system to be vectors—a possibility that we do not consider.

Consider the example of §9.14.2. Let us use MATLAB to examine the system we designed. We examine the performance of the system with the feedback we have designed both when the pole of the system is at its nominal location—at $s = -3$—and when the pole of the system is actually at $s = -1$. We will see that the performance is affected by the change, *but the system's gain at DC is* 1 *in both cases.*

To define the system of Figure 9.5 with its pole at -3, we used the following commands:

```
A1 = [(-3 -4) 12; -1 0];
b1 = [0 1]';
c1 = [1 0];
d1 = [0];
G1 = ss(A1,b1,c1,d1);
```

To define the system with its pole at -1 we added the following commands after the commands given above:

```
A2 = [(-1 -4) 12; -1 0];
G2 = ss(A2,b1,c1,d1);
```

In order to examine the step response of the two systems, we made use of the `step` command—which is happy to work on systems defined using state space definitions. The step response of the two systems is given in Figure 9.6. Note that both systems stabilized on 1—as they should have. However, the poorly modeled system shows oscillation in its output—a sign that its poles are not on the real axis. As the poles were supposed to be at

$s = -3, -4$, this shows that the design is somewhat sensitive to the exact location of the pole.

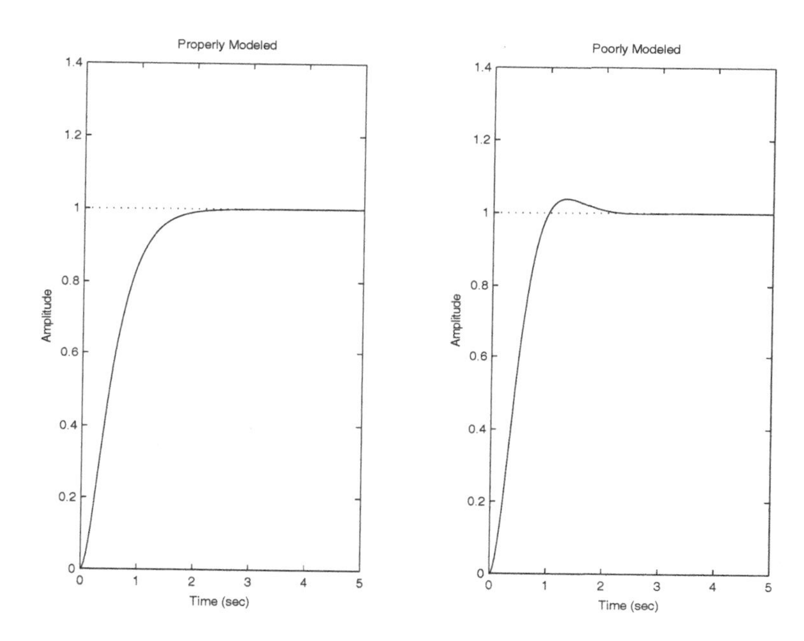

Fig. 9.6 The Step Response of the Properly Modeled System (Left) and the Poorly Modeled System (Right)

9.14.4 *A System that is not Observable*

Let us consider a motor-tachometer system. Such a system can be modeled by a motor followed by a differentiator. (See Figure 9.7.)

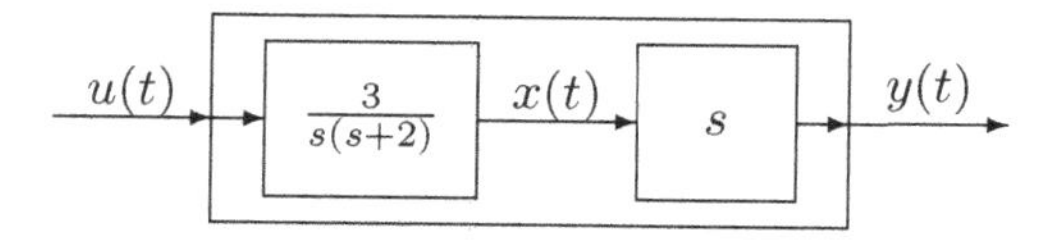

Fig. 9.7 A Motor with a Tachometer

If we are interested in the "true" state of the motor, then we must know the motor shaft's angular position and velocity. Let us define these two quantities as $x(t)$ and $x'(t)$. We find that the equations that define our system are:

$$\begin{bmatrix} x(t) \\ x'(t) \end{bmatrix}' = \begin{pmatrix} 0 & 1 \\ 0 & -2 \end{pmatrix} \begin{bmatrix} x(t) \\ x'(t) \end{bmatrix} + \begin{bmatrix} 0 \\ 3 \end{bmatrix} u(t), \quad y(t) = \begin{bmatrix} 0 & 1 \end{bmatrix} \begin{bmatrix} x(t) \\ x'(t) \end{bmatrix}$$

Note that in this case $y(t)$ is equal to $x'(t)$—as it must be. Let us check to see if this system is observable and/or controllable.

We find that:

$$\text{Obs} = \begin{pmatrix} 0 & 1 \\ 0 & -2 \end{pmatrix}, \quad \text{Con} = \begin{pmatrix} 0 & 3 \\ 3 & -6 \end{pmatrix}.$$

As $\det(\text{Obs}) = 0$ and $\det(\text{Con}) = -9$, we find that this system is controllable, but it is not observable.

That the system is controllable is actually obvious. In §9.14.1 we saw that a system with the same state equations is controllable. The fact that we have changed the output does not change the fact that $u(t)$ can control the state of the system. The system is not observable because the tachometer, whose transfer function is s, cancels a pole of the motor. From the point of view of an observer, the system has only one pole. One could not really expect to see the effect of both states from the outside.

9.14.5 *A Second System that is not Observable*

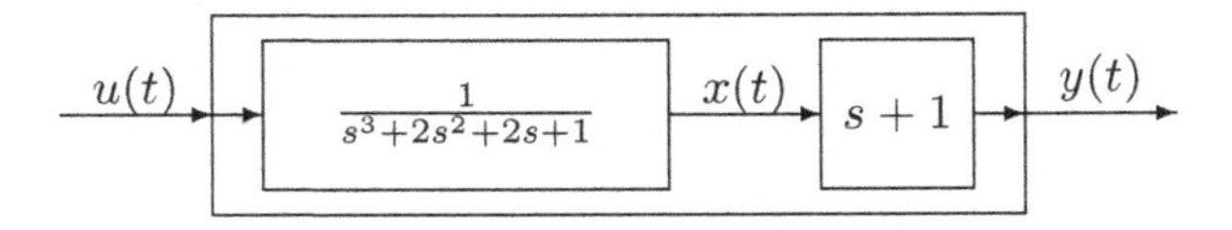

Fig. 9.8 A System that Is not Observable

Consider the system of Figure 9.8. The equations that define this system are:

$$\begin{bmatrix} x(t) \\ x'(t) \\ x''(t) \end{bmatrix}' = \begin{pmatrix} 0 & 1 & 0 \\ 0 & 0 & 1 \\ -1 & -2 & -2 \end{pmatrix} \begin{bmatrix} x(t) \\ x'(t) \\ x''(t) \end{bmatrix} + \begin{bmatrix} 0 \\ 0 \\ 1 \end{bmatrix} u(t), \quad y(t) = \begin{bmatrix} 1 & 1 & 0 \end{bmatrix} \begin{bmatrix} x(t) \\ x'(t) \\ x''(t) \end{bmatrix}.$$

A simple calculation shows that:

$$\text{Obs} = \begin{pmatrix} \vec{c}^T \\ \vec{c}^T\mathbf{A} \\ \vec{c}^T\mathbf{A}^2 \end{pmatrix} = \begin{pmatrix} 1 & 1 & 0 \\ 0 & 1 & 1 \\ -1 & -2 & -1 \end{pmatrix}$$

As the bottom row of the observability matrix is minus the sum of the first two rows, the rows of the matrix are not independent, $\det(\text{Obs}) = 0$, and the system is not observable.

9.14.6 *A System that is neither Observable nor Controllable*

Now consider a system with two identical blocks in parallel—for the sake of this discussion let the blocks both have the transfer function:

$$G(s) = \frac{1}{s+1}.$$

Such a system is shown in Figure 9.9. Clearly $x_1(t)$ and $x_2(t)$ satisfy the equations:

$$\begin{aligned} x_1'(t) &= -x_1(t) + u(t) \\ x_2'(t) &= -x_2(t) + u(t). \end{aligned}$$

We find that the state equations for this system are:

$$\begin{bmatrix} x_1(t) \\ x_2(t) \end{bmatrix}' = \begin{pmatrix} -1 & 0 \\ 0 & -1 \end{pmatrix} \begin{bmatrix} x_1(t) \\ x_2(t) \end{bmatrix} + \begin{bmatrix} 1 \\ 1 \end{bmatrix} u(t), \quad y(t) = [1\ 1] \begin{bmatrix} x_1(t) \\ x_2(2) \end{bmatrix}.$$

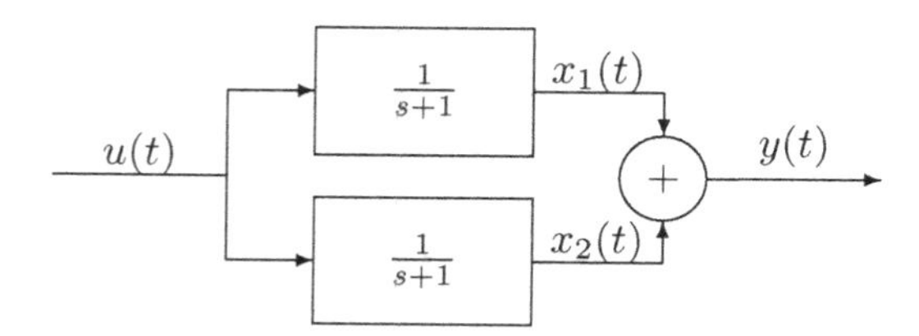

Fig. 9.9 Two Units in Parallel

Calculating the observability and controllability matrices, we find that:

$$\text{Obs} = \begin{pmatrix} 1 & 1 \\ -1 & -1 \end{pmatrix}, \quad \text{Con} = \begin{pmatrix} 1 & -1 \\ 1 & -1 \end{pmatrix}.$$

Clearly:

$$\det(\text{Obs}) = \det(\text{Con}) = 0.$$

The system is neither observable nor is it controllable.

It is clear that the system cannot be controllable. Because the two parallel blocks of which the system is composed are identical, any effect that $u(t)$ has on one block it will have on the other. Thus, if the two subsystems start with different initial conditions, the input cannot possibly bring *both* outputs to zero in finite time. Additionally, because the two subsystems are identical there is no way that someone viewing the system from the outside can hope to separate the contribution of the two internal blocks. Thus, the system cannot be observable either.

9.14.7 *Designing an Observer*

Let us design an observer for a first order system whose transfer function is:

$$\frac{Y(s)}{U(s)} = T(s) = \frac{1}{s+1}$$

and whose state is its output, $y(t)$. (In such a case, one does not really need an observer as the output is available, but this case is easy to analyze and provides a simple example.) We would like the state of the observer to converge to that of the system within (approximately) 30 ms.

We find that the state equations of the system are:

$$[y(t)]' = [-1][y(t)] + [1]u(t), \qquad y(t) = [1][y(t)].$$

As we are working in a one dimensional space,

$$\text{Obs} = [\vec{c}^T] = [1]$$

and the system is observable.

As we are dealing with a first order system, $\vec{l} = [l_1]$. In order to make the observer's state converge to that of the system under observation within

30 ms, we must cause the observer's time constant to be 10 ms. That is, the eigenvalue of:

$$(\mathbf{A} - \vec{l}\vec{c}^T) = (-1 - l_1)$$

must be located at -100. As the eigenvalue of a 1×1 matrix is just the value of the matrix, we find that $l_1 = 99$. Thus, the equations that characterize the observer are:

$$[r(t)]' = [-1][r(t)] + [1]u(t) + [99](y(t) - s(t)), \qquad s(t) = [1][r(t)].$$

9.15 Some Technical Results about Series of Matrices

To show that the series in (9.4) converges in a reasonable fashion, it is easiest to define convergence for sequences of matrices. If one lets the norm of a matrix be the square root of the sum of the squares of the elements of the matrix—if one lets:

$$\|\mathbf{A}\| = \sqrt{\sum_{i,j=1}^{N} |a_{ij}|^2},$$

then (by thinking of the matrices as "vectors") it is clear that:

$$\|\mathbf{A} + \mathbf{B}\| \leq \|\mathbf{A}\| + \|\mathbf{B}\|.$$

Clearly,

$$\|\mathbf{A}\| = 0 \Leftrightarrow \mathbf{A} = \mathbf{0}.$$

Finally, if c is a scalar it is easy to see that:

$$\|c\mathbf{A}\| = |c|\mathbf{A}.$$

Thus, $\|\cdot\|$ is indeed a norm on matrices.

Consider $\|\mathbf{AB}\|$. Denote the ith *row* of $\mathbf{A}$ by $\vec{a}_i^T$, the jth *column* of B by $\vec{b}_j$, the dot product of two vectors by $(\cdot, \cdot)_V$, and the norm of a vector by $\|\cdot\|_V$. We find that:

$$\begin{aligned}
\|\mathbf{AB}\|^2 &= \sum_{i,j=1}^{N} \left|\vec{a}_i^T \vec{b}_j\right|^2 \\
&= \sum_{i,j=1}^{N} |(\vec{a}_i, \vec{b}_j)_V|^2 \\
&\leq \sum_{i,j=1}^{N} \|\vec{a}_i\|_V^2 \|\vec{b}_j\|_V^2 \\
&\leq \sqrt{\sum_{i,j=1}^{N} \|\vec{a}_i\|_V^4} \sqrt{\sum_{i,j=1}^{N} \|\vec{b}_j\|_V^4} \\
&= \sqrt{N \sum_{i=1}^{N} \|\vec{a}_i\|_V^4} \sqrt{N \sum_{j=1}^{N} \|\vec{b}_j\|_V^4} \\
&\leq \sqrt{N \left(\sum_{i=1}^{N} \|\vec{a}_i\|_V^2\right)^2} \sqrt{N \left(\sum_{j=1}^{N} \|\vec{b}_j\|_V^2\right)^2} \\
&= \sqrt{N}\|\mathbf{A}\|^2 \sqrt{N}\|\mathbf{B}\|^2 \\
&= N\|\mathbf{A}\|^2\|\mathbf{B}\|^2.
\end{aligned}$$

We find that:

$$\|\mathbf{AB}\| \leq \sqrt{N}\|\mathbf{A}\|\|\mathbf{B}\|.$$

In particular, we find that:

$$\|\mathbf{A}^n\| \leq \sqrt{N}\|\mathbf{A}\|\|\mathbf{A}^{n-1}\| \leq \cdots \leq \sqrt{N}^{n-1}\|\mathbf{A}\|^n, \quad n \geq 2.$$

If an upper-bound on the norms of a sequence of matrices satisfies the ratio test, then the series that corresponds to the sequence of matrices converges in the matrix norm. Let us consider the series for the matrix exponential—the series whose n^{th} term is $\mathbf{A}^n/n!$. We find that:

$$\left\|\frac{\mathbf{A}^n}{n!}\right\| \leq u_n \equiv \frac{\sqrt{N}^{n-1}\|A\|^n}{n!}.$$

Applying the ratio test to u_n, we find that:

$$\lim_{n\to\infty} \frac{u_{n+1}}{u_n} = \lim_{n\to\infty} \frac{\sqrt{N}\|\mathbf{A}\|}{n+1} = 0.$$

Thus, we see that the series:

$$\mathbf{I} + \mathbf{A} + \cdots + \frac{\mathbf{A}^n}{n!} + \cdots$$

converges to a limit in the matrix norm.

Clearly the value of any given term in the matrix $\mathbf{A}$, say a_{ij}, is less in absolute value than the norm of the matrix. Thus, the convergence of the matrices in the matrix norm also guarantees the convergence of the terms of the matrix in the ordinary sense.

9.16 Exercises

(1) Find the eigenvalues and eigenvectors of the matrix:

$$A = \begin{pmatrix} 0 & 1 \\ 0 & -1 \end{pmatrix}.$$

(2) With A as above, find e^{At} in terms of M, M^{-1}, and Λ. What are the exponents of the exponentials in $e^{\Lambda t}$?

(3) Let:

$$A = \begin{pmatrix} 1 & 0 \\ 1 & 1 \end{pmatrix}.$$

(a) Show that A satisfies its own characteristic equation.
(b) Find A^3 as a linear combination of I and A.

(4) Consider a block whose transfer function is:

$$T(s) = \frac{3}{s^2 + 3s + 2}.$$

Let $y(t)$ be the output of the block and let $u(t)$ be the input to the block. Thus:

$$\frac{Y(s)}{U(s)} = T(s).$$

(a) Find the ordinary differential equation that $y(t)$ satisfies.

(b) Let the states of the system be $y(t)$ and $y'(t)$. Find the equations that define the system.
(c) Is the system controllable? Explain.
(d) Is the system observable? Explain.

(5) Using state feedback place the closed loop poles of the system of the previous exercise at $s = -2$ and $s = -4$. Please provide a sketch of the completed system with all values filled in.
(6) Consider a block whose transfer function is:

$$T(s) = \frac{s+1}{s^2+3s+2}.$$

Let $y(t)$ be the output of the block, and let $u(t)$ be the input to the block. Let $x(t)$ and $x'(t)$ be the states of the block. Let the various signals "in" the block satisfy:

$$\frac{X(s)}{U(s)} = \frac{1}{s^2+3s+2}, \quad \frac{Y(s)}{X(s)} = s+1.$$

Thus:

$$\frac{Y(s)}{U(s)} = \frac{Y(s)}{X(s)}\frac{X(s)}{U(s)} = T(s).$$

(a) Find the ordinary differential equation that $x(t)$ satisfies.
(b) Let the states of the system by $x(t)$ and $x'(t)$. Find the equations that define the system.
(c) Is the system controllable? Explain.
(d) Is the system observable? Explain.

(7) Using state feedback place the closed loop poles of the system of the previous exercise at $s = -2$ and $s = -4$. Please provide a sketch of the completed system with all values filled in.
(8) Consider a block whose transfer function is:

$$T(s) = \frac{s+1}{s^2+5s+6}.$$

Let $y(t)$ be the output of the block, and let $u(t)$ be the input to the block. Let $x(t)$ and $x'(t)$ be the states of the block. Let the various signals "in" the block satisfy:

$$\frac{X(s)}{U(s)} = \frac{1}{s^2+5s+6}, \quad \frac{Y(s)}{X(s)} = s+1.$$

Thus:

$$\frac{Y(s)}{U(s)} = \frac{Y(s)}{X(s)} \frac{X(s)}{U(s)} = T(s).$$

(a) Find the ordinary differential equation that $x(t)$ satisfies.
(b) Let the states of the system by $x(t)$ and $x'(t)$. Find the equations that define the system.
(c) Is the system controllable? Explain.
(d) Is the system observable? Explain.

(9) Using state feedback place the closed loop poles of the system of the previous exercise at $s = -10$ and $s = -11$. Please provide a sketch of the completed system with all values filled in.

(10) In this exercise we show that the system of Figure 9.9 is not controllable by showing that the state of the system cannot generally be brought to zero in finite time. Throughout this problem we make use of the terminology and definitions of §9.14.6.

(a) Define $r(t) = x_1(t) - x_2(t)$. Find the first order ordinary differential equation (ODE) satisfied by $r(t)$.
(b) Find the initial condition, $r(0)$, in terms of $x_1(0)$ and $x_2(0)$.
(c) Find the solution of the ODE.
(d) Show that if $x_1(0) \neq x_2(0)$, then the solution is non-zero for all $t \geq 0$.
(e) Use the results of the preceding section to show that the system is not controllable.

(11) Consider a block whose transfer function is:

$$T(s) = \frac{1}{s+a}, a \in \mathcal{R}.$$

Let $y(t)$ be the output of the block and let $u(t)$ be the input to the block.

(a) Find the ordinary differential equation satisfied by $y(t)$.
(b) Let the state of the equation be $y(t)$. Find the equations that define the state of the system.
(c) Is the system controllable?
(d) Is the system observable?
(e) Explain why for first order systems of this type of "state feedback" is (nearly) identical to "proportional control."

(12) Given a system whose transfer function is

$$G(s) = \frac{s+1}{s^2+3s}$$

(a) Find the state equations of the system.
(b) Show that the system is both observable and controllable.
(c) Design an observer whose time constants are 0.01 and 0.005.
(d) Sketch the system with the observer showing all important connections and parameter values.

(13) Consider the system of Example 12 again. By making a single change to the numerator of the transfer function, cause the system not to be observable.

(14) Consider a system whose transfer function is given by

$$G_p(s) = \frac{s+1}{s^2+s+1}.$$

(a) Find the state equations of the system.
(b) Show that the system is observable.
(c) Design an observer whose poles are located at $s = -10, -11$. What values of l_1 and l_2 must be chosen?

(15) Show that the system of Figure 9.8 *is* controllable.

Chapter 10

Control of Hybrid Systems

10.1 Introduction

Most interesting plants are what can be termed "fully analog." A DC motor, for example, takes a continuous-time voltage as its input and produces in return a shaft-angle—also a continuous-time signal. Controllers, however, are another story. Today control algorithms are often implemented by microcontrollers or microprocessors. These devices process their input at fixed times and change their output at fixed times.

We have learned how to describe systems whose input and output change continuously. For this task we use the Laplace transform and transfer functions. Now we are going to learn how to model systems that process discrete-time signals, and we are going to learn how to models *hybrid systems*—systems that have some parts that work in continuous time (like a motor) and some parts that work in discrete-time (like a microcontroller). The z-transform is the correct tool for modeling many discrete-time systems, so we start by defining it and describing some of its properties. We also give a fairly complete description of the bilinear transform and its uses in the control of hybrid systems. Next, we consider the more general problem of designing controllers for hybrid systems. Finally we give a brief description of the modified z-transform and its uses.

10.2 The Definition of the Z-Transform

Consider a sequence of numbers:

$$\{e(0), e(1), \ldots\}.$$

The z-transform of the sequence, $\mathcal{Z}(\{e(0), e(1), \ldots\})(z)$, is defined as:

$$\mathcal{Z}(\{e(0), e(1), \ldots\})(z) \equiv \sum_{k=0}^{\infty} e(k) z^{-k}.$$

Often the z-transform of the sequence $\{e(0), e(1), \ldots\}$ is denoted by $E(z)$.

10.3 Some Examples

In calculating z-transforms, it is often necessary to sum a geometric series. Recall that the geometric series is the sum:

$$1 + r + r^2 + \cdots = \sum_{k=0}^{\infty} r^k,$$

and its sum is:

$$\sum_{k=0}^{\infty} r^k = \frac{1}{1-r}, \qquad |r| < 1.$$

Let us consider the z-transform of several sequences.

(1) Let

$$e(k) = u(k) = \begin{cases} 1 & k \geq 0 \\ 0 & k < 0 \end{cases}.$$

We find that:

$$E(z) = \sum_{k=0}^{\infty} 1 \cdot z^{-k} = \frac{1}{1 - z^{-1}} = \frac{z}{z-1}, \qquad |z| > 1.$$

(2) Let $e(k) = a^k$. Then we find that:

$$E(z) = \sum_{k=0}^{\infty} a^k z^{-k} = \sum_{k=0}^{\infty} (a/z)^k = \frac{1}{1 - a/z} = \frac{z}{z-a}, \qquad |z| > |a|.$$

(3) Similarly, if we let $e(k) = e^{-ak}$, then we find that:

$$\begin{aligned} E(z) &= \sum_{k=0}^{\infty} e^{-ak} z^{-k} \\ &= \sum_{k=0}^{\infty} (e^{-a} z^{-1})^k \\ &= \frac{1}{1 - e^{-a} z^{-1}} \\ &= \frac{z}{z - e^{-a}}, \qquad |z| > |e^{-a}|. \end{aligned}$$

(4) Let $e(k) = \sin(\omega k)$. Rewriting $\sin(\omega k)$ as:

$$\sin(\omega k) = \frac{e^{j\omega k} - e^{-j\omega k}}{2j},$$

we find that:

$$\begin{aligned} E(z) &= \frac{1}{2j} \left(\frac{z}{z - e^{j\omega}} - \frac{z}{z - e^{-j\omega}} \right) \\ &= \frac{1}{2j} \frac{z(e^{j\omega} - e^{-j\omega})}{z^2 - 2(e^{j\omega} + e^{-j\omega}) + 1} \\ &= \frac{z \sin(\omega)}{z^2 - 2z\cos(\omega) + 1}, \qquad |z| > 1. \end{aligned}$$

10.4 Properties of the Z-Transform

(1) Linearity. If $E(z)$ is the z-transform of the sequence $\{e(0), e(1), \ldots\}$ and $F(z)$ is the z-transform of the sequence $\{f(0), f(1), \ldots\}$, then $aE(z) + bF(z)$ is the z-transform of the sequence $\{ae(0) + bf(0), ae(1) + bf(1), \ldots\}$. The proof is a simple application of the definition of the z-transform.

(2) Final Value. If the sequence $e(k)$ has a limit as $k \to \infty$, then:

$$\lim_{k \to \infty} e(k) = \lim_{z \to 1^+} (z - 1)E(z).$$

The proof of this result is not trivial. Let us start with a sequence, $a(k)$, that converges to zero. That is, for every ϵ there exists an N for which $|a(k)| < \epsilon$ for every $k \geq N$. Let us consider the z-transform of

$a(k)$. We find that:

$$\begin{aligned}
|A(z)| &= \left|\sum_{k=0}^{\infty} a(k)z^{-k}\right| \\
&\leq \left|\sum_{k=0}^{N-1} a(k)z^{-k}\right| + \left|\sum_{k=N}^{\infty} a(k)z^{-k}\right| \\
&\overset{|z|>1}{\leq} \sum_{k=0}^{N-1} |a(k)| + \epsilon \sum_{k=N}^{\infty} |z|^{-k} \\
&\overset{|z|>1}{\leq} \sum_{k=0}^{N-1} |a(k)| + \epsilon \frac{|z|^{-N}}{1-|z|^{-1}} \\
&\overset{z\geq 1}{=} \sum_{k=0}^{N-1} |a(k)| + \epsilon \frac{z^{-N}}{1-z^{-1}} \\
&= \sum_{k=0}^{N-1} |a(k)| + \epsilon \frac{z^{-N+1}}{z-1}.
\end{aligned}$$

We find that:

$$\lim_{z\to 1^+} |(z-1)A(z)| \leq \lim_{z\to 1^+} |(z-1)| \left| \left(\sum_{k=0}^{N-1} |a(k)| + \epsilon \frac{z^{-N+1}}{z-1} \right) \right| = \epsilon.$$

But ϵ can be made arbitrarily small and the limit must be non-negative, so we find that:

$$\lim_{z\to 1^+} |(z-1)A(z)| = 0.$$

From this it is clear that:

$$\lim_{z\to 1^+} (z-1)A(z) = 0.$$

if the sequence $a(k)$ tends to zero.
Suppose that $e(k) \to c$. Then write $e(k)$ as:

$$e(k) = c + (e(k) - c) \equiv c + a(k)$$

where $a(k) \to 0$ as $k \to \infty$. The z-transform of $e(k)$ is then:

$$E(z) = \mathcal{Z}(\{c, c, \ldots\})(z) + A(z) = c\frac{z}{z-1} + A(z).$$

We find that:

$$\lim_{z \to 1^+} (z-1)E(z) = c + 0 = c$$

just as it should.

(3) Initial Value. It is easy to see that:

$$e(0) = \lim_{z \to \infty} E(z).$$

By definition $E(z)$ is:

$$E(z) = e(0) + \sum_{k=1}^{\infty} e(k)z^{-k}.$$

Assuming that the sequence does not grow more than exponentially quickly, we know that:

$$|e(k)| \leq Mc^k, \qquad M, c > 0.$$

We find that:

$$\begin{aligned} \left| \sum_{k=1}^{\infty} e(k)z^{-k} \right| &\leq \sum_{k=1}^{\infty} |e(k)||z^{-k}| \\ &\leq \sum_{k=1}^{\infty} Mc^k|z|^{-k} \\ &= M\sum_{k=1}^{\infty}(c/|z|)^k \\ &\overset{|z|>c}{=} M\frac{c/|z|}{1-c/|z|} \\ &= M\frac{c}{|z|-c}. \end{aligned}$$

Clearly, as $z \to \infty$ this tends to zero. We find that:

$$\lim_{z \to \infty} E(z) = e(0).$$

(4) Translation. Let $E(z)$ be the z-transform of $\{e(0), e(1), \ldots\}$. What is the z-transform of:

$$\{f(0), f(1), \ldots\} = \{\underbrace{0, 0, \ldots, 0}_{n \text{ times}}, e(0), e(1), \ldots\}?$$

We find that:

$$F(z) = \sum_{k=0}^{\infty} f(k)z^{-k} = \sum_{k=n}^{\infty} e(k-n)z^{-k} = z^{-n}\sum_{k=0}^{\infty} e(k)z^{-k} = z^{-n}E(z).$$

The importance of this relationship cannot be overemphasized.

(5) Multiplication by k. Suppose that the z-transform of $\{e(0), e(1), \ldots\}$ is $E(z)$. Then the z-transform of $\{0 \cdot e(0), 1 \cdot e(1), \ldots, ke(k), \ldots\}$ is:

$$-z\frac{d}{dz}E(z).$$

The simplest proof of this property is to consider $-z$ times the derivative of the z-transform. It is:

$$\begin{aligned} -z\frac{d}{dz}\left(e(0) + e(1)z^{-1} + e(2)z^{-2} + \cdots\right) \\ &= -z\left(0e(0) - 1e(1)z^{-2} - 2e(2)z^{-3} + \cdots\right) \\ &= 0e(0) + 1e(1)z^{-1} + 2e(2)z^{-2} + \cdots. \end{aligned}$$

This is the z-transform of the sequence $\{ke(k)\}$.

The Sequence $\{0, 1/1, 1/2, \ldots\}$—An Example

Let us calculate the z-transform of the sequence:

$$e(k) = \begin{cases} 0 & k = 0 \\ 1/k & k > 0 \end{cases}.$$

Denote the z-transform of $e(k)$ by $E(z)$, and note that:

$$\{ke(k)\} = \{0, 1, 1, 1, \ldots\}.$$

Denote the new sequence by $g(k)$. As $g(k)$ is just a shifted unit step we find that:

$$G(z) = z^{-1}\frac{z}{z-1} = \frac{1}{z-1}.$$

From the properties of the z-transform, we know that:

$$-z\frac{d}{dz}E(z) = G(z).$$

Thus, we find that:

$$\begin{aligned}
E(z) &= -\int G(z)/z\,dz + C \\
&= -\int \frac{1}{z(z-1)}\,dz + C \\
&= -\int \left(-\frac{1}{z} + \frac{1}{z-1}\right) dz \\
&= \ln(z) - \ln(z-1) + C \\
&= \ln\left(\frac{z}{z-1}\right) + C.
\end{aligned}$$

All that remains is to evaluate C. As we know that $e(0) = 0$, we know that $\lim_{z\to\infty} E(z) = 0$. As $\ln(z)$ is a continuous function of z at $z = 1$, we find that:

$$\lim_{z\to\infty} E(z) = 0 = \lim_{z\to\infty}\left(\ln\left(\frac{z}{z-1}\right)\right) + C = \ln(1) + C = C.$$

Thus:

$$E(z) = \ln\left(\frac{z}{z-1}\right).$$

Note that this z-transform allows us to calculate an interesting sum[1]. Let us consider $E(-1)$. We find that:

$$\begin{aligned}
E(-1) &= -\ln(2) \\
&= 0 + (1/1)(-1)^{-1} + (1/2)(-1)^{-2} + \cdots \\
&= -\frac{1}{1} + \frac{1}{2} - \frac{1}{3} + \cdots + (-1)^k\frac{1}{k}.
\end{aligned}$$

The sum is just minus the alternating harmonic series. We see that the sum of the alternating harmonic series is $\ln(2)$.

10.5 Sampled-data Systems

One of the main reasons to study the z-transform is to use it to help analyze systems that "run on" samples of data rather than continuously arriving data. In such systems the form of the sequence with which one works is

[1]To justify this calculation is a little bit tricky as the region of convergence of this z-transform is $|z| > 1$, but the calculation can be justified.

often $e(k) = f(kT)$. That is, the elements of the sequence are actually samples of a function from which a sample is acquired every T seconds.

Suppose that one samples the unit step function. That is, suppose that $e(k) = u(kT)$. As this sequence is all ones (for non-negative values of k) its z-transform is:

$$E(z) = \frac{z}{z-1}.$$

Suppose that the function that is being sampled is $5\sin(\omega t)$. Then the sequence whose z-transform we must find is $e(k) = 5\sin(\omega Tk)$. Using the transforms we have already calculated, we find that:

$$E(z) = \frac{5z\sin(\omega Tk)}{z^2 - 2z\cos(\omega Tk) + 1}.$$

Suppose that the function being sampled is the ramp. That is, let $e(t) = tu(t)$. The samples are :

$$\{T \cdot 0, T \cdot 1, \ldots, Tk, \ldots\} = T\{1 \cdot 0, 1 \cdot 1, \ldots, 1 \cdot k, \ldots\}.$$

Thus, the z-transform of the the samples must be $-zT$ times the derivative of the z-transform of the unit step. That is, the z-transform is:

$$\mathcal{Z}(\{0, T, 2T, \ldots, kT, \ldots\})(z) = -zT\frac{d}{dz}\frac{z}{z-1} = \frac{Tz}{(z-1)^2}.$$

10.6 The Sample-and-Hold Element

Because most of the systems in which we are interested have both digital and analog parts, we need a circuit element which takes continuous-time signals and "converts" them to discrete-time signals. The sample-and-hold circuit fills this need. An ideal sample-and-hold element that samples at the rate T samples per second has as its input a signal $r(t)$ and as its output $c(t)$—a signal that satisfies:

$$\begin{aligned} c(t) &= r(0)(u(t) - u(t-T)) + r(T)(u(t-T) - u(t-2T)) \\ &\quad + \cdots + r(kT)(u(t-kT) - u(t-(k+1)T)) + \cdots. \end{aligned} \tag{10.1}$$

(See Figure 10.1 for an example of a sine wave that has been passed through a sample-and-hold circuit.) A signal that has been processed by a sample-and-hold circuit only changes values at discrete times.

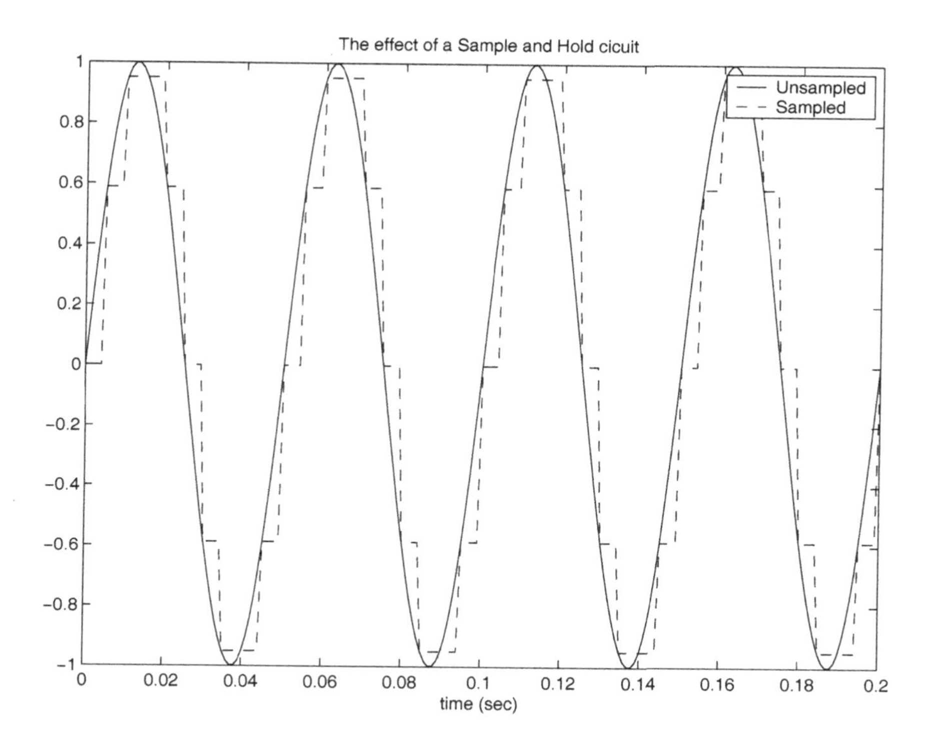

Fig. 10.1 A Sampled-and-Held Sinewave (Dashed Lines) and the Unsampled Sinewave (Solid Lines). The Frequency of the Sinewave is 20 Hz, and 200 Samples Per Second Are Taken.

Taking the Laplace transform of both sides of (10.1), we find that:

$$\begin{aligned} C(s) &= r(0)(1/s - e^{-Ts}/s) + r(T)(e^{-Ts}/s - e^{-2Ts}/2) + \cdots \\ &\quad + r(kT)(e^{-kTs}/s - e^{-k+1Ts}/2) + \cdots . \end{aligned}$$

Inspecting $C(s)$ carefully, we find that it can be written:

$$C(s) = \frac{1-e^{-Ts}}{s} \sum_{k=0}^{\infty} r(kT)e^{-kTs} = \frac{1-e^{-Ts}}{s} \mathcal{Z}(\{r(0), r(T), r(2T), \ldots\})(z)\bigg|_{z=e^{Ts}} .$$

Defining the "star-transform" of the sequence $\{r(0), r(T), r(2T), \ldots\}$ by:

$$R^*(s) \equiv \mathcal{Z}(\{r(0), r(T), r(2T), \ldots\})|_{z=e^{Ts}} ,$$

we find that the Laplace transform of the output of the sample-and-hold circuit is:

$$\frac{1-e^{-Ts}}{s}R^*(s).$$

In the next sections we will see that $R^*(s)$ corresponds to samples of $r(t)$ made by an "ideal sampler" and that $\frac{1-e^{-Ts}}{s}$ is the transfer function of a "zero-order hold."

10.7 The Delta Function and its Laplace Transform

In what follows it will be useful to consider the delta function (or the Dirac delta function). The delta function is the function[2] that is zero at all points other than 0, that is infinite at zero, and whose integral over the real line is one. The delta function is denoted by $\delta(t)$.

Let us consider the family of functions $\delta_h(t)$ defined by:

$$\delta_h(t) \equiv \begin{cases} \frac{1}{h} & 0 \le t \le h \\ 0 & \text{otherwise} \end{cases}.$$

We can say that:

$$\delta(t) = \lim_{h \to 0^+} \delta_h(t).$$

Note that as the integral of $\delta_h(t)$ is one for any h, the integral of $\delta(t)$ should be one as well.

Making use of the definition of the Laplace transform we find that:

$$\mathcal{L}(\delta_h(t))(s) = \int_0^h e^{-st}\frac{1}{h}\,dt = \frac{1-e^{-sh}}{sh}.$$

As $e^{-x} = 1 - x + x^2/2! + \cdots$, we know that for small values of $|x|$ $e^{-x} \approx 1 - x$. We find that as long as sh is small, the Laplace transform of $\delta_h(t)$ is approximately equal to one. As $\delta(t) = \lim_{h\to 0^+} \delta_h(t)$, we find that the Laplace transform of $\delta(t)$ should be:

$$\mathcal{L}(\delta(t)) = \lim_{h \to 0^+} \frac{1-e^{-sh}}{sh} = 1.$$

The Laplace transform of $\delta(t)$ should be, and is, 1.

[2]Strictly speaking the delta function is not quite a function, but we will not worry about the technical problems that surround the delta function's definition.

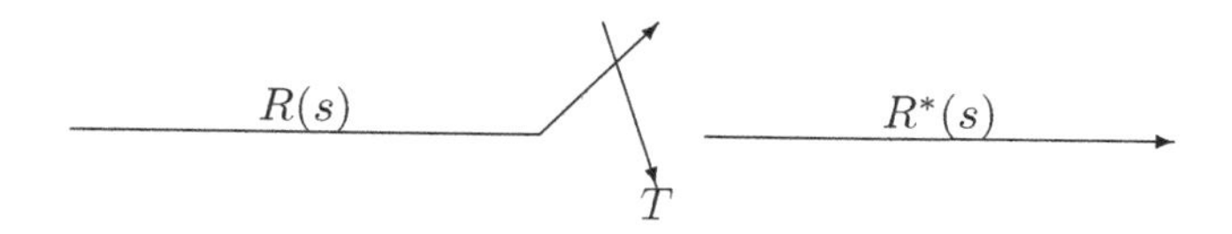

Fig. 10.2 The Symbol for the Ideal Sampler

10.8 The Ideal Sampler

Let us consider the inverse transform of the function e^{-kTs}. We know that the inverse transform of $e^{-Ts}F(s)$ is $f(t-T)u(t-T)$. In our case, $F(s) = 1$. Thus, $f(t) = \delta(t)$. Thus, the inverse transform of e^{-kTs} is $\delta(t - kT)$. We find that the inverse transform of $R^*(s)$, which we denote by $r^*(t)$, is:

$$r^*(t) = r(0)\delta(t) + r(T)\delta(t - T) + \cdots = \sum_{K=0}^{\infty} r(kT)\delta(t - kT).$$

That is, $r^*(t)$ is composed of the samples of $r(t)$ multiplied by delta functions located at the time at which the sample was made. The symbol for an *ideal sampler* is given in Figure 10.2.

10.9 The Zero-Order Hold

For our purposes, what is important about the discrete-time part of our system is not the precise representation of the discrete-time signal. What we need to know is the value of the signal at a particular sample. For this reason, we use either the star-transform or the z-transform to represent our signal in the discrete-time region. To convert the signal from discrete-time back to *steps* in continuous time, we use the zero-order hold block. This block has the transfer function:

$$T(s) = \frac{1 - e^{-Ts}}{s}.$$

Suppose that one inputs a delta function of strength h at time τ to this block. Then the Laplace transform of the output of the block, $C(s)$, is just:

$$C(s) = T(s)he^{-\tau s}1 = h\frac{e^{-\tau s}}{s} - h\frac{e^{-(T+\tau)s}}{s}.$$

Thus, the output as a function of time is:

$$c(t) = h(u(t - \tau) - u(t - \tau - T)).$$

this function is zero until $t = \tau$ and after $t = \tau + T$. Between these times it is a pulse whose height is h.

The zero-order hold takes a delta function of strength h that arrives at time τ and converts it into a pulse of height h that starts at the time at which the delta function arrived and ends T seconds later. (See Figure 10.3.)

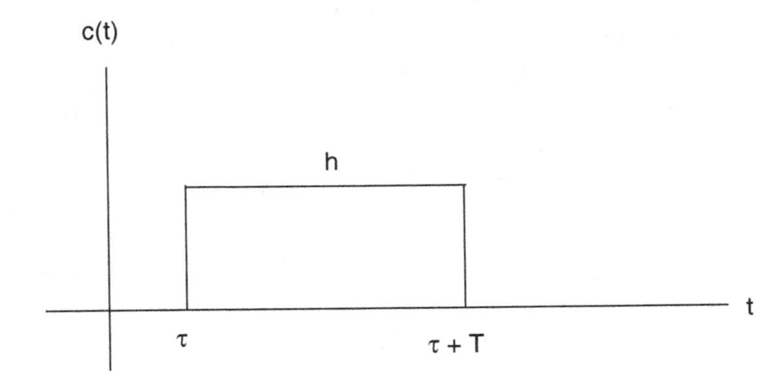

Fig. 10.3 A Pulse of Height h and Duration T Starting at $t = \tau$

10.10 Calculating the Pulse Transfer Function

Let us consider a system like that of Figure 10.4 in which there is a sample and hold circuit followed by an analog element. How can we calculate the output of the analog element?

Consider the input to the zero-order hold. We know that the zero order

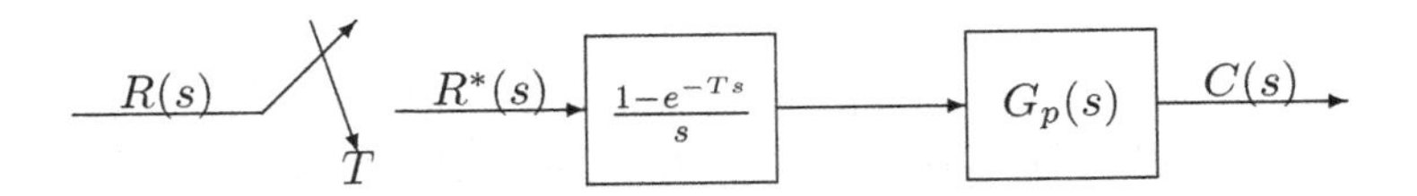

Fig. 10.4 A Very Simple System

hold has a train of delta functions as its input. Let us see what the output of the combination of the zero-order hold and the analog block is when the input to the zero-order hold is just $\delta(t)$.

The Laplace transform of the output is just:

$$\frac{1-e^{-Ts}}{s}G_p(s).$$

To find the output as a function of time one need only consider the inverse Laplace transform of this expression. Because of the time-shifting property of e^{-Ts} the inverse Laplace transform of this expression is actually just:

$$v(t)u(t) - v(t-T)u(t-T)$$

where

$$v(t) \equiv \mathcal{L}^{-1}(G_p(s)/s)(t).$$

(The function $v(t)$ can also be thought of as the impulse response of the block with transfer function $G_p(s)/s$.)

$G_p(s) = 1/s$—An Example

Let $G_p(s) = 1/s$. We find that:

$$v(t) = \mathcal{L}^{-1}(G_p(s)/s)(t) = \mathcal{L}^{-1}(1/s^2)(t) = tu(t).$$

As e^{-Ts} just delays a signal by T, the inverse Laplace transform of $(1-e^{-Ts})(1/s^2)$ is just:

$$tu(t) - (t-T)u(t-T).$$

Let us now consider what sort of data, $r(t)$, would get us a delta function at zero at the output of the sampler. Clearly data that is equal to 1 at $t = 0$ and that is zero at all other sampling instants will do that. If we are willing to limit our interest to the samples of the output, then we can say that when the z-transform of the input is 1, then the output of the system at the sampling times is:

$$v(kT)u(kT) - v(kT-T)u(kT-T).$$

Taking the z-transform of this sequence, we find that the z-transform of the samples of the output when the input has 1 as its z-transform is just:

$$V(z) - z^{-1}V(z) = (1-z^{-1})V(z).$$

Suppose that rather than considering input that has height 1 at $t = 0$ and height zero at all other sampling instants, we consider an input that has height h at $t = kT$ and has zero height at all other sampling instants. Because the system is linear and time-invariant, the output to the system must be the previous output multiplied by the height of the input and shifted by kT seconds. In term of the z-transform of the output, this means multiplication by hz^{-k}. That is, we have found that if the input to a system has z-transform hz^{-k}, then the output must by $hz^{-k}(1-z^{-1})V(z)$. By linearity, we find that if the input to the system is:

$$R(z) = \sum_{k=0}^{\infty} r(kT)z^{-k},$$

then the z-transform of $c(kT)$ must be:

$$\begin{aligned} C(z) &= \sum_{k=0} r(kT)z^{-k}(1 - z^{-1})V(z) \\ &= V(z)(1 - z^{-1}) \sum_{k=0}^{\infty} r(kT)z^{-k} \\ &= V(z)(1 - z^{-1})R(z) \\ &= \mathcal{Z}\left(\frac{G_p(s)}{s}\right)(z)\frac{z-1}{z}R(z) \end{aligned}$$

where:

$$\mathcal{Z}(Y(s))(z) \equiv \mathcal{Z}(\{y(0), y(T), y(2T), \ldots, y(kT), \ldots\})(z)$$

and $y(t)$ is the inverse Laplace transform of $Y(s)$. We now have a nice, simple formula for $C(z)$. The ratio:

$$\frac{C(z)}{R(z)},$$

when it is independent of $R(z)$ is called the *pulse transfer function* (or simply the transfer function) of the system.

Consider the system of Figure 10.5—a system with feedback. Clearly:

$$E(s) = V_{in}(s) - H(s)G_p(s)\frac{1 - e^{-Ts}}{s}E^*(s).$$

Let:

$$w(t) \equiv \mathcal{L}^{-1}(H(s)G_p(s)/s)(t).$$

Following our previous logic, we find that:

$$e(t) = v_{in}(t) - \sum_{k=0}^{\infty} e(k)(w(t - kT) - w(t - (k+1)T)).$$

Sampling these functions at $t = nT$ and taking the z-transform of both sides of this equation we find that:

$$\begin{aligned} E(z) &= V_{in}(z) - \sum_{k=0}^{\infty} e(k)(z^{-k}W(z) - z^{-(k+1)}W(z)) \\ &= V_{in}(z) - E(z)(1 - z^{-1})W(z). \end{aligned}$$

A little bit of algebra shows that:

$$E(z) = \frac{V_{in}(z)}{1 + (1 - z^{-1})W(z)}.$$

To determine the output of the system at sampling times is now simple. We now know what the output of the sampler is. The output of the system is just the output of the hold and $G_p(s)$ when the input is now known to be $E^*(s)$. We have already seen that:

$$V_o(z) = (1 - z^{-1})V(z) \cdot E(z).$$

Thus,

$$V_o(z) = \frac{(1 - z^{-1})V(z)}{1 + (1 - z^{-1})W(z)} V_{in}(z) = \frac{\frac{z-1}{z} \mathcal{Z}\left(\frac{G_p(s)}{s}\right)(z)}{1 + \frac{z-1}{z} \mathcal{Z}\left(\frac{G_p(s)H(s)}{s}\right)(z)}$$

where $w(t)$ and $v(t)$ are as defined above.

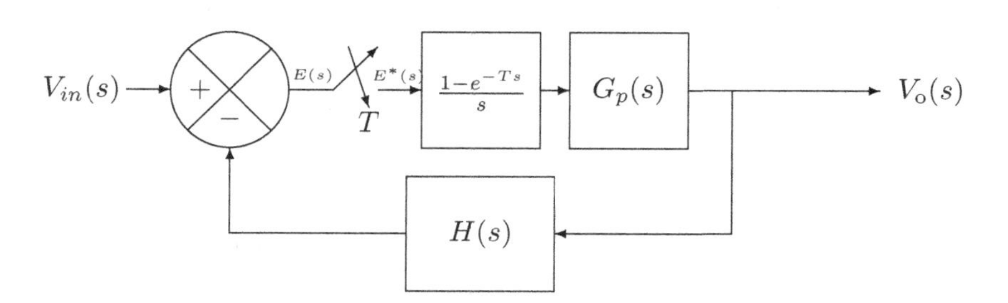

Fig. 10.5 A Simple Feedback System

$G_p(s) = 1/s, H(s) = 1$—An Example

Consider the case in which $H(s) = 1$ and $G_p(s) = 1/s$. Because of the unity-feedback, because $H(s) = 1$,

$$v(t) = w(t) = \mathcal{L}^{-1}(G_p(s)/s)(t).$$

In our case, $v(t) = w(t) = tu(t)$. Thus, we find that:

$$V(z) = W(z) = \frac{Tz}{(z-1)^2}.$$

This shows that:

$$V_o(z) = \frac{\frac{Tz}{(z-1)^2}\frac{z-1}{z}}{1 + \frac{Tz}{(z-1)^2}\frac{z-1}{z}} V_{in}(z) = \frac{T}{z-1+T} V_{in}(z). \qquad (10.2)$$

The pulse transfer function, $T(z)$, is:

$$T(z) = \frac{V_o(z)}{V_{in}(z)} = \frac{T}{z-1+T}.$$

Note that when $T = 1$ we find that $T(z) = z^{-1}$. That is, when $T = 1$ the output of the system at sampling times is just the system's input one sample ago. (Can you explain why this is so?)

10.11 Using MATLAB® to Perform the Calculations

MATLAB has a set of commands that is designed to take care of the calculations that one must perform to calculate a pulse transfer function or to analyze a discrete-time or hybrid system. Let us consider the system of the previous section in which $G_p(s) = 1/s$ and $H(s) = 1$. We found that with these definition the pulse transfer function of the system of Figure 10.5 is:

$$T(z) = \frac{\frac{z-1}{z}\mathcal{Z}(1/s^2)(t)}{1 + \frac{z-1}{z}\mathcal{Z}(1/s^2)(t)}.$$

Let us first define the pulse transfer function $(z-1)/z$. For this purpose, one uses the command `S_H = tf(a, b, Ts)`. (`S_H` is the name of the transfer function object that MATLAB creates.) The array `a` defines the numerator of the transfer function. In our case it is `[1 -1]`. (The polynomials here are represented just as they are in the continuous-time case.) The array `b` defines the denominator of the transfer function. In our case it is

`[1 0]`. Finally, `Ts` is the sample time. If one uses -1 for the sample time, then MATLAB registers the system as discrete-time but it does not pick a sample time for the system. (If no `Ts` is given, then the system is taken to be a continuous-time system.)

Next, we take the Laplace transform $1/s^2$ and "convert" it into a z-transform. MATLAB has a command—`c2d`—whose job it is to perform such *c*ontinuous-time *to* *d*iscrete-time conversions. If we define the transfer function:

```
V = tf([1],[1 0 0])
```

MATLAB will provide us with an object whose "value" is $1/s^2$. To convert this object into a z-transform one uses the command `Gd = c2d(V, Ts, 'imp')`. The first argument of the command is the object one wants transformed. The second argument is the sample time that is desired, and the third argument is the way that MATLAB is being requested to perform the transformation. As we want MATLAB to take samples of the *imp*ulse response, we use the argument `'imp'`. Finally, to calculate $T(z)$ one uses the command:

```
T = S_H * V /(1 + S_H * V)
```

If one gives the following sequence of instruction to MATLAB:

```
V = tf([1],[1 0 0]);
Ts = 1;
S_H = tf([1 -1],[1 0],Ts)
Gd = c2d(V,Ts,'imp')
Td = S_H * Gd / (1 + S_H * Gd);
step(Td,[0:Ts:20]);
```

MATLAB responds with:

```
Transfer function:
z - 1
-----
  z

Sampling time: 1
```

```
Transfer function:
z + 2.719e-016
--------------
z^2 - 2 z + 1

Sampling time: 1
```

and with Figure 10.6 in which the step function has been delayed by $1s$ as predicted in §10.10. Note that the transfer function that corresponds to $V(z)$ is not quite correct. The $+2.719 \times 10^{-16}$ should not be present—it is an artifact of the way MATLAB does the conversion from continuous-time to discrete-time.

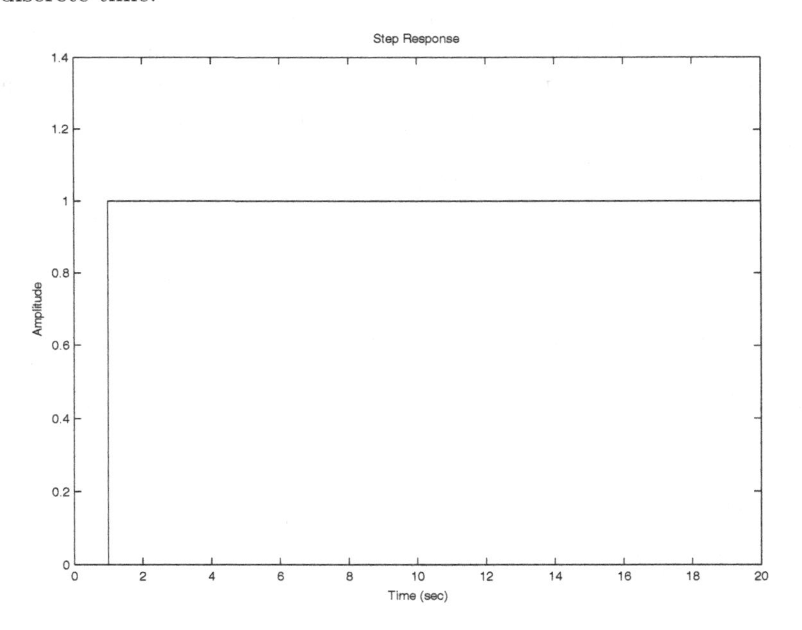

Fig. 10.6 The Step Response of the Hybrid System

10.12 The Transfer Function of a Discrete-Time System

The transfer function of a linear, time-invariant, discrete time system is defined as:

$$T(z) \equiv \frac{V_{out}(z)}{V_{in}(z)}.$$

Let us consider what we can learn from $T(z)$. First of all suppose that the input to the system is 1 when $t = 0$ and zero at all other sample times. Then the input sequence is $\{1, 0, 0, \ldots\}$. (This sequence is sometimes called the *unit pulse.*) The z-transform of this sequence is $V_{in}(z) = 1$. The z-transform of the output of the system with such an input is just $V_{out}(z) = T(z) \cdot 1 = T(z)$. The output of the system with such input will be $t(k) = \mathcal{Z}^{-1}(T(z))(k)$.

Now suppose that we have found that when a unit pulse is input to a system the system's output is $t(k)$ and the z-transform of the output is $T(z)$. Let us consider the output of the system to a generic input $\{a_0, a_1, \ldots, a_n, \ldots\}$. By linearity, we know that the output to this input must be the same as the sum of the outputs of the signals:

$$a_0\{1, 0, 0, \ldots\}, a_1\{0, 1, 0, 0, \ldots\}, a_2\{0, 0, 1, 0, 0, \ldots\}, \cdots.$$

Let us consider the z-transform of the output to the signal. It is clear that the the output to the first term is just $a_0 T(z)$. The next signal is a constant times the first signal delayed by one. As the system is assumed to be linear and time invariant the output must be the same as the output to the unit pulse but scaled and time shifted by one. That is, the output is $a_1 z^{-1} T(z)$. Continuing this way, we find that the z-transform of the output to the signal is:

$$\begin{aligned} V_{out}(z) &= a_0 T(z) + a_1 z^{-1} T(z) + a_2 z^{-2} T(z) + \cdots \\ &= T(z) \sum_{k=0}^{\infty} a_k z^{-k} \\ &= T(z) V_{in}(z). \end{aligned}$$

Thus the transfer function of the system is:

$$\frac{V_{out}(z)}{V_{in}(z)} = T(z).$$

We see that the z-transform of the response to a unit pulse is the (pulse) transfer function and the inverse z-transform of the (pulse) transfer function is the unit pulse response.

10.13 Adding a Digital Compensator

We are now able to consider the system of Figure 10.7—a system that has a digital compensator. Let us calculate the transfer function of such a system.

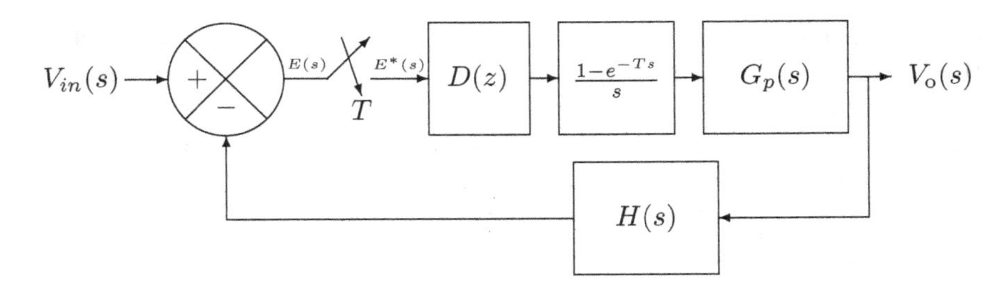

Fig. 10.7 A Simple Feedback System with a Digital Compensator

Before performing the calculation, we must understand *precisely* what the block whose transfer function is $D(z)$ is. $D(z)$ acts on samples of $e(t)$—of the "error." It is the transfer function of the digital compensator. As we saw in the previous section, the z-transform of the sequence at the output of the digital compensator is $D(z)E(z)$. If one would like the star-transform of the sequence rather than the z-transform (because one would like to think of the compensator as processing impulse trains rather than sequences), then one must let $z = e^{Ts}$. Thus, the star-transform of the output of the digital compensator is:

$$D(e^{Ts})\, E(z)|_{z=e^{Ts}} = D(e^{Ts})E^*(s).$$

Let us now calculate $E(s)$. We find that:

$$E(s) = V_{in}(s) - E^*(s)D(e^{Ts})\frac{1-e^{-Ts}}{s}G_p(s)H(s).$$

Let:

$$w(t) = \mathcal{L}^{-1}(G_p(s)H(s)/s)(t)$$

and let:

$$v(t) = \mathcal{L}^{-1}(G_p(s)/s)(t)$$

Then following the logic of §10.10, the transfer function of the system must be:

$$T(z) = \frac{D(z)(1-z^{-1})V(z)}{1 + D(z)(1-z^{-1})W(z)}.$$

10.14 Stability of Discrete-Time Systems

Given that a system's pulse transfer function is $T(z)$, under what conditions can we say that the system is stable? First, let us agree that when we say that a system is stable we mean that for any bounded input, the output of the system remains bounded. I.e. the system is bounded input-bounded output (BIBO) stable. Under what conditions is a system BIBO stable?

Recall that if one has a rational function—a function that can be expressed as the quotient of two polynomials—then the function can be written as a sum of polynomials each of which has a single (possibly complex) pole (that may be repeated several times). (See §1.4 for help with the partial fraction expansion.) Let us consider a generic term of this type and consider the conditions under which the associated sequence remains bounded.

Consider:

$$E(z) = \frac{z}{z-p}.$$

This corresponds to the sequence $e(k) = p^k, k \geq 0$. Clearly this remains bounded if and only if $|p| \leq 1$.

Now consider:

$$-z\frac{d}{dz}E(z) = -z\frac{z-p-z}{(z-p)^2} = \frac{pz}{(z-p)^2}$$

As the transform of $ke(k)$ is $-z$ times $E'(z)$, we find that if the pole in the denominator is of multiplicity two, then the inverse z-transform is a multiple of kp^k. Note that this is bounded as long as $|p| < 1$. If $|p| = 1$, this sequence is unbounded (though it tends to infinity linearly and not exponentially). For higher multiplicities one finds higher powers of k multiplying p^k. This changes nothing.

We conclude that the sequence associated with a term of the form:

$$\frac{z}{(z-p)^n}$$

remains bounded if:

(1) $|p| < 1$, or
(2) $|p| = 1$ and $n = 1$.

Now consider a system whose transfer function is $T(z)$. Under what conditions can we say that the system is stable? If $T(z)$ has any poles outside of the unit circle, then almost no matter what the input to the system is, the partial fractions expansion of the output will have a term with a pole outside of the unit circle and will tend to infinity exponentially fast. Thus, if *any poles* of $T(z)$ are strictly *outside the unit circle*, then the system is *wildly unstable.*

Next, consider a system with poles inside or on the unit circle. First let us suppose that the system has no poles that are greater than or equal to one in absolute value. That is, assume that all poles are strictly inside the unit circle. Then if the input is $R(z)$, the output will be $T(z)R(z)$. Assume that $R(z)$ is a rational function of z. In order for the input to be bounded all of the poles of $R(z)$ must be inside or on the unit circle, and if they are on the unit circle they must be simple poles. Under these conditions, we know that the partial fractions expansion of $T(z)R(z)$ will have terms with poles inside the unit circle that correspond to the poles of $T(z)$. It will have terms inside the unit circle that correspond to the poles of $R(z)$. It may have simple poles on the unit circle that correspond to simple poles of $R(z)$ that are on the unit circles. Since all such poles lead to bounded sequences, the output must be bounded. That is, *if all of the poles of* $T(z)$ *are located inside the unit circle, then the system is stable.*

Suppose now that $T(z)$ has a simple pole on the unit circle. Consider the following scenario—suppose that the input to the block has a simple pole at the same location. Then $T(z)R(z)$ will have a double pole on the unit circle. In the partial fractions expansion of $T(z)R(z)$ there will be a term with a double pole at that point on the unit circle. That will cause the output to run off to infinity in a "linear" fashion. That is, certain bounded inputs lead to unbounded outputs. Thus, even a simple pole on the unit circle leads to an unstable system. A system with a simple pole on the unit circle is said to be "marginally stable." It is almost, but not quite, stable.

A System for which $T(z) = T/(z-1)$—An Example

Suppose that one is given a system for which:

$$T(z) = \frac{T}{z-1}.$$

To demonstrate that this system is unstable, input a signal with a pole in the same location as the system's pole. One can pick as the input the unit step function. With this choice of input $V_{in}(z) = z/(z-1)$. The z-transform of the system's output is $V_{out}(z) = Tz/(z-1)^2$. This corresponds to $v_{out}(kT) = Tku(kT)$.

That is, the output is a sampled ramp function. We see that the output is unbounded even though the input is bounded.

10.15 A Condition for Stability

Much of the time the transfer functions we examine are of the form:

$$T(z) = \frac{P(z)}{Q(z)}$$

where $P(z)$ and $Q(z)$ are polynomials with positive coefficients. We state and prove a theorem that gives a necessary condition that such polynomials have all their zeros inside the unit disk.

We start by proving a related result. Consider the polynomial:

$$Q(z) = a_0 + a_1 z + a_2 z + \cdots a_N z^N, \qquad a_i \in \mathcal{R}.$$

Suppose that $a_0 > a_1 > a_2 > \cdots > a_N > 0$. We show that all of the zeros of $Q(z)$ are located *outside* the unit disk.

Consider the function $(1 - z)Q(z)$. Clearly:

$$(1 - z)Q(z) = a_0 + (a_1 - a_0)z + \cdots + (a_N - a_{N-1})z^N - a_N z^{N+1}. \quad (10.3)$$

From the triangle inequality, we know that $|a(z) + b(z)| \leq |a(z)| + |b(z)|$. Letting $a(z) = f(z) + g(z)$ and $b(z) = -g(z)$, we find that:

$$|f(z)| \leq |f(z) + g(z)| + |-g(z)| = |f(z) + g(z)| + |g(z)|.$$

Rearranging terms, we find that:

$$|f(z) + g(z)| \geq |f(z)| - |g(z)|.$$

Combining this result with (10.3), we find that:

$$\begin{aligned} |(1 - z)Q(z)| &\geq a_0 - |(a_1 - a_0)z + (a_2 - a_1)z^2 \\ &\qquad + \cdots + (a_N - a_{N-1})z^N - a_N z^{N+1}| \\ &\geq a_0 - ((a_0 - a_1)|z| + (a_1 - a_2)|z|^2 \\ &\qquad + \cdots + (a_{N-1} - a_N)|z|^N + a_N|z|^{N+1}). \end{aligned}$$

As we are interested in showing that the polynomial has no roots in the region $|z| \leq 1$, we consider z inside or on the unit circle. Because all of the coefficients in the final form are positive, it is clear that (for points inside or on the unit circle) we are subtracting as much as possible when $|z| = 1$. At this point we find that $|(1-z)Q(z)| \geq a_0 - a_0 = a_0 \geq 0$. For all $|z| < 1$ we find that $|(1-z)Q(z)| > 0$. Thus, $Q(z)$ has no zeros strictly inside the unit circle. If $|z| = 1$, the only way for equality to hold is if $z = 1$. (Because all of the coefficients of the terms being subtracted are positive, if $z \neq 1$ there is some cancellation. Because of the cancellation, $|(1-z)Q(z)| > 0$.) When $z = 1$ it is clear that $(1-z)Q(z) = 0$. However, $Q(1) = a_0 + \cdots + a_N > 0$. Thus $Q(z)$ has no zeros inside or on the unit circle.

Now consider a polynomial $Q(z)$ of the same form but for which $a_N > a_{N-1} > a_{N-2} > \cdots a_0 > 0$. Note that:

$$z^N Q(1/z) = a_0 z^N + \cdots + a_{N-1} z + a_N.$$

This new polynomial satisfies the conditions of our previous result. The new polynomial has all of its zeros outside the unit circle. (The multiplication by z^N does not add zeros at $z = 0$. It removes a removable singularity.) If $a \neq 0$ is a zero of $z^N Q(1/z)$, then $1/a$ is a zero of $Q(z)$. As every root of $Q(1/z)$ is located outside the unit circle, every root of $Q(z)$ is located inside the unit circle.

We have found that a *sufficient* condition for a polynomial $Q(z) = a_0 + a_1 z + a_2 z + \cdots a_N z^N$ to have all of its roots inside the unit circle is that $a_N > a_{N-1} > a_{N-2} > \cdots a_0 > 0$. This result is known as Enestrom's theorem.

A Quick Check Using MATLAB®—An Example

Let us consider the polynomial $Q(z) = z^4 + 0.9z^3 + 0.8z^2 + 0.7z + 0.6$. According to our theorem all of roots of this polynomial must be inside the unit circle. Additionally, let us consider $z^4 Q(1/z) = 0.6z^4 + 0.7z^3 + 0.8z^2 + 0.9z + 1$. The roots of this polynomial should be the reciprocals of the roots of the first polynomial; they should all lie outsize the unit circle. Using MATLAB we find that:

```
>> rts1 = roots([1 0.9 0.8 0.7 0.6])

rts1 =

  -0.7128 + 0.5149i
  -0.7128 - 0.5149i
   0.2628 + 0.8408i
   0.2628 - 0.8408i
```

```
>> abs(rts1)

ans =

    0.8793
    0.8793
    0.8809
    0.8809

>> rts2 = roots([0.6 0.7 0.8 0.9 1])

rts2 =

   0.3386 + 1.0835i
   0.3386 - 1.0835i
  -0.9219 + 0.6659i
  -0.9219 - 0.6659i

>> 1./rts2

ans =

   0.2628 - 0.8408i
   0.2628 + 0.8408i
  -0.7128 - 0.5149i
  -0.7128 + 0.5149i

>> abs(rts2)

ans =

    1.1351
    1.1351
    1.1373
    1.1373
```

(Note that `1./rts2` means divide 1 by each element of `rts2` and produce a new array from the results.) The results of the calculations performed by MATLAB bear out all of our assertions.

10.16 The Frequency Response

Suppose that the input to a sampled-data system is $r(t) = e^{j\omega t}$. Then the z-transform of the sampled input, $r(kT)$, is:

$$R(z) = \frac{z}{z - e^{j\omega T}}.$$

Assume that the transfer function of the system to which this signal is being input is $T(z)$ and that the system is stable—all of its poles are located inside the unit circle. Then the z-transform of the output sequence, $C(z)$, is:

$$\begin{aligned} C(z) &= T(z)\frac{z}{z - e^{j\omega T}} \\ &= \frac{Az}{z - e^{j\omega T}} + \text{terms with poles inside the unit circle.} \end{aligned}$$

Multiplying through by $z - e^{j\omega T}$, and then substituting $z = e^{j\omega T}$, we find that:

$$A = T(e^{j\omega T}).$$

Thus, we find that the output of the system is:

$$c(kT) = T(e^{j\omega T})e^{j\omega T} + \text{terms that decay.}$$

Now let us consider what happens when:

$$r(t) = \cos(\omega t) = \frac{e^{j\omega t} + e^{-j\omega t}}{2}.$$

As $-\omega$ is also an angular frequency, we can use our previous results and linearity to say that:

$$c(kT) = \frac{T(e^{j\omega T})e^{j\omega kT} + T(e^{-j\omega T})e^{-j\omega kT}}{2} + \text{terms that decay.}$$

Assuming that $T(z)$ is a rational function with real coefficients, then it is easy to show that $\overline{T(z)} = T(\overline{z})$. Thus, under this (not terribly restrictive) condition we find that:

$$c(kT) = \mathrm{Re}(T(e^{j\omega T})e^{j\omega kT}) + \text{terms that decay.}$$

Let us write:

$$T(e^{j\omega T}) = |T(e^{j\omega T})|e^{j\angle T(e^{j\omega T})}.$$

Then we find that:

$$c(kT) = |T(e^{j\omega T})|\cos(\omega kT + \angle T(e^{j\omega T})) + \text{terms that decay}.$$

Thus, if the input to a system is a cosine wave at a particular frequency, then the steady-state output of the system is a cosine wave at the same frequency shifted by the phase of $T(e^{j\omega T})$ and amplified by the magnitude of $T(e^{j\omega T})$. The function:

$$T(e^{j\omega T})$$

is called the *frequency response* of the discrete-time system. Note that the frequency response at D.C.—when $\omega = 0$—is just $T(1)$. Thus, the steady-state amplification seen by a D.C. signal is $T(1)$.

A Simple Low-Pass Filter—An Example

Consider a system whose (pulse) transfer function is:

$$T(z) = \frac{z^{-1} + 1}{2}.$$

Let the input sequence to the system be denoted by $x(k)$ and the sequence output from the system be denoted by $y(k)$. We find that:

$$\frac{Y(z)}{X(z)} = \frac{z^{-1} + 1}{2}.$$

Cross-multiplying and converting the resulting equation back to an equation for sequences, we find that:

$$y(k) = \frac{x(k) + x(k+1)}{2}.$$

That is, $y(k)$ is an average of the last two values of $x(k)$.

Now consider the frequency response of the filter. We find that:

$$\begin{aligned} T(e^{j\omega T}) &= \frac{1 + e^{-j\omega T}}{2} \\ &= e^{-j\omega T/2}\frac{e^{j\omega T/2} + e^{-j\omega T/2}}{2} \\ &= e^{-j\omega T/2}\cos(\omega T/2). \end{aligned}$$

Thus, the magnitude of the frequency response is:

$$|T(e^{j\omega T})| = |\cos(\omega T/2)|.$$

Clearly this function has a maximum at $\omega = 0$ and decreases to zero at $\omega = \pi/T$. In this region the system is clearly low-pass. In the next section we discuss why only the angular frequencies between 0 and π/T are of interest to us.

10.17 A Bit about Aliasing

There is something a bit odd about the frequency response, $T(e^{j\omega T})$—it is periodic with period $2\pi/T$. That means that high enough frequencies are affected by our system in the same way low frequencies are.

Actually this behavior is not surprising at all. Suppose that the input to the system is $r(t) = \sin((\omega + 2\pi/T)t)$. The samples of this sine wave are:

$$r(kT) = \sin((\omega + 2\pi/T)kT) = \sin(\omega kT + 2\pi k) = \sin(\omega kT).$$

That is, the samples here are the same as the samples of $\sin(\omega t)$—a lower frequency signal. This is an example of *aliasing*.

As a rule, if one does not sample a sine wave fast enough, one cannot determine the frequency of the sine wave from which the samples were taken. Because we did not sample the sine wave fast enough, the samples that we did take looked like the samples of sine wave at a lower frequency. It can be shown that the frequency beyond which aliasing occurs, the Nyquist[3] frequency, is:

$$\omega_{Nyquist} \equiv \frac{\pi}{T}, \quad f_{Nyquist} \equiv \frac{1}{2T}.$$

This effect is also seen in the frequency response. The frequency response is $2\pi/T$ periodic, but in the range $[-\pi/T, \pi/T)$ half of the frequencies are negative and, physically, are essentially the same as the positive frequencies (for real inputs). The frequency range of interest is $[0, \pi/T)$.

10.18 The Behavior of the System in the Steady-State

Consider the system given in Figure 10.5. If the input to the system is a constant and the output tends to a constant, then the input to the sample and hold element is tending to a constant too. As a sample and hold circuit with a constant input does not change its input at all, the output of the sample and hold circuit is precisely its input—and the sample and hold

[3] Once again, Harry Nyquist of the eponymous plot.

circuit can be ignored. Thus, the gain in the steady state can be calculated by removing the sample and hold elements and then calculating the steady state gain of the remaining analog system.

As an example let us consider the system discussed at the end of §10.10. In (10.2) we found that the transfer function of a feedback system composed of a sample and hold element and an integrator is:

$$T(z) = \frac{T}{z - 1 + T}.$$

It is easily seen that the system without the sample-and-hold element tracks a DC signal perfectly. (To see this, either calculate the transfer function and plug in $s = 0$, or note that the integrator has infinite gain at DC.) Plugging $z = 1$ into the transfer function, we find that $T(1) = 1$. That is, the discrete-time system tracks DC perfectly too—as it should.

10.19 The Bilinear Transform

In §10.15 we showed that a particular class of systems is stable. In §10.14 we determined a necessary and sufficient condition for stability—that the poles of the system lie strictly inside the unit circle. Now we need to develop a general technique that will allow us to determine whether or not the poles of a system all lie inside the unit circle. We *choose* to transform the whole question into a question that we know how to answer.

Suppose that we had a mapping $z = B(w)$ that mapped every point in the right half of the w plane in a one-to-one fashion onto the region outside the unit circle; that mapped every point in the left half of the w plane in a one-to-one fashion onto the region inside the unit circle; and that mapped the imaginary axis of the w plane onto the unit circle of the z plane in a one-to-one fashion.

Suppose that we have a function $F(z)$ and we need to check whether its poles are insider or outside of the unit circle. If we consider $F(B(w))$ instead, the question becomes whether or not the poles of the composite function lie in the left half-plane. The advantage of this question is that it is one that we have already developed many techniques for answering.

Let us consider the mapping:

$$z = B(w) \equiv \frac{2 + Tw}{2 - Tw}.$$

This function maps every point in the w plane except the point $w = 2/T$

into a point in the finite z plane. The point $w = 2/T$ is mapped to $z = \infty$. With this definition of $B(w)$ we find that w as a function of z is:

$$\begin{aligned} z &= \frac{2+Tw}{2-Tw} \\ &\Leftrightarrow 2z - zTw = 2 + Tw \\ &\Leftrightarrow 2z - 2 = Tw + zTw \\ &\Leftrightarrow w = \frac{2}{T}\frac{z-1}{z+1} \\ &= B^{-1}(z). \end{aligned}$$

As there is an inverse for every value except for $z = -1$, we find that $B(w)$ must be a one-to-one mapping onto the z-plane less the point -1. Considering $B(w)$, we find that $B(w)$ maps ∞ to -1. If one considers ∞ to be a single point, then the mapping is indeed one-to-one and onto.

To check that this mapping maps the regions correctly is easy. First of all, if $w = j\omega_w$, then:

$$|z| = \left|\frac{2+Tj\omega_w}{2-Tj\omega_w}\right| = \sqrt{\frac{2+T^2\omega_w^2}{2+T^2\omega_w^2}} = 1.$$

That is, the mapping takes the imaginary axis in the w-plane into the unit circle of the z-plane.

Now let $w = a + jb$. Then we find that:

$$|z| = \left|\frac{2+Ta+Tjb}{2-Ta+Tjb}\right| = \sqrt{\frac{(2+Ta)^2+(Tb)^2}{(2-Ta)^2+(Tb)^2}}.$$

If $a > 0$, then $|z| > 1$. Similarly, if $a < 0$, we find that $|z| < 1$. Thus, we find that the mapping is just the type of mapping we were looking for.

This particular mapping has another nice property. We have seen that the value of the transfer function at $z = e^{j\omega T}$ determines the gain and phase shift that the system imparts to a sine wave input at the frequency ω_z. Let us consider what value corresponds to $z = e^{j\omega_z T}$ under the bilinear transform. We already know that all points on the unit circle in the z-plane are mapped into points on the imaginary axis in the w-plane. Thus, we can call the point into which $z = e^{j\omega_z T}$ is mapped $w = j\omega_w$. We find that:

$$\begin{aligned} j\omega_w &= \frac{2}{T}\frac{z-1}{z+1} \\ &= \frac{2}{T}\frac{e^{j\omega_z T}-1}{e^{j\omega_z T}+1} \\ &= \frac{2}{T}\frac{e^{j\omega_z T/2}-e^{-j\omega_z T/2}}{e^{j\omega_z T/2}+e^{-j\omega_z T/2}} \\ &= j\frac{2}{T}\tan(\omega_z T/2). \end{aligned}$$

Recall that ω_z is the actual frequency under consideration. Currently $j\omega_w$ is just the point on the imaginary axis to which $e^{j\omega_z}$ is mapped under the bilinear transform.

Now suppose that $|\omega_z T|/2 << 1$. Then, $\tan(\omega_z T/2) \approx \omega_z T/2$. In that case, we find that:

$$j\omega_w \approx j\omega_z.$$

That is, under the bilinear transformation angular frequencies are mapped almost correctly to the appropriate point on the imaginary axis in the w-plane. The correspondence is perfect when $\omega_z = 0$, and it is very good when $\omega_z T$ is small. For very large values, it is not good.

A final and very major advantage of this mapping is that it takes rational functions of z into rational functions of w. As most transfer functions that we are interested in are rational functions of z, most of the transformed functions are rational functions of w—and we have many tools for examining such functions.

Let us consider several examples.

(1) In the system of Figure 10.7, let $D(z) = 1$, $H(s) = 1$, and $G_p(s) = 1/s$. As we saw in (10.2), the transfer function of the system is:

$$T(z) = \frac{\frac{T}{z-1}}{1+\frac{T}{z-1}}.$$

Let us use the bilinear transform to convert this to an analog-style function. Making use of the definition of the bilinear transform, we find that:

$$z - 1 \Rightarrow \frac{2+Tw}{2-Tw} - \frac{2-Tw}{2-Tw} = \frac{2Tw}{2-Tw}$$

Thus, we find that the bilinear transform of $T(z)$ is:

$$T(w) = \frac{\frac{2-Tw}{2w}}{1+\frac{2-Tw}{2w}}.$$

Note that the function that plays the role that $G_p(s)$ would play in an analog system—$\frac{2-Tw}{2w}$ has an interesting property. When one lets $T \to 0$, one finds that this function tends to $1/w$. That is it tends to $G_p(s)$ with s replaced by w. That is, the bilinear transform of the z-transform of the analog portion of the system (with the hold circuit) tends to the transfer function of the analog element as $T \to 0$.

(2) Consider the same problem, but with $H(s) = 1$ and:

$$G_p(s) = \frac{s}{s+5}.$$

We find that the inverse Laplace transform of $G_p(s)/s$ is just:

$$v(t) = e^{-5t}u(t).$$

Thus:

$$V(z) = \frac{z}{z - e^{-5T}}.$$

We find that the transfer function of the system is:

$$\begin{aligned} T(z) &= \frac{(1-z^{-1})V(z)}{1+(1-z^{-1})V(z)} \\ &= \frac{\frac{z-1}{z-e^{-5T}}}{1+\frac{z-1}{z-e^{-5T}}} \\ &= \frac{z-1}{z-1+z-e^{-5T}}. \end{aligned}$$

Let us consider the bilinear transform of this function. We find that:

$$z - 1 = \frac{2+Tw}{2-Tw} - \frac{2-Tw}{2-Tw} = \frac{2Tw}{2-Tw}.$$

Also,

$$z - e^{-5T} = \frac{2+Tw}{2-Tw} - e^{-5T}\frac{2-Tw}{2-Tw} = \frac{2(1-e^{-5T}) + Tw(1+e^{-5T})}{2-Tw}.$$

Thus,

$$T(w) = \frac{2Tw}{2(1-e^{-5T}) + Tw(3+e^{-5T})}.$$

Let us consider the value of this function as $T \to 0$. We find that it tends to:

$$T(w) \to \frac{2Tw}{2(5T) + Tw(3+1)} = \frac{w}{5+2w}.$$

Note that in the original system if the sample and hold element is removed, then $T(s) = s/(2s+5)$. Thus, by taking T to zero, we have "recovered" the analog system.

Using MATLAB®—An Example

MATLAB can calculate bilinear transforms as well. The command `d2c` converts *d*iscrete-time transfer functions *to* *c*ontinuous-time transfer functions. If one would like to use `d2c` to transform a discrete-time function to a continuous-time function, the syntax for the command is `d2c(G,'tustin')`. Here `G` is the discrete-time transfer function that is to be transformed and `'tustin'` tells MATLAB that it is to use the bilinear transform (also known as the Tustin transform) to transform the continuous-time function. If, for example, one gives MATLAB the commands:

```
Ts = 0.001
Gdisc = tf([1 -1],[2 -(1+exp(-5*Ts))],Ts)
Gcont = d2c(Gdisc, 'tustin')
```

MATLAB replies with:

```
Ts =

    0.0010

Transfer function:
   z - 1
-----------
2 z - 1.995

Sampling time: 0.001

Transfer function:
0.5006 s
---------
s + 2.497
```

Note how close the final transfer function is to $s/(2s+5)$.

Let us analyze the stability of the system. We have found that:

$$(1-z^{-1})V(z) = \frac{z-1}{z-e^{-5T}}.$$

The bilinear transform of this element—our "$G_p(z)$"—is:

$$G_p(w) = \frac{2Tw}{2(1-e^{-5T}) + Tw(1+w^{-5T})}.$$

Clearly:

$$T(w) = \frac{G_p(w)}{1+G_p(w)}.$$

As the root locus diagram that corresponds to $G_p(w)$ has one branch that travels along the real axis from the single negative pole to the zero at the origin, the system is stable no matter how much gain is inserted. In particular, it is stable for a gain of one. Thus, the system is stable. However, its gain at DC is (plugging in $w=0$) zero. This could well be a problem.

10.20 The Behavior of the Bilinear Transform as $T \to 0$.

We consider the behavior of the bilinear transform of the z-transform of:

$$f(t) = \mathcal{L}^{-1}\left(\frac{1-e^{-Ts}}{s}G(s)\right)(t) = v(t) - v(t-T), \qquad V(s) = \frac{G(s)}{s}.$$

Let us consider $f(t)$. As e^{-Ts} produces a delay of T seconds and multiplication by $1/s$ is integration in time, we find that:

$$f(t) = \int_0^t g(y)\,dy - u(t-T)\int_0^{t-T} g(y)\,dy = \int_{t-T}^t g(y)\,dy, t \geq T.$$

If $g(t)$ is continuous $f(0)=0$. For $t \geq T$, as long as T is small enough and $g(t)$ is differentiable[4] we find that:

$$f(t) = Tg(t) + O(T^2).$$

[4]Differentiability is not actually necessary, but it makes the truth of the conclusion easier to prove.

Thus, the z-transform of $f(t)$ is:

$$\begin{aligned}F(z) &= \sum_{k=0}^{\infty} f(kT)z^{-k}\\ &= \sum_{k=1}^{\infty} \left(g(kT)z^{-k}T + O(T^2)z^{-k}\right)\\ &= \sum_{k=1}^{\infty} g(kT)z^{-k}T + O(T^2)O\left(\frac{z}{z-1}\right).\end{aligned}$$

The bilinear transform of $F(z)$ is:

$$F(w) = \sum_{k=1}^{\infty} g(kT)\left(\frac{2+Tw}{2-Tw}\right)^{-k} T + O(T^2)O(1/(2Tw/(2-Tw))).$$

Note that:

$$\frac{2+Tw}{2-Tw} = \frac{1+(T/2)w}{1-(T/2)w} \approx \frac{e^{(T/2)w}}{e^{-(T/2)w}} = e^{Tw}, \quad |Tw| << 1.$$

Proceeding formally[5], we find that as $T \to 0$:

$$\begin{aligned}F(w) &= \sum_{k=1}^{\infty} g(kT)\left(\frac{2+Tw}{2-Tw}\right)^{-k} T + O(T)\\ &\to \sum_{k=1}^{\infty} g(kT)e^{-wkT}T + O(T)\\ &\to \int_0^{\infty} g(t)e^{-wt}\,dt = G(w).\end{aligned}$$

That is, the bilinear transform tends to the Laplace transform of the analog element from which it is derived—just as we saw previously.

10.21 Digital Compensators

We now show how using the bilinear transform the *design* of a digital compensator can be reduced to the design of an analog compensator—a task we have already mastered—and a few mechanical steps. Let us consider an example.

[5]That is, without making all the estimates and listing all the conditions that are necessary to properly prove the statement.

Suppose that in Figure 10.7 we let $H(s) = 1$ and $G_p(s) = 1/(s+1)$. Then:

$$\begin{aligned} V(z) &= \mathcal{Z}\left(\frac{1}{s(s+1)}\right)(z) \\ &= \mathcal{Z}((1-e^{-t})u(t))(z) \\ &= \frac{z}{z-1} - \frac{z}{z-e^{-T}} \\ &= \frac{(1-e^{-T})z}{(z-1)(z-e^{-T})}. \end{aligned}$$

As:

$$(1-z^{-1})V(z) = \frac{z-1}{z}V(z) = \frac{1-e^{-T}}{z-e^{-T}},$$

we find that:

$$T(z) = \frac{D(z)\frac{1-e^{-T}}{z-e^{-T}}}{1+D(z)\frac{1-e^{-T}}{z-e^{-T}}}.$$

A simple check that we have not made any silly mistakes yet is to check the gain at DC when $D(z) = 1$—when there is no compensator. As we have already seen, the gain at DC must be the same both in the circuit with the sample-and-hold element and in the purely analog circuit. In the purely analog circuit it is easy to see that the gain at DC is $1/(1+1) = 1/2$. In the digital case, let $z = 1$. We find that here too $T(1) = 1/2$.

Suppose that we want to compensate the system in such a way that the gain of the system at DC will be 1. We use the bilinear transform to give us functions of w to which we relate as though the functions represent analog components.

Let:

$$z = \frac{2+Tw}{2-Tw}.$$

Then we find that:

$$z - e^{-T} = \frac{2+Tw}{2-Tw} - e^{-T}\frac{2-Tw}{2-Tw} = \frac{2(1-e^{-T}) + Tw(1+e^{-T})}{2-Tw}.$$

From this we see that the expression $(1-e^{-T})/(z-e^{-T})$—an expression which plays the role of the given plant—goes over to

$$G_p(w) \equiv \frac{(1-e^{-T})(2-Tw)}{2(1-e^{-T}) + Tw(1+e^{-T})}.$$

Thus we find that:

$$T(w) = \frac{D(w)G_p(w)}{1 + D(w)G_p(w)}.$$

Note that if $D(w) = 1$—if there is no compensator—then:

$$T(0) = \frac{D(0)G_p(0)}{1 + D(0)G_p(0)} = \frac{1 \cdot 1}{1 + 1 \cdot 1} = 1/2.$$

This is exactly what we saw when we used the z-transform.

How can we compensate this system so that the gain at DC—when $w = 0$—will be 1? As $G_p(0) = 1$, we find that the gain at DC is just:

$$T(0) = \frac{D(0)}{1 + D(0)}.$$

The only way that this can be 1 is if $D(0) = \infty$. The simplest function for which this holds is $D(w) = K_I/w$. This compensator is, in some sense, an integrator. Let us check the stability of our system with this compensator.

With $D(w) = K_I/w$, we find that the denominator of $T(w)$ is:

$$\begin{aligned}
\text{denominator} &= 1 + \frac{K_I}{w}\frac{(1 - e^{-T})(2 - Tw)}{2(1 - e^{-T}) + Tw(1 + e^{-T})} \\
&= \frac{w(2(1 - e^{-T}) + Tw(1 + e^{-T})) + K_I(1 - e^{-T})(2 - Tw)}{2(1 - e^{-T}) + Tw(1 + e^{-T})}.
\end{aligned}$$

The zeros of the denominator—the poles of the system—are the solutions of:

$$w(2(1 - e^{-T}) + Tw(1 + e^{-T})) + K_I(1 - e^{-T})(2 - Tw) = 0$$

Rewriting this, we find that we are looking for the solutions of:

$$T(1 - e^{-T})w^2 + \left(2(1 - e^{-T}) - K_I T(1 - e^{-T})\right) w + K_I(1 - e^{-T}) = 0.$$

It is well known (and easy to show using the Routh-Hurwitz theorem—see Problem 8a of §4.4) that the condition for all of the roots of a quadratic to be in the left half-plane is that all of the coefficients have the same sign. As the coefficient of w^2 is positive the other coefficients must be positive too. Thus, $K_I > 0$, and $TK_I < 2$. These then are the conditions for stability.

Let us now convert our compensator back into a function of z. As:

$$w = \frac{2}{T}\frac{z - 1}{z + 1},$$

we find that:

$$D(z) = \frac{T}{2}\frac{z+1}{z-1}.$$

What recursion relation do the input, $x(k)$, and the output, $y(k)$, satisfy? By the definition of the transfer function,

$$\frac{Y(z)}{X(z)} = \frac{T}{2}\frac{z+1}{z-1} = \frac{T}{2}\frac{1+z^{-1}}{1-z^{-1}}.$$

Cross-multiplying, we find that:

$$\frac{Y(z) - z^{-1}Y(z)}{T} = \frac{X(z) + z^{-1}X(z)}{2}.$$

Inverse transforming we find that:

$$\frac{y(k) - y(k-1)}{T} = \frac{x(k) + x(k-1)}{2}.$$

If T is small, then this looks very much like "the derivative of y is equal to x." That is, y is the integral of x—just as we planned in the w-plane.

If one would like to write this equation in a way that is reasonably simple to implement, one writes:

$$y(k) = y(k-1) + (T/2)x(k) + (T/2)x(k-1).$$

(This equation can be implemented in a couple of lines of code on a simple microcontroller.)

10.22 When Is There No Pulse Transfer Function?

Consider a linear time invariant system that has a sample-and-hold element in it. Under what conditions will there *not* be a pulse transfer function? The answer to this question is essentially simple. There will be no pulse transfer function when the output of the system at sampling intervals is not only a function of the input at sampling intervals but also of the input between sampling intervals.

Consider the system of Figure 10.8. Let us show that this system has no transfer function. Let:

$$r(t) = \begin{cases} 0 \; t < T/4 \\ 1 \; T/4 \leq t < 3T/4 \\ 0 \; t \geq 3T/4. \end{cases}$$

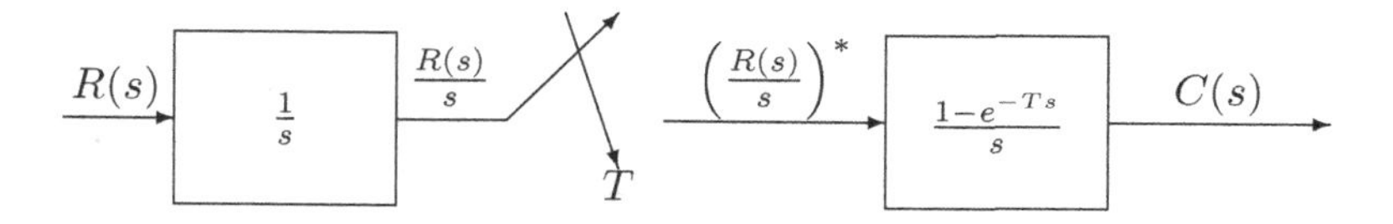

Fig. 10.8 A System Without a Transfer Function

Note that the z-transform of the input—the z-transform of $r(kT)$—is zero. The input is always zero at the sampling instants. Let us consider what the output of the system is. Integrating this pulse gives us the function:

$$\int_0^t r(y)\,dy = \begin{cases} 0 & t < T/4 \\ (t - T/4) & T/4 \leq t < 3T/4 \\ T/2 & t \geq 3T/4 \end{cases}.$$

Thus the output of the system at sample times is $\{0, T/2, T/2, \ldots\}$. The z-transform of the output is:

$$\frac{T}{2}\frac{1}{z-1}.$$

Now suppose that the system has a transfer function, $T(z)$. Then as the z-transform of the input is 0, the z-transform of the output must be:

$$V_{out}(z) = T(z) \cdot 0 = 0.$$

But the z-transform of the output is not zero. Thus the system has no transfer function. The *reason* that there is no transfer function is that the input to the sample-and-hold element is not dependent on the values of the input only at the sample instants. The input to the sample-and-hold is the integral of the input to the system. Hence it depends on the input to the system at all times. Thus there can be no pulse transfer function.

10.23 An Introduction to The Modified Z-Transform

Until now we have assumed that in a system with a sample-and-hold circuit one cannot calculate the value of the output between sampling times. This is not true. By working a little harder than one does to find out the value of the output at sampling times, one can find the value between sampling times too.

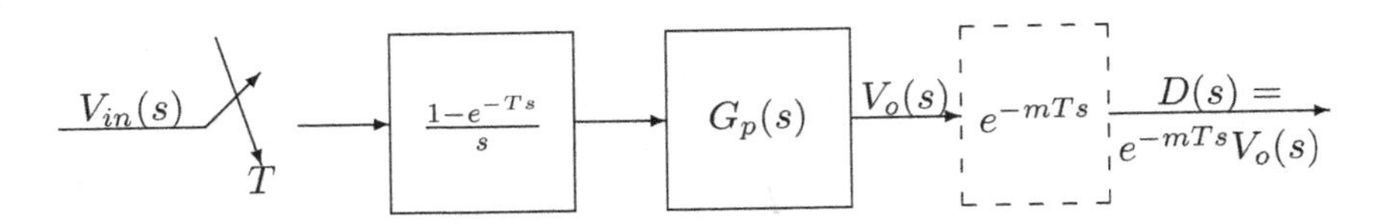

Fig. 10.9 A Generic System for Use with the Modified Z-transform

There are several ways of seeing what one can do. One way is to consider the system of Figure 10.9. In this figure the final element is a "virtual element" that delays the output by mT seconds. Note that we require that $0 < m < 1$. The delay is *always* a fraction of one sampling period. The actual output of the system is $V_o(s)$. However, the output of the virtual element is $e^{-mTs}V_o(s)$. The inverse z-transform of samples of the output of this virtual element is:

$$v_o(kT - mT).$$

As m is between 0 and 1, these samples are samples of the output *between* sampling intervals.

Let us consider an example. Let $G_p(s) = 1/s$ and let $V_{in}(s) = 1/s$. We proceed to analyze the system including the virtual element. We find that the transfer function of the system is:

$$(1 - z^{-1})\mathcal{Z}(e^{-mTs}/s^2)(z).$$

The inverse Laplace transform of $1/s^2$ is $tu(t)$. The time shifted version of this function is $(t-mT)u(t-mT)$. This equals $tu(t-mT) - mTu(t-mT)$. Sampling at $t = kT$, we find that the samples are:

$$kTu(kT - mT) - mTu(kT - mT).$$

Clearly the z-transform of the samples is just:

$$\mathcal{Z}(e^{-mTs}/s^2)(z) = \frac{Tz}{(z-1)^2} - \frac{mT}{z-1}.$$

Now let us consider the samples of the output of the system when the input to the system is the unit step function. Clearly, $V_{in}(z) = z/(z-1)$.

Thus,

$$D(z) = T(z)V_{in}(z) = \frac{z-1}{z}\left(\frac{Tz}{(z-1)^2} - \frac{mT}{z-1}\right)\frac{z}{z-1} = \frac{Tz}{(z-1)^2} - \frac{mT}{z-1}.$$

The inverse z-transform of this function is:

$$d(kT) = kTu(kT) - mTu((k-1)T).$$

As at zero the k in the first term forces the term to zero, this function can be written:

$$d(kT) = v_o(kT - mT) = (k - mT)u((k-1)T).$$

Let us check to see if our answer is correct. If one passes a unit step through a sample-and-hold element, the unit step is output by the sample-and-hold without any changes. Thus, the input to the integrator is a unit step function. The output of the integrator is $tu(t)$. The output of the delay is $(t - mT)u(t - mT)$. After sampling the output, we find that $d(kT) = (kT - mT)u(kT - mT)$. Since we are only interested in integer values of k this is equal to:

$$d(kT) = (kT - mT)u((k-1)T).$$

This is precisely the answer we got using the pulse transfer function.

To treat systems with feedback, one must add a virtual delay in the forward path and a virtual "predictor" (something that sees into the future—whose transfer function is e^{+mTs}) in the feedback. The "predictor" precisely cancels the effects of the delay. This allows one to calculate the output of a system with feedback at non-sample times.

The z-transforms that one calculates when e^{-mTs} is part of the transfer function are called modified z-transforms. Tables of these transforms are available. (See [Jur58], for example.)

10.24 Exercises

(1) Find the z-transform of the sequence whose elements are:

(a) $e(k) = \cos(\omega k)$.
(b) $e(k) = k^2$.
(c) $e(k) = k\sin(\omega k)$.

(2) Can one use the final value theorem to calculate:

$$\lim_{k\to\infty} e(k)$$

when $e(k) = \sin(k)$? Why or why not?

(3) Use the initial value theorem to show that if:

$$E(z) = \frac{z}{z-p},$$

then $e(0) = 1$, while if:

$$E(z) = \frac{1}{z-p},$$

then $e(0) = 0$.

(4) Making use of translation property and the z-transform of the sequence $e(k) = a^k$, find the inverse z-transform of:

$$\frac{1}{z-p}.$$

How does this explain the results of the previous question?

(5) Find the star-transform of the sequence whose elements are $e(kT) = e^{-kT}$.

(6) Let:

$$T(z) = \frac{z}{z^2+1}.$$

(a) Show that a block with transfer function $T(z)$ is marginally stable.

(b) Produce a bounded input for which the output of the block is unbounded.

(c) Calculate the output of the block for the input given in the previous section.

(7) Let:

$$T(z) = \frac{z}{z-a}, \qquad 0 < a < 1.$$

(a) What is the frequency response of the block?

(b) What kind of filter does this block represent?

(8) In Figure 10.5 let $G_p(s) = 1/(s+1)$ and let $H(s) = 1$.

(a) Find $T(z)$.

(b) Find $T(w)$.

(c) Show that as $T \to 0$, $T(w)$ approaches the transfer function of the system without the sample-and-hold circuit.

(9) In Figure 10.7 let $G_p(s) = 1/(s+1)$ and $H(s) = 1$.

(a) Design a compensator, $D(w)$, that will cause the system to track a DC input in the steady-state.
(b) Show that the system with the compensator is stable.
(c) Find the recurrence relation satisfied by the input and the output of the compensator.

(10) In Figure 10.5 let $G_p(s) = 1/s^2$ and let $H(s) = 1$.

(a) Find $G_p(z) = (1 - z^{-1})V(z)$.
(b) Find $G_p(w)$.
(c) Use the Routh-Hurwitz criterion to show that the system is unstable for all $T > 0$.

(11) In the system of Figure 10.8 replace the integrator block (the block whose transfer function is $1/s$) with a differentiator block (a block whose transfer function is s). Show that the new system has no pulse transfer function. Explain why this is so.
(12) Find the z-transform of $u(t) - u(t - T)$.
(13) In Figure 10.9 let $T = 0.1$ and $G_p(s) = 1/(s+1)$. Let $v_{in}(t) = u(t) - u(t-T)$. Find $v_o(kT - mT)$.
(14) In Figure 10.9 let $T = 0.1$ and $G_p(s) = 1/(s+1)$. Let $v_{in}(t) = u(t)$. Find $v_o(kT - mT)$.
(15) Show that if $E(1)$ is well defined, then:

$$E(1) = \sum_{k=0}^{\infty} e(k).$$

(16) (a) Please calculate the pulse transfer function of the system of Figure 10.5 when $G_p(s) = s/(s+1)$ and $H(s) = 1$.
(b) What is $v_o(kT)$ when the input to the system is $v_{in}(t) = u(t)$?
(17) (a) Please calculate the pulse transfer function of the system of Figure 10.7 when $G_p(s) = 1/s$, $H(s) = 1$, and $D(z) = (1 - z^{-1})/T$.
(b) What is $v_o(kT)$ when the input to the system is $v_{in}(t) = u(t)$?
(18) In Figure 10.7 let $G_p(s) = 1/(s+3)$ and $H(s) = 1$.

(a) Design a compensator, $D(w)$, that will cause the system to track a DC input in the steady-state.
(b) Determine the values of T for which the system with the compensator is stable.

(c) Find recurrence relation satisfied by the input and the output of the compensator.

(19) In Figure 10.7 let $G_p(s) = 1/s^2$, and let $H(s) = 1$. Use a PD compensator in the w-plane to stabilize the system.

(a) Determine the range of values of K_P and K_D that stabilize the system. Note that the range of values of K_P and K_D that stabilizes the system may be a function of T.

(b) If $x(k)$ is the input to the compensator and $y(k)$ is its output, what equation do $x(k)$ and $y(k)$ satisfy?

Chapter 11

Answers to Selected Exercises

11.1 Chapter 1

11.1.1 *Exercise 1*

1.a. On page 5 we found that:

$$\mathcal{L}(\cos(t))(s) = \frac{s}{s^2+1}.$$

Using the dilation property of the Laplace transform, we find that:

$$\mathcal{L}(\cos(\omega t))(s) = \frac{1}{\omega}\frac{s/\omega}{(s/\omega)^2+1} = \frac{s}{s^2+\omega^2}.$$

1.b. We find that:

$$\begin{aligned}
\mathcal{L}(te^{-t}\cos(t))(s) &= \mathcal{L}(t\cos(t))(s+1) \\
&= -\frac{d}{ds}\mathcal{L}(\cos(t))(s)\bigg|_{s\to s+1} \\
&= -\frac{d}{ds}\frac{s}{s^2+1}\bigg|_{s\to s+1} \\
&= \frac{s^2-1}{(s^2+1)^2}\bigg|_{s\to s+1} \\
&= \frac{(s+1)^2-1}{((s+1)^2+1)^2}.
\end{aligned}$$

1.c. We find that:

$$\begin{aligned}\mathcal{L}(t^2e^{-t})(s) &= \mathcal{L}(t^2)(s+1)\\ &= \left.\frac{2}{s^3}\right|_{s\to s+1}\\ &= \frac{2}{(s+1)^3}.\end{aligned}$$

1.d. We find that:

$$\begin{aligned}\mathcal{L}(e^{-t}\sin(2t))(s) &= \mathcal{L}(\sin(2t))(s+1)\\ &= \left.\frac{2}{s^2+4}\right|_{s\to s+1}\\ &= \frac{2}{(s+1)^2+4}.\end{aligned}$$

11.1.2 *Exercise 3*

3. We must solve the ODE:

$$y''(t) + 5y'(t) + 4y(t) = 1$$

subject to the initial conditions $y(0) = y'(0) = 0$. Taking the Laplace transform of both sides (and noting that the Laplace transform of 1 and of $u(t)$ must be the same because the functions are identical when $t > 0$), making use of the initial conditions, and denoting the Laplace transform of $y(t)$ by $Y(s)$, we find that:

$$s^2Y(s) + 5sY(s) + 4Y(s) = \frac{1}{s}.$$

Thus, we find that:

$$\begin{aligned}Y(s) &= \frac{1}{s(s^2+5s+4)}\\ &= \frac{1}{s(s+1)(s+4)}\\ &\overset{\text{partial fractions}}{=} \frac{A}{s} + \frac{B}{s+1} + \frac{C}{s+4}.\end{aligned}$$

After clearing denominators we find that:

$$1 = A(s+1)(s+4) + Bs(s+4) + Cs(s+1).$$

Rewriting this, we find that:

$$1 = (A + B + C)s^2 + (5A + 4B + C)s + 4A.$$

Equating coefficients, we find that:

$$\begin{aligned} A + B + C &= 0 \\ 5A + 4B + C &= 0 \\ 4A &= 1. \end{aligned}$$

We find that $A = 1/4, B = -1/3$, and $C = 1/12$. Thus:

$$Y(s) = \frac{1/4}{s} - \frac{1/3}{s+1} + \frac{1/12}{s+4}.$$

We see that:

$$y(t) = \left(\frac{1}{4} - \frac{1}{3}e^{-t} + \frac{1}{12}e^{-4t}\right) u(t).$$

11.1.3 *Exercise 5*

5. We must solve the integral equation:

$$\int_0^t y(r)\, dr = -y(t) + \sin(\omega t)u(t).$$

Taking Laplace transforms and denoting the Laplace transform of $y(t)$ by $Y(s)$, we find that:

$$\frac{Y(s)}{s} = -Y(s) + \frac{\omega}{s^2 + \omega^2}.$$

Rearranging the terms, we find that:

$$sY(s) + Y(s) = \frac{s\omega}{s^2 + \omega^2}.$$

We find that:

$$Y(s) = \frac{s\omega}{(s+1)(s^2 + \omega^2)} = \frac{A}{s+1} + \frac{Bs + C}{s^2 + \omega^2}.$$

Clearing the denominators, we find that:

$$A(s^2 + \omega^2) + (Bs + C)(s + 1) = s\omega.$$

Rewriting this, we find that:

$$(A + B)s^2 + (B + C)s + (A\omega^2 + C) = s\omega.$$

Equating coefficients we find that:

$$\begin{aligned} A + B &= 0 \\ B + C &= \omega \\ A\omega^2 + C &= 0. \end{aligned}$$

Solving these equations we find that:

$$A = -\frac{\omega}{1+\omega^2}, \quad B = \frac{\omega}{1+\omega^2}, \quad C = \frac{\omega^3}{1+\omega^2}.$$

We find that:

$$Y(s) = \frac{\omega}{1+\omega^2}\left(\frac{-1}{s+1} + \frac{s+\omega^2}{s^2+\omega^2}\right).$$

By inspection, the inverse transform of this is just:

$$y(t) = \frac{\omega}{1+\omega^2}\left(-e^{-t} + \cos(\omega t) + \omega\sin(\omega t)\right) u(t).$$

11.1.4 *Exercise 7*

7. Following the instructions given in the exercise, we find that:

$$\begin{aligned} \mathcal{L}\left(\frac{1}{\sqrt{t}}\right)(s) &= \int_0^\infty e^{-st}\frac{1}{\sqrt{t}}\,dt \\ &\overset{y=\sqrt{st},\ dy=\frac{\sqrt{s}}{2\sqrt{t}}}{=} \frac{2}{\sqrt{s}}\int_0^\infty e^{-y^2}\,dy \\ &= \frac{2}{\sqrt{s}}\frac{\sqrt{\pi}}{2} \\ &= \sqrt{\frac{\pi}{s}}. \end{aligned}$$

11.1.5 *Exercise 9*

9.a. The only time the difference is non-zero is when $0 \le t < T$, and when $0 \le t < T$, the difference is just $g(t)$. The result then follows from the definition of the Laplace transform.

9.b. Recall that $\mathcal{L}(g(t-T))(s) = e^{-sT}$. Making use of linearity, the result follows immediately.

9.c. This result follows immediately from the results of the preceding two sections.

9.d. For small values of s, $1 - e^{-sT} = sT + O(s^2)$. Thus:

$$\lim_{s \to 0} \frac{s}{1 - e^{-sT}} = \frac{1}{T}.$$

Also, as $s \to 0$, $e^{-st} \to 1$. Assuming that $g(t)$ is a "reasonable" function:

$$\lim_{s \to 0} \int_0^T e^{-st} g(t)\, dt = \int_0^T g(t)\, dt.$$

Combining the two results leads us to the conclusion that:

$$\lim_{s \to 0} sG(s) = \frac{1}{T} \int_0^T g(t)\, dt = \text{average}(g(t)).$$

11.1.6 *Exercise 11*

11.a.

$$Y(s) = \frac{1}{s} - \frac{1}{(s+2)^2 + 1^2} = \frac{1}{s} - \frac{1}{s^2 + 4s + 5}.$$

11.b.

$$y(\infty) = \lim_{s \to 0^+} sY(s) = \lim_{s \to 0^+} \left(1 - \frac{s}{s^2 + 4s + 5}\right) = 1.$$

11.c.

$$y(0^+) = \lim_{s \to \infty} sY(s) = \lim_{s \to \infty} \left(1 - \frac{s}{s^2 + 4s + 5}\right) = 1.$$

11.d. As the initial value of the function is 1, the Laplace transform of the derivative is:

$$sY(s) - 1 = -\frac{s}{s^2 + 4s + 5}.$$

11.e. We find that:

$$y'(0^+) = \lim_{s \to \infty} s\left(-\frac{s}{s^2 + 4s + 5}\right) = -1.$$

11.f. As we are only interested in $t > 0$, we can ignore $u(t)$. Differentiating what is left of $y(t)$, we find that:

$$y'(t) = 2e^{-2t}\sin(t) - e^{-2t}\cos(t), \qquad t > 0.$$

Clearly, $y'(0^+) = -1$.

11.2 Chapter 2

11.2.1 *Exercise 1*

1.a. We must find the transfer function of the system whose input, $x(t)$, and output, $y(t)$, satisfy the ODE:

$$y''(t) + y'(t) + y(t) = x(t).$$

Recall that in finding a transfer function we assume that all initial conditions are zero. Letting $X(s)$ and $Y(s)$ be the Laplace transform of $x(t)$ and $y(t)$ respectively, we find that:

$$s^2Y(s) + sY(s) + Y(s) = X(s).$$

Rearranging terms, we find that:

$$\frac{Y(s)}{X(s)} = \frac{1}{s^2 + s + 1}.$$

This is the transfer function.

1.b. To get MATLAB to print the Bode plots that correspond to this transfer function, proceed as follows. First type:

```
T = tf([1],[1 1 1])
```

MATLAB responds with:

```
Transfer function:
    1
---------
 2
s + s + 1
```

Then type `bode(T)`. MATLAB will respond with the Bode plots.

11.2.2 *Exercise 3*

3. In Exercise 2c we are asked to evaluate the step response of the system whose transfer function is:

$$T(s) = \frac{5}{s^2 + 10s + 50}.$$

Let us estimate this step response. First of all, we find that $T(0) = 0.1$. Thus, the steady state amplitude of the output will be 0.1. We find that $\omega_n = \sqrt{50}$ and that $2\zeta\omega_n = 10$. Thus, $\omega_n = 7.01$ and $\zeta = 0.70$. From the value of ζ (using (2.4)) we calculate that the percent overshoot is 4.3%—i.e. there is not much overshoot. The rise time of the output is about $3/(\zeta\omega_n) = 0.6$. Assuming that the rise is more or less linear, we find that the slope of the rise in this 0.6 second interval is approximately $0.1/0.6 = 0.17$. Note that the response we are asked to consider should be the derivative of the response of the system of Exercise 2c. Assuming that the response of the first figure is linear until $t = 0.6$ and zero afterwards, we find that the output should be approximately 0.17 until $t = 0.6$ and zero afterwards. The actual response is given in Figure 11.1. Though the response does not quite agree with our estimate, note that the time-frame is correct and the values given are not too far from the correct values.

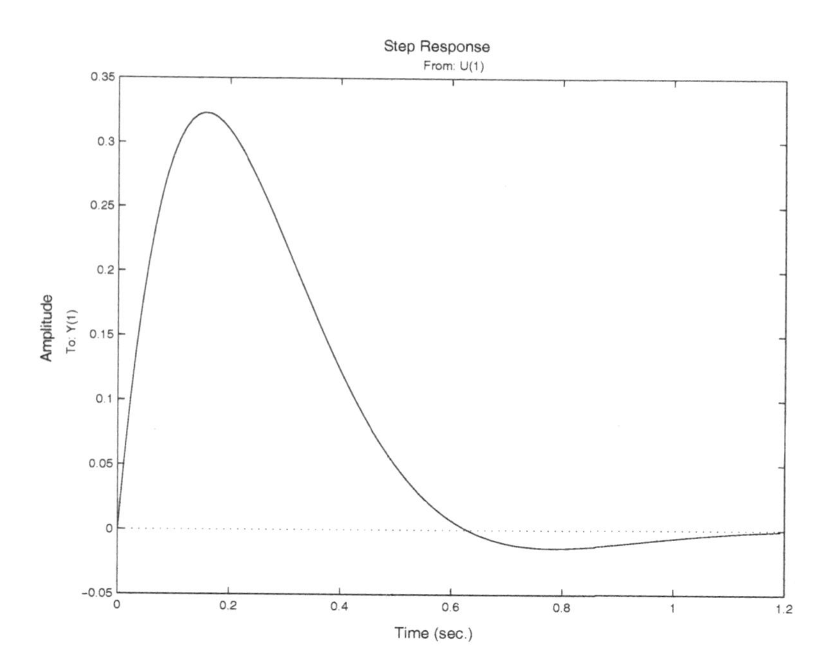

Fig. 11.1 The Step Response

11.2.3 *Exercise 5*

5. We are asked to consider the system described by:

$$T(s) = \frac{\omega_n^2}{s^2 + 2\zeta\omega_n s + \omega_n^2}.$$

Note that the poles of the transfer function are:

$$s = -\zeta\omega_n \pm \omega_n\sqrt{\zeta^2 - 1} = \omega_n\zeta\left(-1 \pm j\frac{\sqrt{1-\zeta^2}}{\zeta}\right). \tag{11.1}$$

For this system, the rise time is approximately $3/(\zeta\omega_n)$ and the percent overshoot is given by:

$$\text{percent overshoot} = e^{-\frac{\pi\zeta}{\sqrt{1-\zeta^2}}} \times 100\%.$$

The two conditions on the output translate into:

$$\frac{3}{\zeta\omega_n} \leq T$$
$$e^{-\frac{\pi\zeta}{\sqrt{1-\zeta^2}}} \leq O.$$

Rearranging the conditions, we find that they can be written:

$$\zeta\omega_n \geq 3/T$$
$$\frac{\sqrt{1-\zeta^2}}{\zeta} \leq -\frac{\pi}{\ln(O)}.$$

We find that the second condition limits the angle of

$$\mu \equiv -1 \pm j\frac{\sqrt{1-\zeta^2}}{\zeta}.$$

In fact, considering the pole with positive imaginary part, we find that it can have imaginary part no larger than $-\pi/\ln(O)$. Thus its angle must be between 180° and

$$180^\circ - \tan^{-1}\left(-\frac{\pi}{\ln(O)}\right).$$

By symmetry, we find that the angle of μ must be in the interval:

$$\left[180^o - \tan^{-1}\left(-\frac{\pi}{\ln(O)}\right), 180^o + \tan^{-1}\left(-\frac{\pi}{\ln(O)}\right)\right].$$

As the angle of μ is equal to the angle of the pole, the poles in our region must have angles in this interval too. Furthermore, as $\mathcal{R}(\text{pole}) = -\zeta\omega_n$, the other condition says that the real part of the pole must be less than or equal to $-3/T$. See Figure 11.2 to get an idea of what the region looks like.

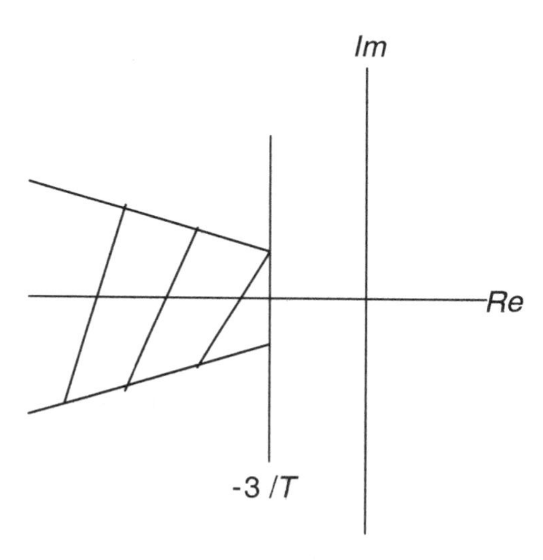

Fig. 11.2 The Region

11.2.4 *Exercise 7*

7. In the analysis of Exercise 5 we found that the poles of this system are $-\omega_n\zeta \pm j\omega_n\sqrt{1-\zeta^2}$. If $\zeta = 0$, then there are two poles on the imaginary axis—the system is marginally stable (which means that the system is just barely unstable). If $\zeta > 0$, then it is not hard to show that both of the poles of the system are in the left half-plane. We consider the cases $\zeta < 1$ and $\zeta \geq 1$ separately.

If $\zeta < 1$, then the poles are:

$$-\omega_n\zeta \pm j\omega_n\sqrt{1-\zeta^2}.$$

Clearly both poles have as their real part $-\zeta\omega_n < 0$.

If $\zeta \geq 1$, then the poles are both real and their value is:

$$-\omega_n\zeta \pm \omega_n\sqrt{\zeta^2 - 1}.$$

As $\sqrt{\zeta^2 - 1} < \zeta$, it is clear that the poles are both negative and the system is stable.

11.2.5 *Exercise 9*

9.a. The Laplace transformof the step response, $Y(s)$, is:

$$Y(s) = \frac{1}{s}\frac{2-s}{(s+1)(s+3)}.$$

The partial fractions expansion of $Y(s)$ is:

$$Y(s) = \frac{1}{s} + \frac{2}{s+2} - \frac{3}{s+1}.$$

Thus,

$$y(t) = (1 + 2e^{-2t} - 3e^{-t})u(t).$$

Using MATLAB to plot both the step response of the system and the $y(t)$ we just calculated brings us to Figure 11.3—where we find that the system suffers from undershoot.

9.b. The step response suffers from undershoot.

9.c. The transfer function's positive zero.

11.2.6 *Exercise 11*

11.a. As the system is stable:

$$\lim_{t\to\infty} y(t) = \lim_{s\to 0+} s\frac{1}{s}T(s) = T(0) = 1/6.$$

11.b.

$$y(0^+) = \lim_{s\to\infty} s\frac{1}{s}T(s) = 0.$$

11.c. As $y(0^+) = 0$, we find that:

$$y'(0^+) = \lim_{s\to\infty} s\frac{1}{s}sT(s) = -1.$$

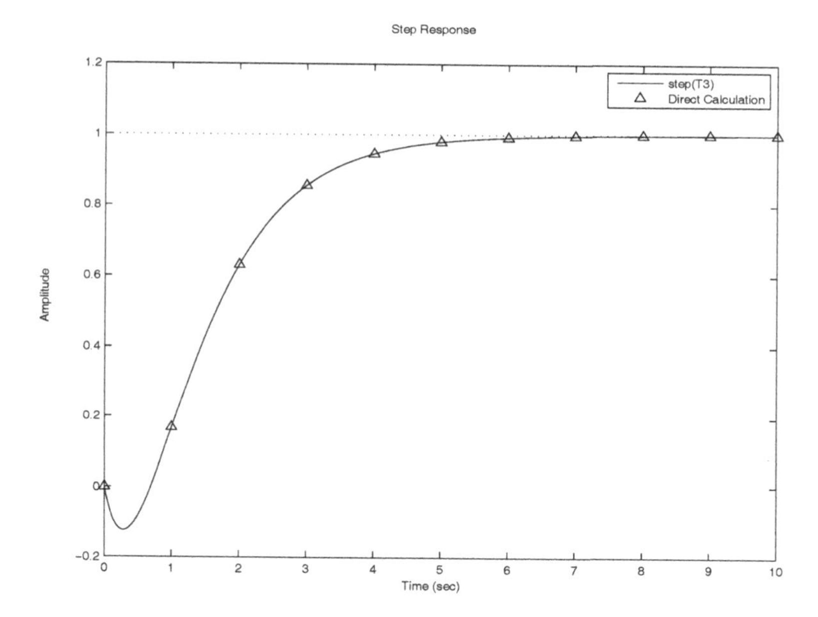

Fig. 11.3 The Step Response of the System Calculated in Two Different Ways

11.d. The partial fractions expansion of $y(t)$ is:

$$\frac{2-s}{s(s+3)(s+4)} = \frac{1/6}{s} - \frac{5/3}{s+3} + \frac{3/2}{s+4}.$$

Thus, $y(t) = (1/6)(1 - 10e^{-3t} + 9e^{-4t})u(t)$. Clearly, $y(0^+) = 0$. For $t > 0$, $y'(t) = 5e^{-3t} - 6e^{4t}$. Thus, $y'(0^+) = -1$—just as it should.

Clearly, this system suffers from undershoot.

11.3 Chapter 3

11.3.1 *Exercise 1*

1.a. We find that:

$$\begin{aligned} V_o(s) &= \frac{A}{s\tau + 1}(V_+(s) - V_-(s)) \\ &= \frac{A}{s\tau + 1}(V_{in}(s) - V_0(s)). \end{aligned}$$

Rearranging the terms we find that:

$$V_o(s)\left(1+\frac{A}{s\tau+1}\right)=\frac{A}{s\tau+1}V_{in}(s).$$

That is:

$$\begin{aligned}V_o(s)&=\frac{\frac{A}{s\tau+1}}{1+\frac{A}{s\tau+1}}V_{in}(s)\\&=\frac{A}{s\tau+1+A}V_{in}(s).\end{aligned}$$

Thus, the transfer function is:

$$T(s)=\frac{A}{s\tau+1+A}.$$

1.b. We find that:

$$T(0)=\frac{A}{A+1}.$$

As A is generally very large this number is generally *very* close to one.

1.c. We find that the sensitivity is:

$$\begin{aligned}S_A^{T(s)}&=\frac{\partial T(s)}{\partial A}\frac{A}{T(s)}\\&=\frac{(s\tau+1+A)-A}{(s\tau+1+A)^2}\frac{A}{T(s)}\\&=\frac{s\tau+1}{s\tau+1+A}.\end{aligned}$$

1.d. We find that when $s=0$:

$$\left.S_A^{T(s)}\right|_{s=0}=\frac{1}{1+A}\approx 0.$$

11.3.2 *Exercise 3*

3.a. In case i we find:

$$T(s)=\frac{1/1}{1+1/1}=\frac{1}{2}.$$

This is certainly stable.

In case ii we find that:

$$T(s)=\frac{1/s}{1+1/s}=\frac{1}{s+1}.$$

As this has a single negative pole, this too must be stable.

In case iii we find that:

$$T(s) = \frac{1/s^2}{1+1/s^2} = \frac{1}{s^2+1}.$$

Here the two poles are $\pm j$. As these are on the imaginary axis this system is (just barely) unstable.

3.b. If a system is unstable then using the final value theorem can lead to *wrong* answers.

11.3.3 *Exercise 5*

5.a. Considering Figure 3.13 we find that:

$$V_o(s) = D(s) + G_p(s)(V_{in}(s) - V_o(s)).$$

Rewriting this we find that:

$$V_o(s)(1 + G_p(s)) = G_p(s)V_{in}(s) + D(s).$$

We see that:

$$V_o(s) = \frac{G_p(s)}{1+G_p(s)}V_{in}(s) + \frac{1}{1+G_p(s)}D(s).$$

Finally, if we let $G_p(s) = K$, then we find that:

$$V_o(s) = \frac{K}{1+K}V_{in}(s) + \frac{1}{1+K}D(s).$$

5.b. The disturbance does not affect the output very much.

5.c. The disturbance arrives at the output with very little attenuation.

11.3.4 *Exercise 7*

7. Note that the gain provided by $1/s$ is very large when $|s|$ is small and small when $|s|$ is large. Thus, low frequency disturbances should have practically no effect on the output whereas high frequency disturbances suffer practically no change.

11.3.5 *Exercise 9*

9.a. As:

$$T(s) = \frac{G_c(s)G_p(s)}{1 + G_c(s)G_p(s)} = \frac{3}{(s+2)(s-1)+3} = \frac{3}{s^2+s+1},$$

and all of the roots of the denominator are in the left half-plane, the system is BIBO stable.
9.b. As there is no cancellation of a pole in the right half-plane by a zero in the right half-plane, the system is also internally stable.

11.3.6 *Exercise 11*

11.a.

$$T(s) = \frac{C}{s^2 + as + C}.$$

11.b.

$$\begin{aligned} S_a^{T(s)} &= \frac{\partial T(s)}{\partial a}\frac{a}{T(s)} \\ &= -\frac{Cs}{(s^2+as+C)^2}\frac{a}{T(s)} \\ &= -\frac{as}{s^2+as+C}. \end{aligned}$$

11.c. Clearly as $C \to \infty$, the sensitivity tends to zero. Large gains, C, make the system less sensitive to changes in a.
11.d. $S_a^{T(0)} = 0$.
11.e. As $T(0) = 1$ for all $a > 0$, it should come as no surprise that the sensitivity of $T(0)$ to changes in a is zero.

11.3.7 *Exercise 13*

We prove the result by induction. For $m = 0$, we need to show that:

$$\mathcal{L}(u(t))(s) = \frac{1}{s}.$$

This has already been proved. (See p. 2.) Now assume that for m the result holds—$\mathcal{L}(t^m u(t))(s) = m!/s^{m+1}$. As multiplication in the time domain is equivalent to calculating minus the derivative in the s domain, we find that:

$$\mathcal{L}(t^{m+1}u(t))(s) = -\frac{d}{ds}\frac{m!}{s^{m+1}} = \frac{(m+1)!}{s^{m+2}}$$

which is just what we needed to show.

11.4 Chapter 4

11.4.1 *Exercise 1*

1.a. The Routh array corresponding to the polynomial is:

$$\begin{matrix} 1 & 1 \\ 3 & 0 \\ 1 & 0 \end{matrix}.$$

As there are no sign changes in the first column, there are no RHP zeros.

1.b. The Routh array corresponding to the polynomial is:

$$\begin{matrix} 1 & 0 \\ 0 & 1 \end{matrix}.$$

As the array terminates prematurely there must be some RHP zeros. As one of the zeros of our polynomial s^3+1 is -1, we know that s^3+1 is divisible by $s+1$. In fact:

$$\frac{s^3+1}{s+1} = s^2 - s + 1.$$

Let us determine how many zeros s^2-s+1 has in the RHP. The Routh array that corresponds to this polynomial is:

$$\begin{matrix} 1 & 1 \\ -1 & 0 \\ 1 & 0 \end{matrix}.$$

As there are two sign changes, this polynomial has two zeros in the RHP. As $s^3+1 = (s+1)(s^2-s+1)$, we find that s^3+1 has two zeros in the RHP.

1.c. The Routh array that corresponds to the polynomial is:

$$\begin{matrix} 1 & 3 & 1 \\ 0 & 0 & 0 \end{matrix}.$$

As the Routh array terminates prematurely there are RHP zeros.

1.d. In this case, we find that the Routh array is:

$$\begin{matrix} 1 & 1 & 1 \\ 1 & 1 & 0 \\ 0 & 1 & 0 \end{matrix}.$$

As the Routh array terminates prematurely, we know that there is at least one RHP zero.

1.e. Here the Routh array is:

$$\begin{matrix} 1 & 0 & 1 \\ 1 & 0 & 0 \\ 0 & 1 & 0 \end{matrix},$$

and once again we see that the array terminates prematurely. As there are no sign changes before the array terminates, we can look at the polynomial that corresponds to the last two rows to get some more information about the number of RHP zeros. The number of RHP zeros of our original polynomial must equal the number of RHP zeros of this polynomial. The polynomial that corresponds to the last two rows is:

$$s^3 + 1.$$

Rather than using the analysis we used in 1.b, we find the solutions of $s^3 + 1 = 0$ by using some facts from complex variables. The three solution of $s^3 + 1 = 0$ are

$$s = -1, s = e^{j\pi/3} = \frac{1}{2} + j\frac{\sqrt{3}}{2}, s = e^{-j\pi/3} = \frac{1}{2} - j\frac{\sqrt{3}}{2}.$$

We find that there are two zeros in the RHP.

11.4.2 *Exercise 3*

3. If one defines $x = s^2$, then one finds that our polynomial can be written as:

$$x^2 + 3x + 1.$$

The zeros of the original fourth order polynomial must satisfy the equation:

$$s = \pm\sqrt{\text{solutions of } x^2 + 3x + 1 = 0}.$$

Consider what the solutions of this equations must be. If the square root is in the LHP, the minus the square root is in the RHP. If the square root is in the RHP not including the imaginary axis, then minus the square root is in the LHP. If the square root is imaginary—and hence (just barely) in the RHP, then minus the square root is also (just barely) in the RHP. Thus each solution leads to either one or two zeros in the RHP. We find that on general grounds, this type of fourth order polynomial must have either two or four zeros in the RHP.

11.4.3 *Exercise 5*

5. Here we find that:

$$\begin{aligned} T(s) &= \frac{\frac{as^2+b}{s^3}}{1+\frac{as^2+b}{s^3}} \\ &= \frac{as^2+b}{s^3+as^2+b}. \end{aligned}$$

The Routh array that corresponds to the denominator is:

$$\begin{array}{cc} 1 & 0 \\ a & b \\ -b/a & 0 \\ b & 0 \end{array}.$$

As there are two sign changes in the Routh array, we see that our system has two poles in the RHP and is certainly not stable.

11.4.4 *Exercise 7*

7.a. With:

$$G_p(s) = \frac{(s-a)(s+3)}{s^2+3s+2},$$

we find that:

$$\begin{aligned} T(s) &= \frac{K(s-a)(s+3)}{s^2+3s+2+K(s-a)(s+3)} \\ &= \frac{K(s-a)(s+3)}{(1+K)s^2+(3+K(3-a))s+2-3Ka}. \end{aligned}$$

The Routh array corresponding to the denominator is:

$$\begin{array}{cc} 1+K & 2-3Ka \\ 3+K(3-a) & 0 \\ 2-3Ka & 0 \end{array}.$$

As $K > 0$, we find that for all the elements of the first column to have the same sign all of the elements must be positive (as the first element is definitely positive). This leads to the two conditions:

$$\begin{aligned} 3 + K(3 - a) &> 0 \\ 2 - 3Ka &> 0. \end{aligned}$$

The second condition leads us to the condition that:

$$K < \frac{2}{3a}.$$

The first condition is certainly true if $a \leq 3$. When $a > 3$, it leads to the condition that:

$$K < \frac{3}{a-3}.$$

Let us compare the sizes of these conditions when $a > 3$. We find that:

$$\frac{3}{a-3} - \frac{2}{3a} = \frac{9a - 2(a-3)}{3a(a-3)} = \frac{7a+6}{3a(a-3)} > 0, \quad a > 3.$$

Thus, the condition:

$$K < \frac{2}{3a}$$

is always the stricter condition and is the only one we need to consider.
7.b. Consider the Routh array when $a = 0$. It is all positive if $K > 0$. Thus the system (with $a = 0$) is stable for all $K > 0$.
7.c. We see that zeros in the right half plane tend to make a system unstable at relatively high gains.

11.4.5 *Exercise 9*

9. Here we find that:

$$\begin{aligned} T(s) &= \frac{K(s^2 + 201s + 10100)}{s(s^2 + s + 1) + K(s^2 + 201s + 10100)} \\ &= \frac{K(s^2 + 201s + 10100)}{s^3 + (1+K)s^2 + (1 + 201K)s + 10100K}. \end{aligned}$$

The Routh array that corresponds to the denominator is:

$$\begin{array}{rr} 1 & 1+201K \\ 1+K & 10100K \\ 1+201K-\frac{10100K}{1+K} & 0 \\ 10100K & 0 \end{array}.$$

In order for the system to be stable all of the elements in the first column must have the same sign. We find that:

$$\begin{aligned} 1+K &> 0 \\ 1+201K-\frac{10100K}{1+K} &> 0 \\ 10100K &> 0. \end{aligned}$$

The first inequality requires that $K > -1$, and the last requires that $K > 0$. As we know that for stability $K > 0$, we can clear the denominator of the second condition. This leads to the condition that:

$$(1+201K)(1+K) - 10100K = 201K^2 + (202 - 10100)K + 1 > 0.$$

This is a quadratic equation in K. It is clear that for large $|K|$ this quadratic is positive. Thus there are certainly some values of K for which the system is stable. Let us find the zeros of the quadratic. We find that:

$$K = \frac{(10100-202) \pm \sqrt{(10100-202)^2 - 4 \cdot 201}}{2 \cdot 201}.$$

Evaluating this expression, we find that the two roots are:

$$K_1 = 1.01 \times 10^{-4}, \quad K_2 = 49.2.$$

As we know that for large values of $|K|$ the system is stable, we see that the system is unstable only when:

$$K_1 \geq K \geq K_2.$$

11.4.6 *Exercise 11*

11.a. Consider the partial products, $(s - r_1)$, $(s - r_1)(s - r_2), \ldots$. We claim that all of the coefficients of each of the partial products are positive, and the coefficient of the highest power is one. We proceed by induction.

Start with $s - r_1$. The coefficient of s is 1, and $-r_1$ is positive. Now assume that for some k all of the coefficients of $R(s) = (s - r_1) \cdots (s - r_k)$ are positive and the coefficient of the highest power is 1. After multiplying by $(s - r_{k+1})$, we have $sR(s) - r_k R(s)$. Clearly the coefficient of the highest power is once again one, and, as $-r_k > 0$ and all of the coefficients of $R(s)$ are positive, the coefficients of the new polynomial are also all positive.

11.b. This part is trivial.

$$(s - \alpha - j\beta)(s - \alpha + j\beta) = s^2 - 2\alpha s + \alpha^2 + \beta^2.$$

As $\alpha < 0$, all the coefficients are positive.

11.c. In this more general case, all the factors are of the form $s^2 - 2\alpha_k + \alpha_k^2 + \beta_k^2$ or $s - r_k$ where all the coefficients are positive and the coefficient of the highest power is one. The same basic induction as the one used in 11.a. shows that all the coefficients of the original polynomial must have been positive.

11.d. Dividing by a constant cannot change the zeros of a polynomial. Thus the zeros of $P(s)$ and of $A(s) \equiv P(s)/a_0$ are the same. If all of the zeros of $P(s)$ are in the left half-plane, so are all of the zeros of $A(s)$. As the coefficient of the highest power of $A(s)$ is one, we know that all of the coefficients of $A(s)$ are positive. Thus, all of the coefficients of $P(s) = a_0 A(s)$ are of the same sign as a_0.

11.4.7 *Exercise 13*

The transfer function of the system is:

$$T(s) = \frac{K}{s^3 + 5s^2 + K}.$$

As the denominator "skips over" the first order term, the denominator has at least one zero in the right half-plane, and the system is not stable.

11.5 Chapter 5

11.5.1 *Exercise 1*

1.a. We must consider how the unit circle is mapped by $1/z^2$. In this case it is clear that the circle is mapped into two circles about the origin but whose direction is the opposite of the direction of the original circle. Clearly $1/z^2$ has no zeros in the unit circle—its numerator cannot be zero. We find that $E = -2$ and $N = 0$. Thus $P = 2$. (Clearly $1/z^2$ does have two poles inside the unit circle—it has two poles at the origin.)

1.b. Consider Figure 11.4. Our region is the unit circle. It is plotted in the left-hand diagram. The mapping of the boundary under the map

$$f(z) = 1/(z(z-2))$$

is shown in the right-hand diagram.

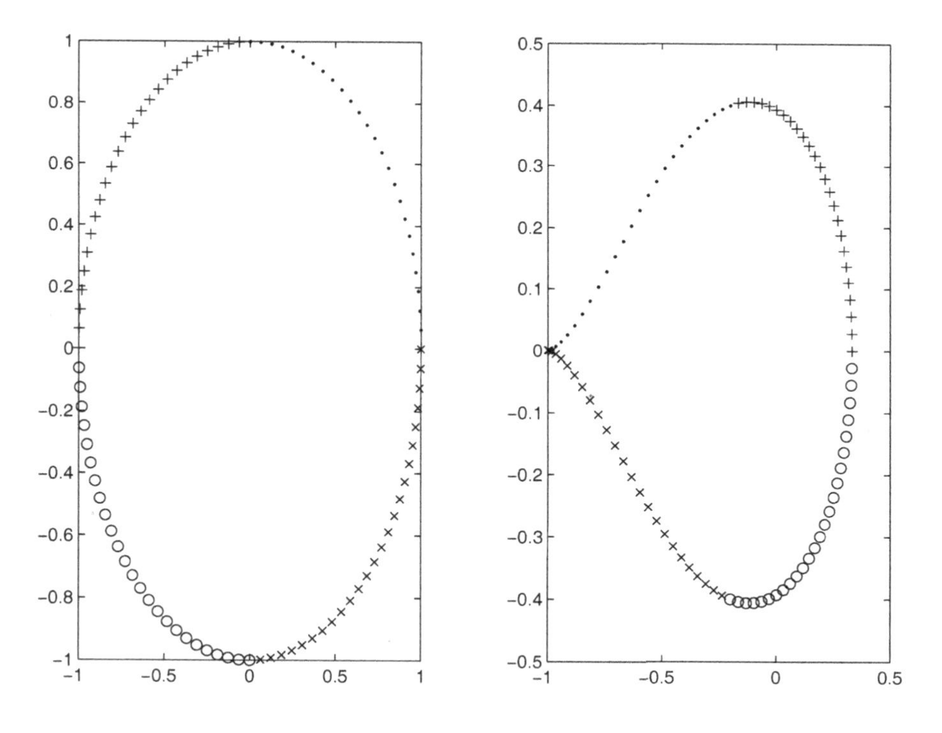

Fig. 11.4 $f(z) = \frac{1}{z(z-2)}$

In order to make it possible to determine the direction of each plot, we have used dots, o's, pluses and then x's in that order in the plots. We see that the curve in the right-hand diagram encircles the origin once in the direction opposite that taken by the curve in the left-hand diagram—which is the boundary of the region of interest. As it is clear that $f(z)$ has no zeros in the region of interest, we find that:

$$E = -1, N = 0.$$

Thus, we see that $P = 1$.

1.c. Consider the plots in Figure 11.5. We find that here the mapping of the boundary of region of interest circles the origin once in the positive direction. As it is clear that $f(z) = \frac{z}{z^2-4}$ has one zero in the region, we find that:

$$N = 1, E = 1.$$

Thus, we find that $P = 0$—as it should be.

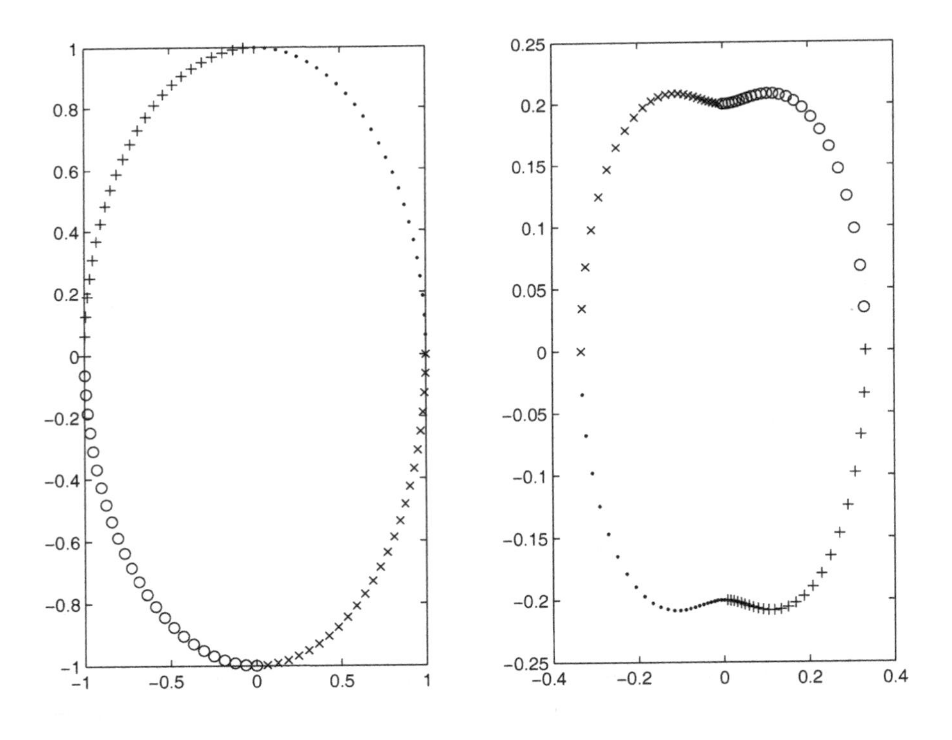

Fig. 11.5 $f(z) = \frac{z}{z^2-4}$

11.5.2 *Exercise 3*

3. We plot the Nyquist plot in Figure 11.6. We see that because of the negative phase contributed by the term:

$$\frac{K}{s+1},$$

and because of the fact that the term $1/s^2$ contributes a (at least almost) full circle when it maps the small semi-circle that skirts the origin, the Nyquist plot of:

$$G_p(s) = \frac{K}{s^2(s+1)}$$

encircles -1 twice for all positive K. Thus, the system is unstable for all positive K.

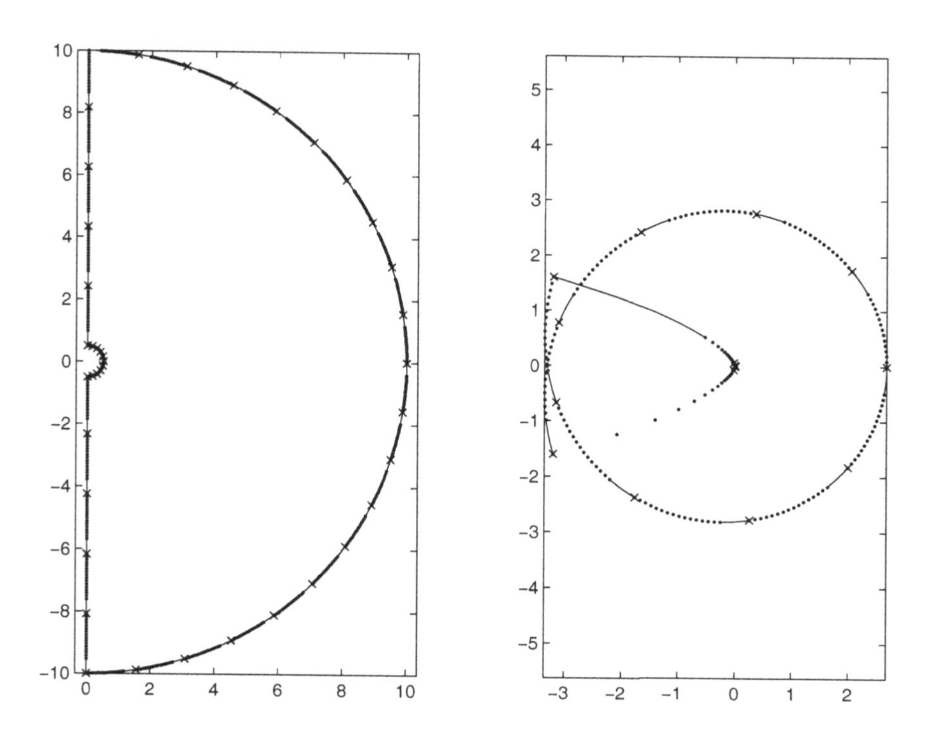

Fig. 11.6 $G_p(s) = \frac{K}{s^2(s+1)}$

11.5.3 *Exercise 5*

5. This exercise is very similar to Exercise 3. Here too we make use of the fact that $1/s^2$ is borderline unstable and can tolerate no addition of negative phase. We use this to show that a term like e^{-Ts}—which always adds negative phase—is guaranteed to destabilize our system. To show this in a graphical way, consider the Nyquist plot that corresponds to the system with $T = 0.5$ given in Figure 11.7. We see that -1 is encircled twice here—showing that our system is unstable.

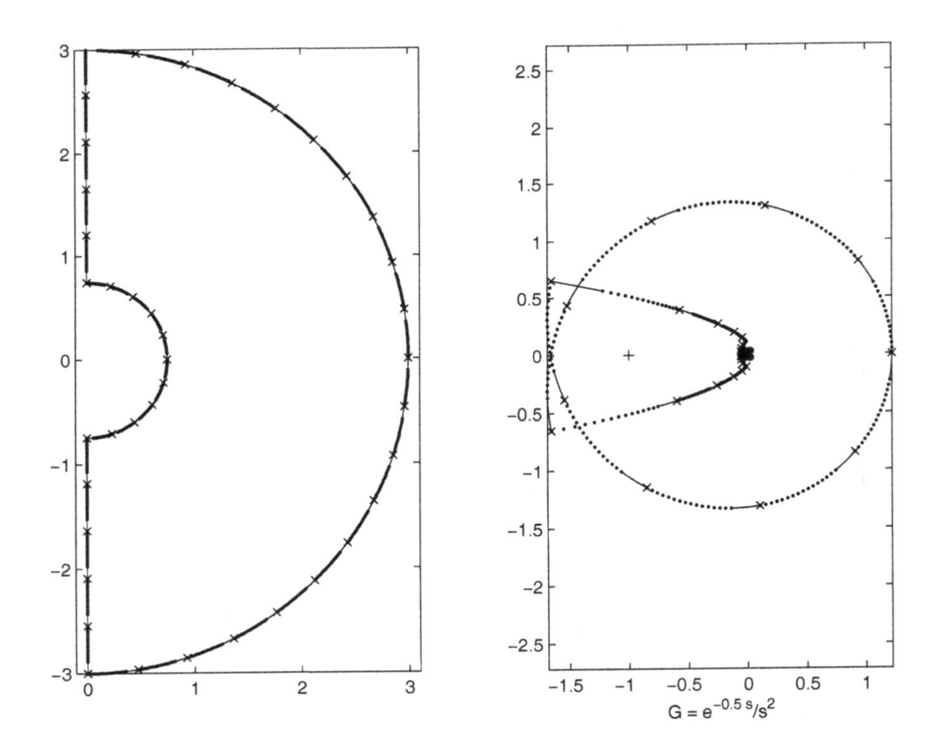

Fig. 11.7 $G_p(s) = \frac{e^{-0.5s}}{s^2}, \rho = 0.75, R = 3$

11.5.4 *Exercise 7*

7.a. The gain plot is a line that is always decreasing at 20dB/dec. The phase plot is always decreasing and its value is:

$$-180^o - T\omega 180^o/\pi.$$

As the ω axis is scaled logarithmically, this will *not* appear as a straight line.

7.b. As we have already shown in Exercise 5, this system is *absolutely unstable* when $T > 0$. To calculate the gain margin—which should be negative—one must find the point at which the Nyquist plot crosses the negative real axis. In this case, that crossing is caused by the "infinitesimal" semi-circle that skirts the origin. Thus, the only gain that leads to a stable system is 0. One over that gain is ∞. Thus, the gain margin is $-\infty$dB. To find the phase margin we must determine when $|G_p(j\omega)| = 1$. Clearly this occurs when $\omega = 1$. At this point,

$$\angle G_p(j\omega) = -180^o - T\omega 180^o/\pi.$$

We find that the phase margin is $-T\omega 180^o/\pi$.

7.c. The system is quite unstable.

7.d. To say that it would be unwise to use such a system is to massively understate the case. One should stay *far far* away from a system that promises to be as poorly behaved as this system.

11.5.5 *Exercise 9*

9. Near zero the transfer function of a system with n poles at the origin will look like:

$$G_p(s) \approx \frac{K}{s^n}.$$

Thus, at low frequencies we will find that:

$$G_p(j\omega) \approx \frac{K}{j^n \omega^n}.$$

Thus, the amplitude of the frequency response (in dB) will be:

$$20\log_{10}\left(\frac{K}{\omega^n}\right) = -20n\log_{10}(\omega) + 20\log_{10}(K).$$

Clearly this is falling at a rate of $20n$ dB per decade. That is, if at low frequencies the frequency response falls off at the rate of $20n$ dB per decade then the transfer function has n poles at the origin.

11.5.6 *Exercise 11*

11.a. Using the first approximation, we find that the poles of the system are (approximately) the solutions of the equation:

$$1 + \frac{1 - Ts}{s} = 0.$$

The lone solution of this equation is:

$$s = \frac{1}{T - 1}.$$

As this is negative until $T = 1$, we find that the set of values for which the system is stable is $T < 1$.

11.b. Using the second approximation, we find that the poles of the system are (approximately) the solutions of the equation:

$$1 + \frac{1 - Ts + (T^2/2)s^2}{s} = 0.$$

We find that the poles of the system are (approximately) the solutions of the quadratic equation:

$$(T^2/2)s^2 + (1 - T)s + 1 = 0.$$

We have found that a necessary and sufficient condition for all the roots of this equation to be in the left half-plane is that all of the coefficients have the same sign. Here that translates into the condition $T < 1$ yet again.

11.c. Using the third approximation, we find that the poles of the system are (approximately) the solutions of the equation:

$$1 + \frac{1 - Ts + (T^2/2)s^2 - (T^3/6)s^3}{s} = 0.$$

We find that the poles of the system are (approximately) the solutions of the cubic equation:

$$-(T^3/6)s^3 + (T^2/2)s^2 + (1 - T)s + 1 = 0.$$

Let us consider the Routh matrix that corresponds to this polynomial. We find that it is:

$$\begin{matrix} -(T^3/6) & 1-T \\ T^2/2 & 1 \\ 1-\frac{2T}{3} & 0 \\ 1 & 0 \end{matrix}.$$

As always, we require that $T > 0$—that our delay be a delay. We find that the first column can never be of one sign. Thus, this approximation gives an unstable system for all delays—a result that we know is very wrong.

11.d. Note that the system give in the previous section is not at all low-pass in nature. In fact the degree of the numerator is two greater than the degree of the denominator. That leads the large semi-circle to be mapped into a large circle. This circle leads to an encirclement of -1 for all values of T—and this is why the approximation does not give reasonable results.

11.5.7 *Exercise 13*

13.a. As must be the case for a system for which the denominator of $G_c(s)G_p(s)$ is of higher order than its numerator and that has a pole at zero, the magnitude of the frequency response tends to zero at high frequencies and to infinity at low frequencies. Let's consider the phase of the frequency response. We find that:

$$\angle G_c(j\omega)G_p(j\omega) = \tan^{-1}(\omega) - \pi/2 - \tan^{-1}(\omega/3) - \tan^{-1}(\omega/5).$$

As the arctangent function is an increasing function, it is easy to see that for any positive frequency, the phase of the frequency response is never as negative as $-\pi$. Also, for small values of ω, $\tan^{-1}(\omega) \approx \omega$. Thus, for small values of ω we find that:

$$\angle G_c(j\omega)G_p(j\omega) \approx \omega - \pi/2 - \omega/3 - \omega/5 > -\pi/2.$$

That is, for low frequencies, the plot will be located in the fourth quadrant! At higher frequencies, the phase indicates that the plot will enter the third quadrant and remain there.

Let's consider the asymptotes. When s is small,

$$\begin{aligned}
G_c(s)G_p(s) &= \frac{s+1}{s^3+8s^2+15s} \\
&\approx \frac{s+1}{15s+8s^2} \\
&= \frac{1}{15s}\frac{s+1}{1+8s/15} \\
&\approx \frac{1}{15s}(s+1)(1-8s/15) \\
&\approx \frac{1}{15s}(1+s-8s/15) \\
&= \frac{1}{15s}+7/225.
\end{aligned}$$

We find that when $s = j\omega$ and ω is small the real part of $G_c(s)G_p(s)$ is positive and its imaginary part is negative. That is, the asymptote associated with small positive values of ω lies in the fourth quadrant—as we saw previously by considering the phase.

As usual, the small semi-circle about the origin is mapped into a large semi-circle. The Nyquist plot is given in Figure 11.8.

13.b and c. The Bode plots as plotted by the `margin` command are given in Figure 11.9.

13.d. The system is *very* stable.

11.5.8 *Exercise 15*

15.a. For $\omega > 0$, $|G_p(j\omega)|$ is monotonically decreasing in ω. Additionally, for $\omega > 0$:

$$\angle G_p(j\omega) = -\pi/2 - 6\tan^{-1}(\omega).$$

Thus, for small values of ω the phase is near -90°, and as $\omega \to \infty$, the phase tends monotonically to -630°. Clearly there should be two points at which the plot crosses the negative real axis—when the phase is -180° and when the phase is -540°. To find these two points, we look for the points at which $G_p(s)$ is negative. At these points, it must be real, so we must find the zeros of $-j\omega^7 + j15\omega^5 - 15j\omega^3 + j\omega$. Using MATLAB we find that the three positive zeros are at $\omega \approx 3.732, 1, 0.267$. (In fact, if one is careful enough, one can avoid using MATLAB.) The associated value of $G_p(j\omega)$ are approximately $-8 \times 10^{-5}, 1/8$, and -3.03.

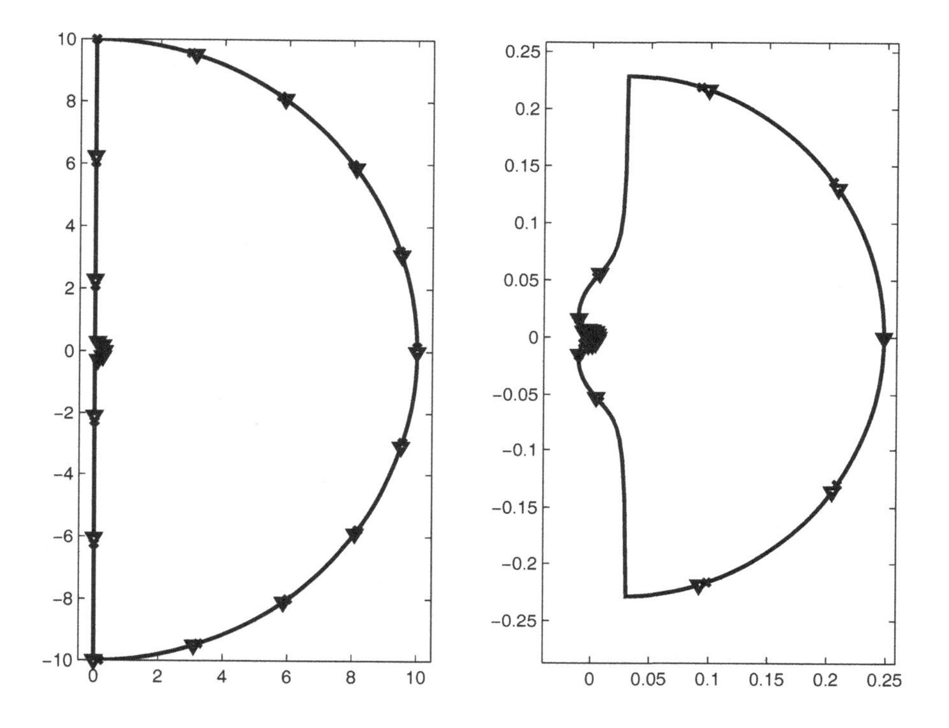

Fig. 11.8 $G_p(s) = \frac{s+1}{s(s+3)(s+5)}, \rho = 0.3, R = 10$

The Nyquist plot is given in Figure 11.10, and an enlarged version of the region near the origin is given in Figure 11.11.

15.b. As $P = 0$ and $E = 2$, we find that $N = 2$ and the system is unstable.

15.c. From the Nyquist plot, it is clear that the system is stable for $K < 1/3.03$.

11.5.9 *Exercise 17*

17.a-c. Looking at $G_p(s)$, it is clear that as $s = j\omega \to j0$ the system has a vertical asymptote. Additionally, as $s = j\omega \to j\infty$, $G_p(s)$ approaches zero and its phase tends to -90°. For $\omega > 0$,

$$\angle G_p(j\omega) = \angle(a^2 - \omega^2) - \pi/2 - 2\tan^{-1}(\omega).$$

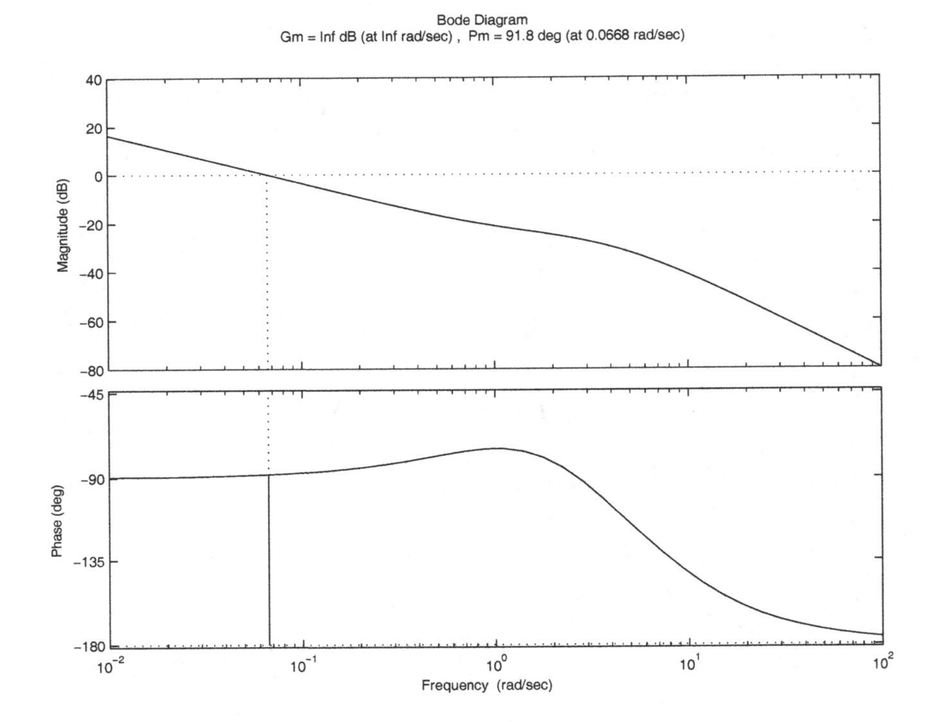

Fig. 11.9 The System's Bode Plots

When $\omega = a$, $G_p(ja) = 0$. As ω goes from less than a to more than a, the phase gains 180°.

Near $s = 0$, $G_p(s) \sim a^2/s$, so the small semi-circle we use to avoid the origin in the region of interest becomes a large semi-circle taken in the counter clockwise direction in the Nyquist plot.

17.a. Consider $s = j\omega, 0 < \omega < a$. When ω is near zero, the phase of $G_p(j\omega)$ is near -90°. For $a < 1$, as $\omega \to a^-$, the phase of the denominator is less than 180^0, so as s tends to ja, $G_p(j\omega)$ lies in the third quadrant. For $\omega > a$, the phase of the numerator is 180°. Thus for ω slightly larger than a, $G_p(j\omega)$ is located in the first quadrant. As ω increases the phase of the denominator hits 180° (when $\omega = 1$). At that point $G_p(j\omega) = (-1 + a^2)/(-2) > 0$. As ω increases, the phase of the denominator increases to 270°, so that as $\omega \to \infty$ the phase of $G_p(j\omega) \to -90^o$. When sketching the Nyquist plot, we *start* from ω near infinity, so we find that the Nyquist plot starts from the

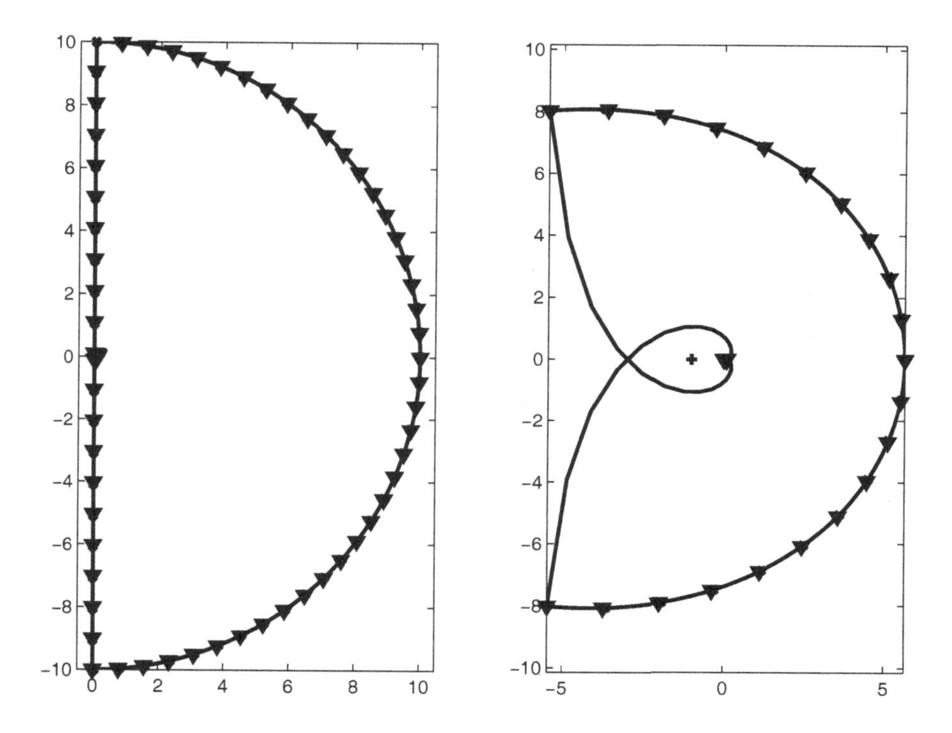

Fig. 11.10 The Nyquist plot. $G_p(s) = 1/(s(s+1)^6)$.

origin, leaves at an angle -90°, proceeds to a real positive value, loops back towards the origin, and finally continues to an asymptote. The full Nyquist plot is given in Figure 11.12 and a blow-up of the region near the origin is given in 11.13. As $P = 0$ and we see that $E = 0$, we find that $N = 0$, and the system is stable. In fact, since the Nyquist plot does not intersect the negative real-axis, the system is stable for all positive gains—the system's gain margin is infinite.

17.b. In this case, for $0 < \omega < a = 1$, the plot is located in the third quadrant. When $\omega = a = 1$, $G_p(j\omega) = 0$, and when $\omega > 1$, $-90^o < \angle G_p(j\omega) < 0^o$. As we start plotting the Nyquist plot from $\omega \to \infty$, we should find that the plot leaves the origin at -90°, proceeds through the fourth quadrant to the origin, is tangent to the real axis at the origin, and then proceeds to the third quadrant where it has an asymptote. The Nyquist plot corresponding to the case $a = 1$ is given in Figure 11.14. The

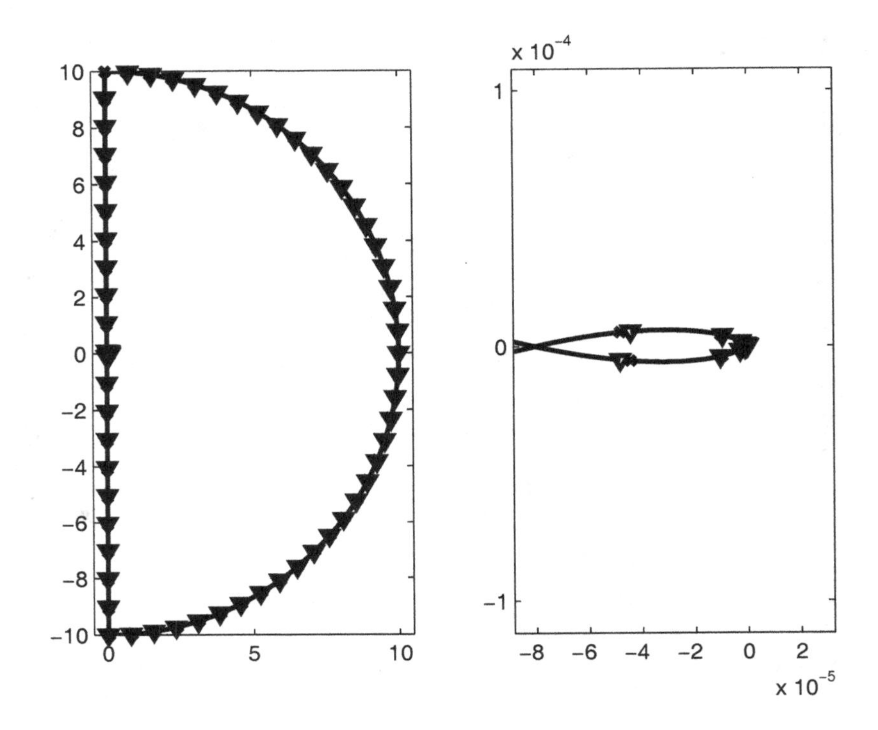

Fig. 11.11 The Nyquist plot in the neighborhood of the origin. $G_p(s) = 1/(s(s+1)^6)$.

blowup is given in Figure 11.15. We know that $P = 0$. From the plots, it is clear that $E = 0$. Thus, $N = 0$, and the system is closed-loop stable. In fact, looking at the Nyquist plot, it is clear that the system's gain margin is infinite.

17.c. When $a > 1$, we find that while $0 < \omega < 1$, the phase of $G_p(j\omega)$ is near -90° for small ω and it gets more negative as ω increases. When $\omega = 1$, the phase is precisely -180°—that is, $G_p(j \cdot 1) = (-1 + a^2)/-2 < 0$. For $1 < \omega < a$, $G_p(j\omega)$ is located in the second quadrant. When $\omega = a$, $G_p(j\omega) = 0$. For $\omega > a$, $-90o < \angle G_p(j\omega) < 0$, and as $\omega \to \infty$, the phase tends to -90°. As here the Nyquist plot intersects the negative real axis, we expect the system to have a finite gain margin.

The Nyquist plot corresponding to the case $a = \sqrt{2}$ is given in Figure 11.16. The blowup is given in Figure 11.17. Here, $P = 0$, $E = 0$, and, therefor, $N = 0$ and the system is stable. Considering the Nyquist plot, it

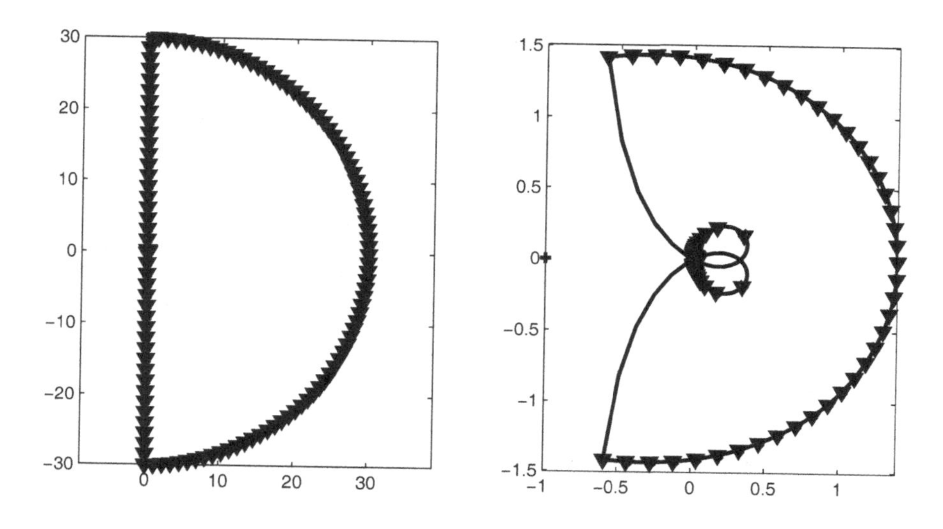

Fig. 11.12 The Nyquist plot. $G_p(s) = (s^2 + a^2)/(s(s+1)^2), a < 1$.

is clear that the maximum gain for which the system is stable is 2.

17.d. The Routh array for our system is:

$$\begin{array}{ll} 1 & 1 \\ 2+K & Ka^2 \\ 1-(K/(2+K))a^2 & 0 \\ Ka^2 & 0 \end{array}.$$

Thus, as long as $a \leq 1$, the system is stable for all $K > 0$—its gain margin is infinite. When $a > 1$, the system's gain margin is:

$$\text{GM} = 20\log_{10}\left(\frac{2}{a^2-1}\right).$$

In 17.a and 17.b, where $a \leq 1$, we found that the gain margin is infinite. In 17.c, where $a > 1$, we found that the gain margin is finite—and that is just what we found by using the Routh array.

11.5.10 *Exercise 19*

19.a. As there is no pole at the origin, there is no need to avoid the origin when plotting the Nyquist plot. Clearly for both large *and* small values of

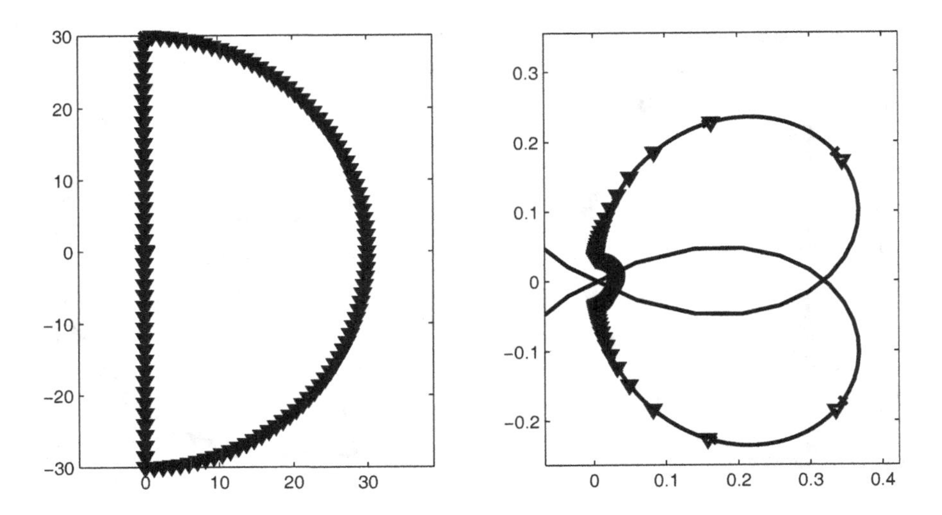

Fig. 11.13 The Nyquist plot in the neighborhood of the Origin. $G_p(s) = (s^2+a^2)/(s(s+1)^2), a < 1$. The small semi-circle near the origin is, of course, the image of the large semi-circle in the region of interest.

s, $|G_p(s)| \to 0$. Additionally, the phase of the frequency response is:

$$\angle G_p(j\omega) = \pi/2 - \tan^{-1}(\omega) - \tan^{-1}(\omega/2) - \tan^{-1}(\omega/3).$$

Thus, for large values of ω, the phase is near -180° (but never hits -180°). As ω tends to zero, the phase increases towards $+90^{\circ}$. For $\omega > 0$, the Nyquist plot starts at the origin (for $\omega \to \infty$), leaves the origin at -180°, passes through the third and fourth quadrants, enters the first quadrant, and, as $\omega \to 0$, tends to zero with an angle of $+90^{\circ}$. Negative frequencies can be taken care of by symmetry. See Figure 11.18. As the plot has no intersection with the negative real axis, the gain margin is seen to be infinite.

19.b. The transfer function of the closed-loop system with $G_c(s) = K$ is:

$$T(s) = \frac{s}{s^3 + 6s^2 + (11 + K)s + 6}.$$

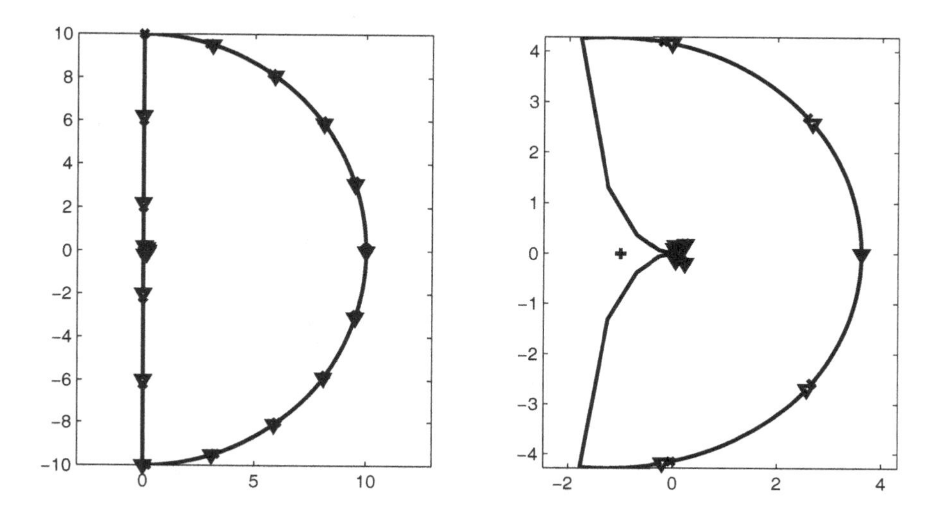

Fig. 11.14 The Nyquist plot. $G_p(s) = (s^2 + a^2)/(s(s+1)^2), a = 1$.

The Routh array that corresponds to the denominator is:

$$\begin{array}{ll} 1 & 11+K \\ 6 & 6 \\ 10+K & 0 \\ 6 & 0 \end{array}$$

The system is stable for all $K > -10$, and we see that its gain margin is infinite.

11.5.11 *Exercise 21*

21.a. Considering $s = j\omega$, we find that for large ω the magnitude of $G_p(j\omega)$ is small and the phase is near -270°, and for small ω, its magnitude is large and its phase is near -90°. Clearly,

$$\angle(G_p(j\omega)) = -\pi/2 - \tan^{-1}(\omega/2) - \tan^{-1}(\omega/4).$$

Thus, as ω goes from nearly infinity towards zero, the phase increases monotonically from very close to -270° to very close to -90°. In particular, the Nyquist plot crosses the negative real axis at only one point. Considering $G_p(j\omega)$, we find that at that point, $-j\omega^3 + 8j\omega = 0$. As our region is built to avoid $\omega = 0$, the solutions of interest to us are the solution of $\omega^2 = 8$.

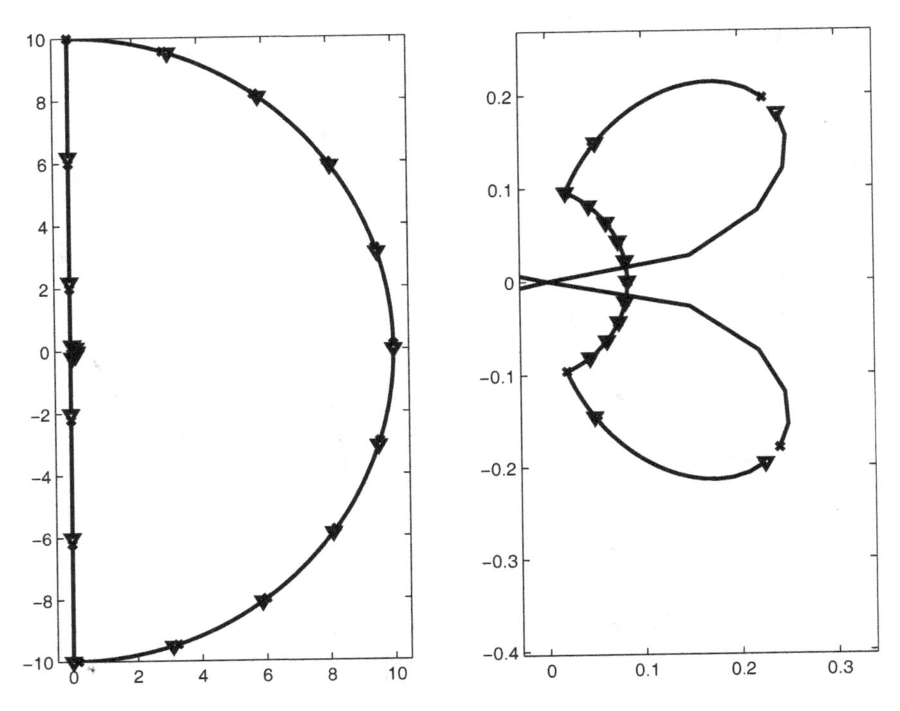

Fig. 11.15 The Nyquist plot in the neighborhood of the origin. $G_p(s) = (s^2+a^2)/(s(s+1)^2), a = 1$. The small semi-circle near the origin is, of course, the image of the large semi-circle in the region of interest.

As this point, $G_p(j\sqrt{8}) = -1/48$.

Considering $G_p(s)$ for small values of s, we find that:

$$\begin{aligned}
G_p(s) &= \frac{1}{s(s+2)(s+4)} \\
&= \frac{1}{8s}\frac{1}{1+(6/8)s+s^2/8} \\
&\approx \frac{1}{8s}\frac{1}{1+(3/4)s} \\
&\approx \frac{1}{8s}(1-(3/4)s) \\
&= \frac{1}{8s} - \frac{3}{32}.
\end{aligned}$$

Thus, the asymptote is the line whose real part is $-3/32$—which is less

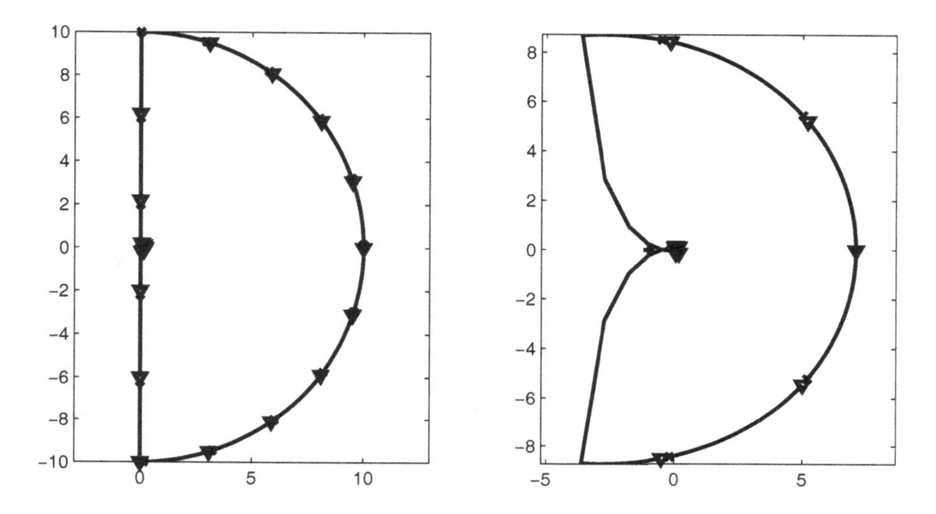

Fig. 11.16 The Nyquist plot. $G_p(s) = (s^2 + a^2)/(s(s+1)^2), a = \sqrt{2}$.

than one tenth in absolute value.

21.b. The gain margin is $20\log_{10}(48) = 33.6\,\text{dB}$.

21.c. Letting $G_c(s) = K$, we find that the relevant Routh array is:

$$\begin{array}{ll} 1 & 8 \\ 6 & K \\ 8 - K/6 & 0 \\ K & 0 \end{array}$$

We find that for stability $K < 48$, and the gain margin is 33.6 dB.

21.d. Given the nature of the asymptote, the phase margin should be near 90°.

21.e. Yes, the system is stable enough to use. One might even say that the system is too stable. Some of the "excess" in the margins could possibly have been used to improve the system's performance.

11.5.12 *Exercise 23*

23.a. As $G_p(j\omega) = j\omega/(-\omega^2 - 1)$, for all $\omega > 0$, the phase is -90°. The Nyquist plot lies on the imaginary axis. Thus, $E = 0$. As $G_p(s)$ has one pole in the right half-plane, $P = 1$, and $N = 1$. Thus, the system is not stable. To check this, consider the transfer function of the closed-loop

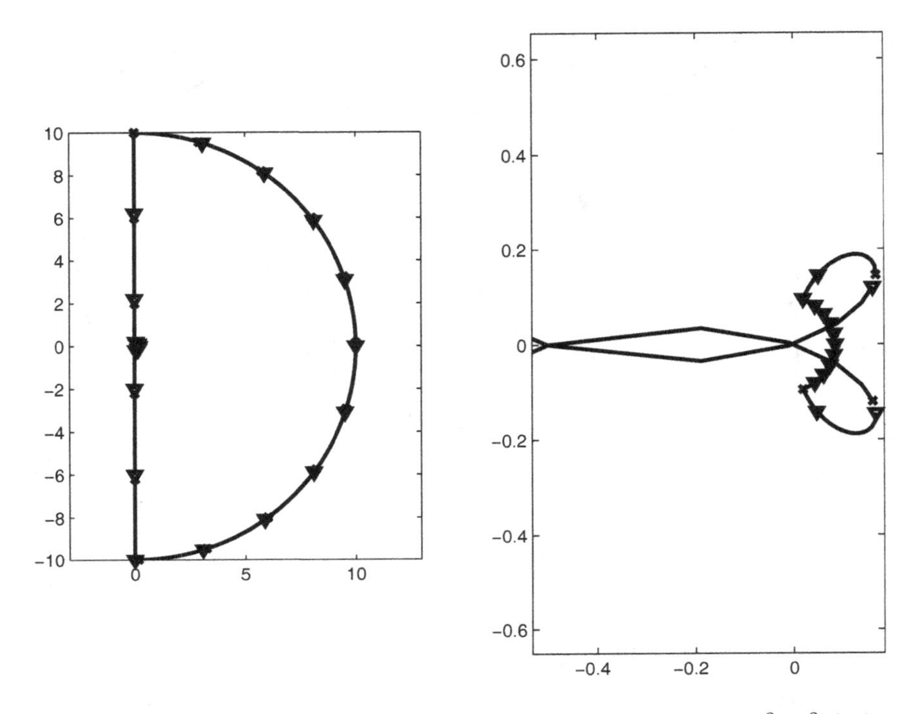

Fig. 11.17 The Nyquist plot in the neighborhood of the origin. $G_p(s) = (s^2+a^2)/(s(s+1)^2)$, $a = \sqrt{2}$. The small semi-circle near the origin is, of course, the image of the large semi-circle in the region of interest.

system. It is:

$$T(s) = \frac{s}{s^2+s-1}.$$

As the coefficients of the denominator do not all have the same sign, the system must be unstable.

23.b. We find that:

$$G_p(j\omega) = \frac{j\omega+2}{-\omega^2-1}.$$

Clearly,

$$\angle G_p(j\omega) = \tan^{-1}(\omega/2) - \pi.$$

For large values of ω the phase tends to -90° When $\omega = 0$, the phase is -180°. In fact, when $\omega = 0$, we find that $G_p(0) = -2$. Thus, the Nyquist plot encircles -1 once, but the encirclement is in the clockwise direction.

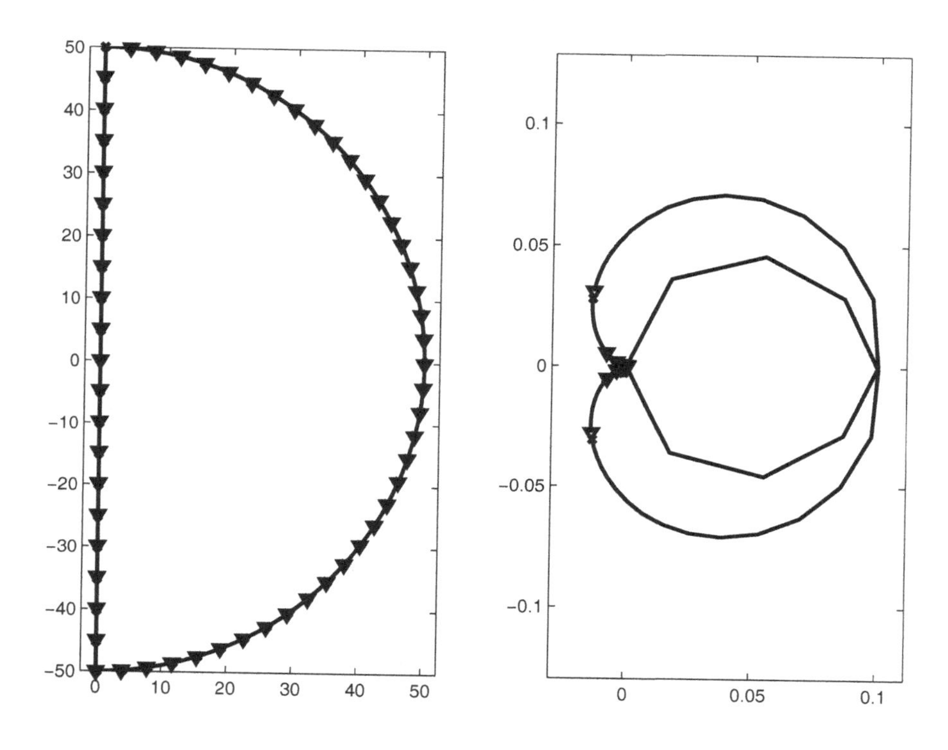

Fig. 11.18 The Nyquist plot corresponding to $G_p(s) = s/((s+1)(s+2)(s+3))$

Here $P = 1$, $E = -1$, and $N = 0$. That is, the system is stable. It is simple to check this result by considering the closed-loop transfer function of the system.

11.6 Chapter 6

11.6.1 *Exercise 1*

1.a. The root locus has one branch because the degree of the denominator (and the numerator) is one. From rule 2, we see that the branch connects $s = 0$ to $s = -1$ and lies on the real axis. See Figure 11.19 for the root locus diagram.

1.b. There are two branches; from rule 2 we see that one connects 0 to -1 and the other connects -2 to $-\infty$. See Figure 11.20 for the root locus diagram.

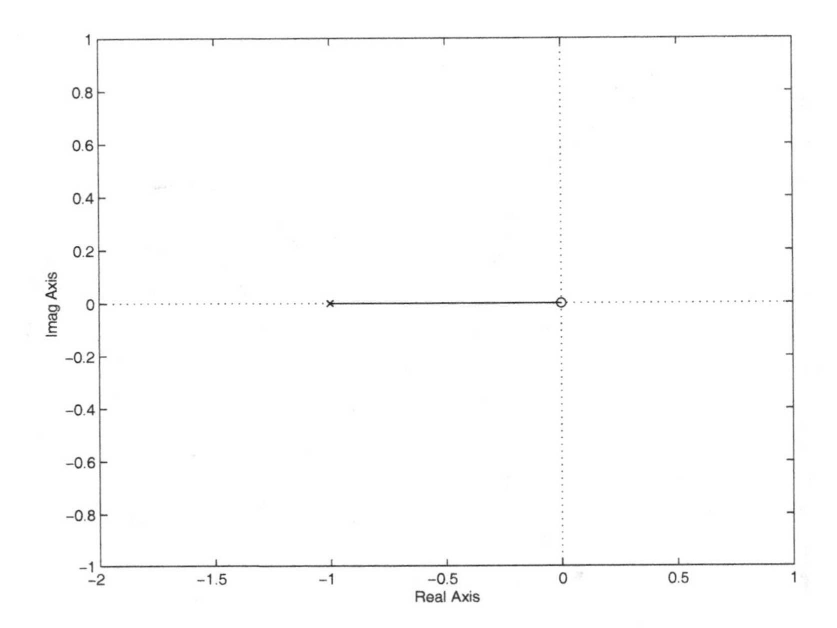

Fig. 11.19 The Root Locus Diagram for $\frac{s}{s+1}$

1.c. From rule 3 we see that three branches leave from the origin. From rule 4 we see that one branch leaves the pole at zero and connects to the "zero" at $-\infty$. The other two connect 0 to the zeros of the numerator—$\pm j$. From rule 10 we see that near the origin the root locus must behave like the root locus of $1/s^3$. Thus, two poles must leave the origin and enter the RHP while one leaves the origin and moves into the LHP. See Figure 11.21 for the root locus diagram.

11.6.2 *Exercise 3*

3. The root locus diagram corresponding to $G_1(s)$ is given in Figure 11.22. As the denominator of $G_1(s)$ is a fourth order polynomial (and the degree of the numerator is smaller than four), the root locus has four branches. One of the branches connects the pole at zero to the zero at -0.1. The next branch connects the pole at -0.2 to the pole at -10. As this is a connection between two poles, we know that the branch must leave the real axis between the two poles. As the degree of the denominator is three greater than the degree of the numerator, we find that the sum of all the

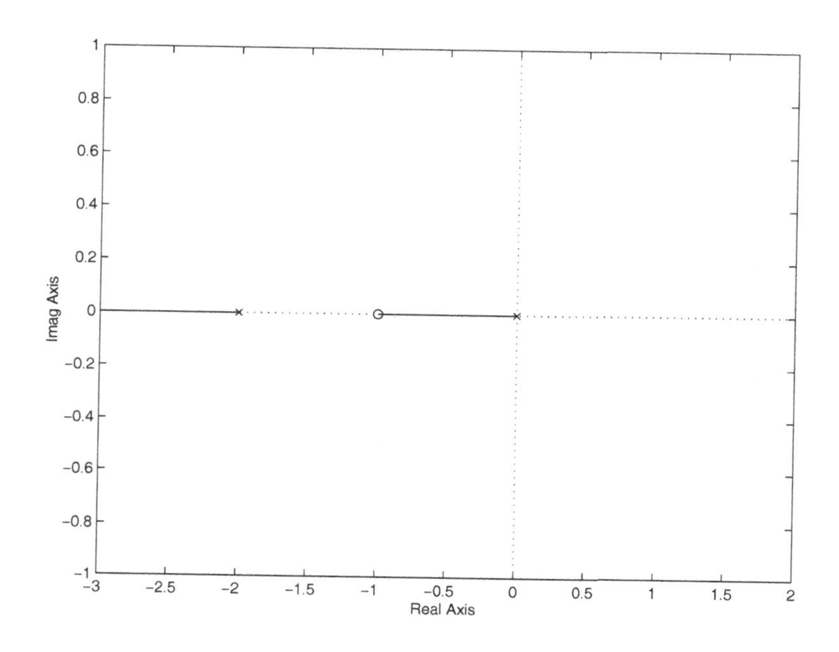

Fig. 11.20 The Root Locus Diagram for $\frac{s+1}{s(s+2)}$

poles must remain constant. As the last branch connects -20 to $-\infty$, we find that the two branches must tend towards $+\infty$.

The root locus diagram that corresponds to $G_2(s)$ is given in Figure 11.23. As the degree of the denominator is two, the root locus diagram has two branches. One branch connects the pole at zero to the zero at -0.1 while the second branch connects the pole at -0.2 to the zero at $-\infty$.

The reason that the two plots are similar near the origin is that the difference between the two transfer functions is two poles that are far from the origin. Such poles do not change the root locus diagram very much near the origin.

11.6.3 *Exercise 5*

5. The root locus diagram that corresponds to $G_p(s)$ is given in Figure 11.24. We see that two branches of the root locus break away between $s = 0$ and $s = -1$. To calculate where they break away we must find the

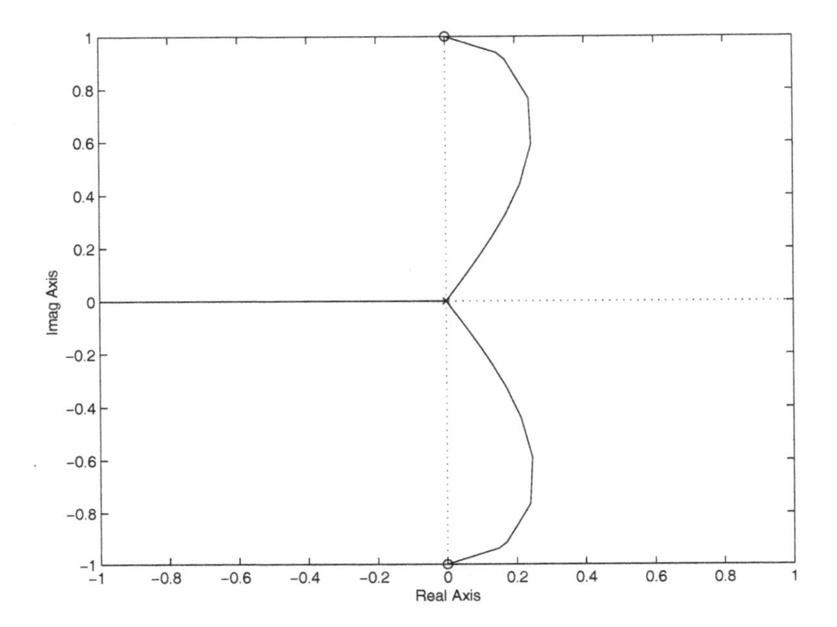

Fig. 11.21 The Root Locus Diagram for $G_p(s) = \frac{s^2+1}{s^3}$

solutions of:

$$\begin{aligned}
\frac{d}{ds}G_p(s) &= \frac{d}{ds}\frac{s+5}{s(s+1)(s+2)} \\
&= \frac{d}{ds}\frac{s+5}{s^3+3s^2+2s} \\
&= \frac{(s^3+3s^2+2s)-(s+5)(3s^2+6s+2)}{(s^3+3s^2+2s)^2} \\
&= \frac{-2s^3-18s^2-30s-10}{(s^3+3s^2+2s)^2} \\
&= 0.
\end{aligned}$$

The zeros of this function are the zeros of its numerator. Thus, we must calculate the zeros of:

$$-2s^3 - 18s^2 - 30s - 10 = 0.$$

The solution of this equation that lies between 0 and -1 is approximately -0.45 (as found by a simple numerical calculation). At $s_0 = -0.45$ we find

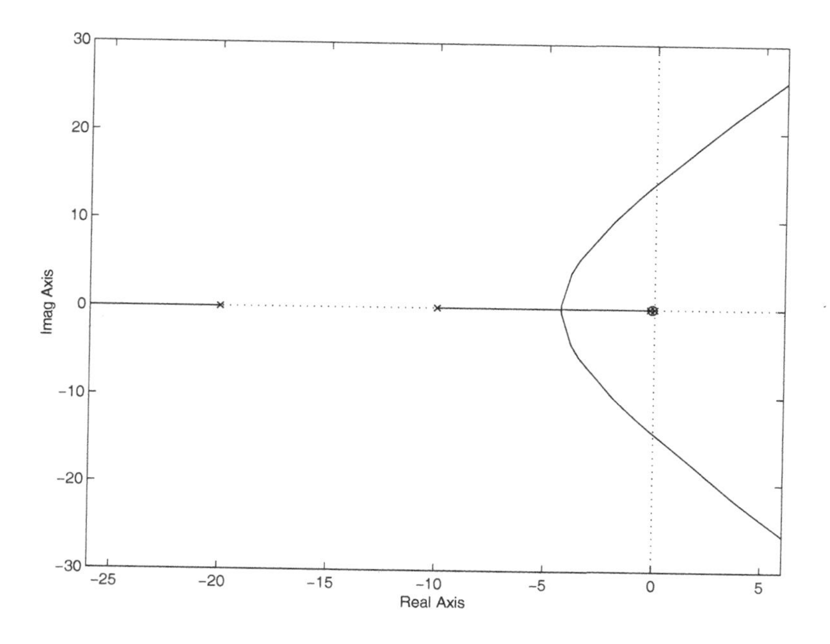

Fig. 11.22 The Root Locus Diagram of $G_1(s) = \frac{s+0.1}{s(s+0.2)(s+10)(s+20)}$.

that $G_p(s_0) = -11.9$. In order that $1 + KG_p(s_0) = 0$, as it must on points of the root locus, we find that $K = 0.084$. We see that for all K greater than (approximately) 0.084 the system has complex poles.

11.6.4 *Exercise 7*

7. In this exercise, since the degree of the denominator is greater than the degree of the numerator by 2, we know that the sum of the poles is constant. Clearly, one branch of the root locus goes from the pole at -5 to the zero at -4. Thus, the two symmetric branches that leave from $\pm j$ must move from $\pm j$ to values whose real part is more negative than zero. That is, the branches must move *into* the left half-plane. The system must, therefore, be stable for all $K > 0$. For a plot of the root locus diagram that corresponds to this system, see Figure 11.25.

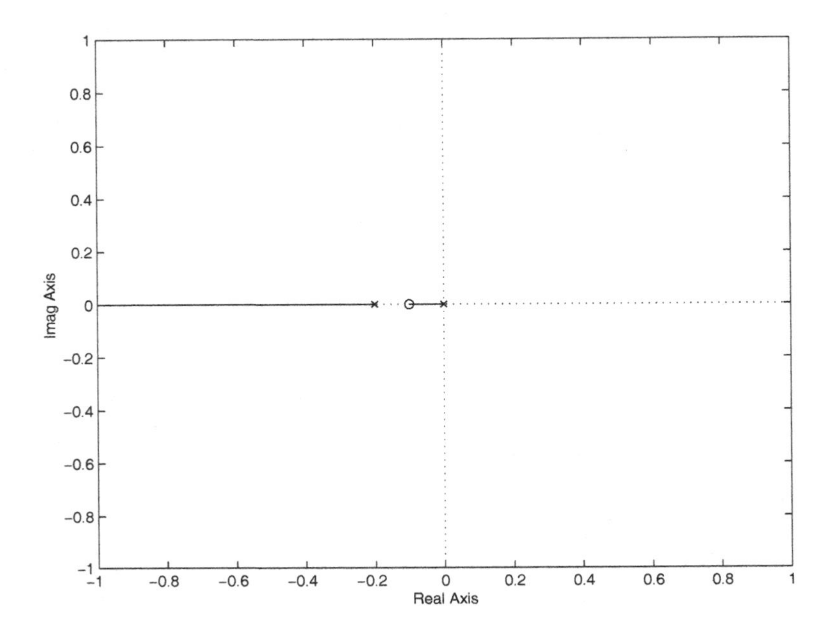

Fig. 11.23 The Root Locus Diagram of $G_2(s) = \frac{s+0.1}{s(s+0.2)}$

11.6.5 *Exercise 9*

9.a. As $G_p(s)H(s)$ has two (real) poles and two (imaginary) zeros, we find that the root locus must start at the poles and end on the zeros. From rule 2, we see that the interval $[-1, 0]$ is on the root locus. Differentiating $G_p(s)H(s)$, we find that:

$$\begin{aligned}\frac{d}{ds}\frac{s^2+1}{s(s+1)} &= \frac{2s \cdot s(s+1) - (2s+1)(s^2+1)}{s^2(s+1)^2} \\ &= \frac{2s^3+2s^2-2s^3-s^2-2s-1}{s^2(s+1)^2} \\ &= \frac{s^2-2s-1}{s^2(s+1)^2}.\end{aligned}$$

We find that the derivative equals zero when:

$$s^2 - 2s - 1 = (s-1)^2 - 2 = 0$$

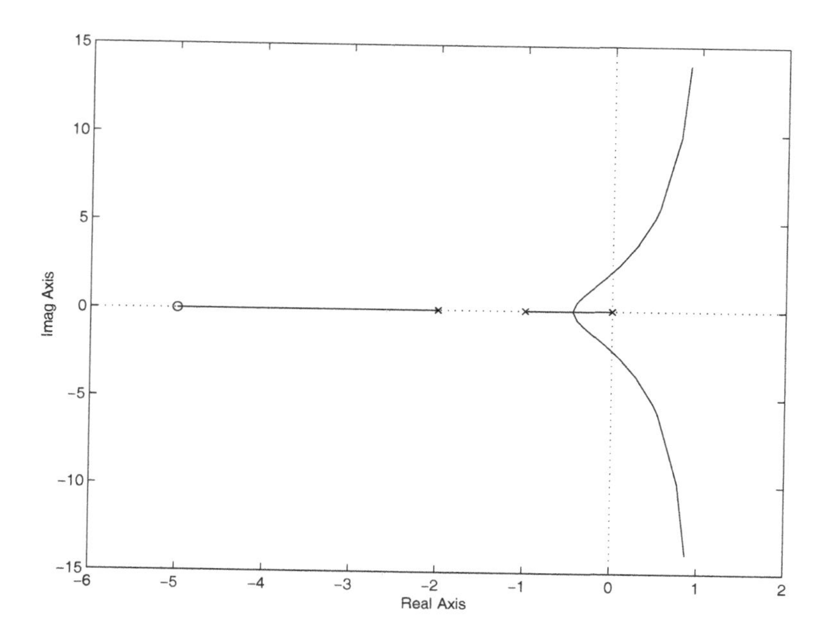

Fig. 11.24 The Root Locus Diagram of $G_p(s) = \frac{s+5}{s(s+1)(s+2)}$

Thus, the points at which the derivative equals zero are:

$$s = 1 \pm \sqrt{2} = -0.4142, 2.4142.$$

As only the first of these numbers is located on the root locus diagram we see that the point at which the branches leave the real axis is just $s = -0.4142$. We plot the root locus in Figure 11.26.

9.b. The system's zeros are just the zeros of $G_p(s)$. Thus, the zeros are $\pm j$.

9.c. The transfer function of the system is:

$$T(s) = \frac{KG_p(s)}{1 + KG_p(s)} = \frac{K(s^2+1)}{s^2 + s + K(s^2+1)}.$$

9.d. For very large K the system response will be approximately one at any point that is not very near the zeros of $\pm j$. When $s = \pm j$, the system response will be zero. That is the system passes all frequencies except for frequencies very near $\omega = 1$.

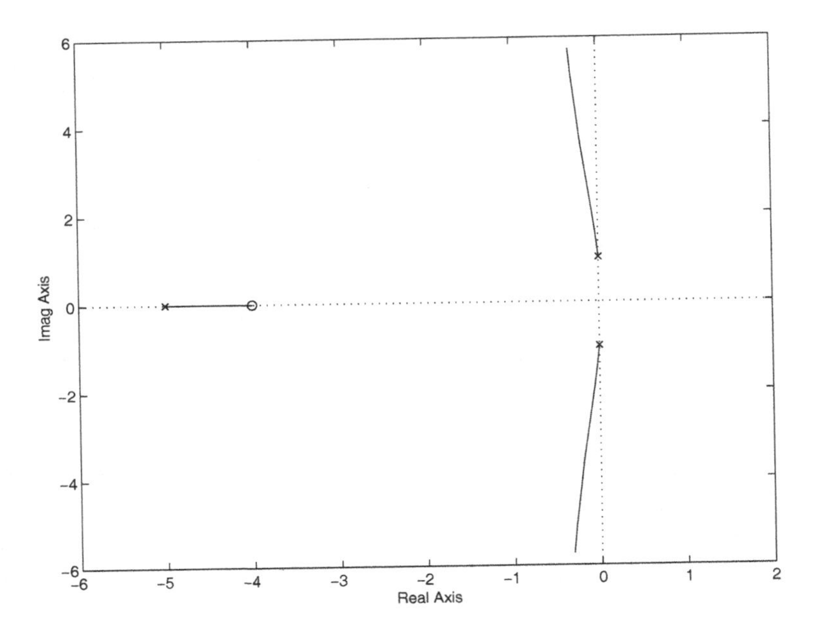

Fig. 11.25 The Root Locus Diagram of $G_p(s) = \frac{s+4}{(s+5)(s^2+1)}$

9.e. This type of system can be used to stop one frequency. (In our case the system stops $\omega = 1$ while passing all frequencies that are "reasonably far" from $\omega = 1$.)

11.6.6 *Exercise 11*

11.a Rearranging $G_c(s)G_p(s)$ we find that:

$$G_c(s)G_p(s) = \frac{3s^2 + 3s + 1}{s^3}.$$

This function has a third order pole at zero and zeros at $-1/2 \pm j/(2\sqrt{3})$. Thus, the root locus contains the entire negative real axis, all three branches start at zero, and two branches tend to the complex open loop zeros. To see where branches enter or leave the negative real axis, we consider the zeros of the derivative of $G_c(s)G_p(s)$. We find that the derivative of the

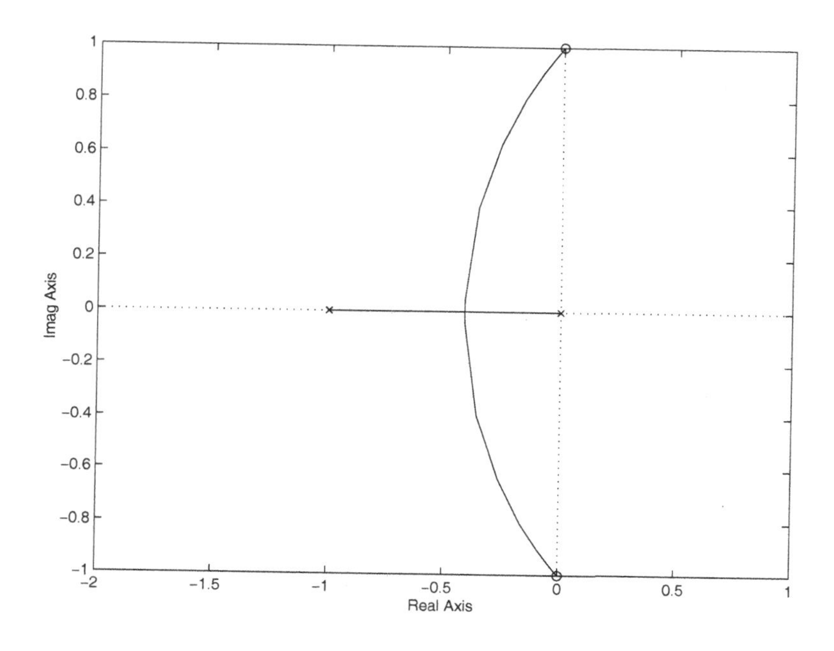

Fig. 11.26 The Root Locus Diagram of $G_p(s) = \frac{s^2+1}{s(s+1)}$

open loop transfer function is:

$$\begin{aligned}\frac{d}{ds}G_c(s)G_p(s) &= \frac{3(s+1)^2s^3 - 3s^2(s+1)^3}{s^6} \\ &= \frac{(s+1)^2(3s - 3(s+1))}{s^4} \\ &= -\frac{3(s+1)^2}{s^4}.\end{aligned}$$

Clearly, the only zero of the derivative is located at $s = -1$. Considering the behavior of the root locus diagram at $s = -1 + \epsilon$, near $s = -1$, we find that:

$$Gc(s)G_p(s) = -1 + \frac{\epsilon^3}{s^3}.$$

It is easy to see that when ϵ is very small the condition for $G_c(s)G_p(s)$ to be real is, essentially, that ϵ^3 be real. That happens when $\epsilon = |\epsilon|\exp(n\pi j/3)$, $n = 0, \ldots, 5$. Thus, all three branches must enter the negative real axis at -1, and they all leave from there.

One more point is worth noting. There is a triple pole at zero. Near zero, we expect the root locus to look like the root locus of $1/s^3$—we expect to see to branches heading off into the right half-plane. That is, for small gains, we expect the closed loop system to be unstable.

The root locus diagram is given in Figure 11.27.

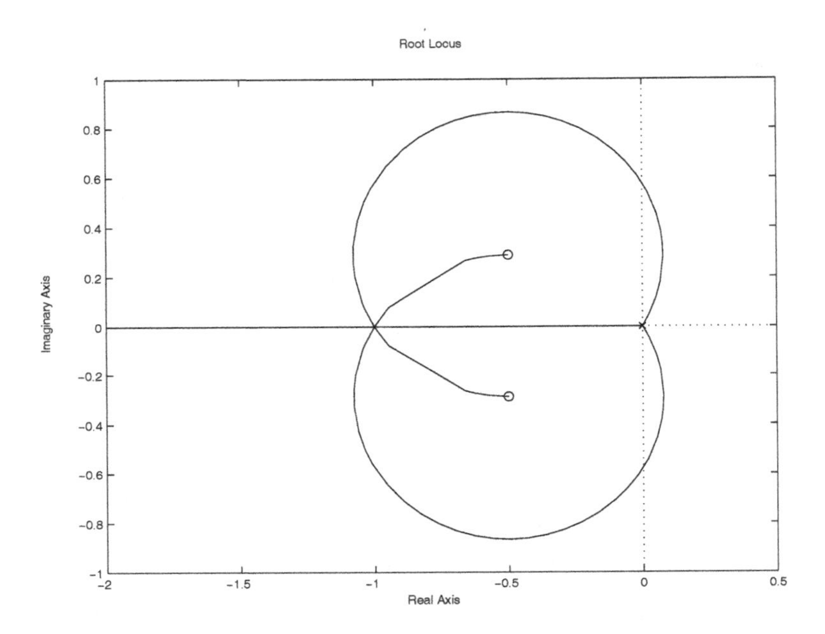

Fig. 11.27 The Root Locus Diagram of $G_c(s)G_p(s) = \frac{3s^2+3s+1}{s^3}$

11.b. The polynomial whose zeros we must check is $s^3 + 3Ks^2 + 3Ks + K$. The Routh array is:

$$\begin{array}{lr} 1 & 3K \\ 3K & K \\ 3K - 1/3 & 0 \\ K & 0 \end{array}$$

Thus, the system is stable for all $K > 1/3$.

11.c. Yes, the system is unstable at low gains and stable at high gains, and that comes through clearly from the root locus diagram and from the Routh array.

11.d. Not really. The gain margin tells us up to what gain a system is stable. Here the system is only stable *beyond* a certain gain.

11.e. Here $X(s) = 2/s^3$ and the closed loop transfer function is $T(s) = ((s+1)^3 - s^3)/(s+1)^3$. Thus, the output of the system, $y(t)$, satisfies:

$$Y(s) = \frac{2}{s^3} - \frac{2}{(s+1)^3}.$$

We find that $y(t) = t^2(1 - e^{-t})u(t)$. Clearly $y(t)$ tracks the input.

11.6.7 *Exercise 13*

13.a. As the degree of the denominator is seven higher than that of the numerator, there are seven branches that tend to infinity. The angles of the asymptotes must be $-180^\circ/7 + n360^\circ/7$. That is, the angles of the asymptotes are approximately $-26^\circ, 26^\circ, 77^\circ, 129^\circ, 180^\circ, 231^\circ$, and 283°. Thus there are four asymptotes in the right half-plane and three in the left half-plane.

The entire negative half of the real axis is part of the root locus. One branch must start at $s = 0$ and six at $s = -1$. Let's consider the other point(s) from which branches enter and/or leave the axis.

We must solve :

$$\begin{aligned}\frac{d}{ds}G_p(s) &= -\frac{1}{s^2(s+1)^{12}}((s+1)^6 + 6s(s+1)^5)\\ &= -\frac{1}{s^2(s+1)^7}(7s+1) = 0.\end{aligned}$$

Clearly, $s = -1/7$ is a point at which branches enter or leave the real axis. As the point is located between *poles*, branches must leave the axis at that point.

Putting all the information together, we find that the root locus must look like the root locus of Figure 11.28.

13.b. As the system is unstable for high enough gains, its gain margin is finite.

11.6.8 *Exercise 15*

15.a. The system has three poles—at $s = 0, -1 + j$ and $-1 - j$, and it has no zeros. The root locus has one branch that starts at the origin and continues out to $-\infty$. The other two branches start at the other two poles and must be complex conjugates of one another. Additionally, because

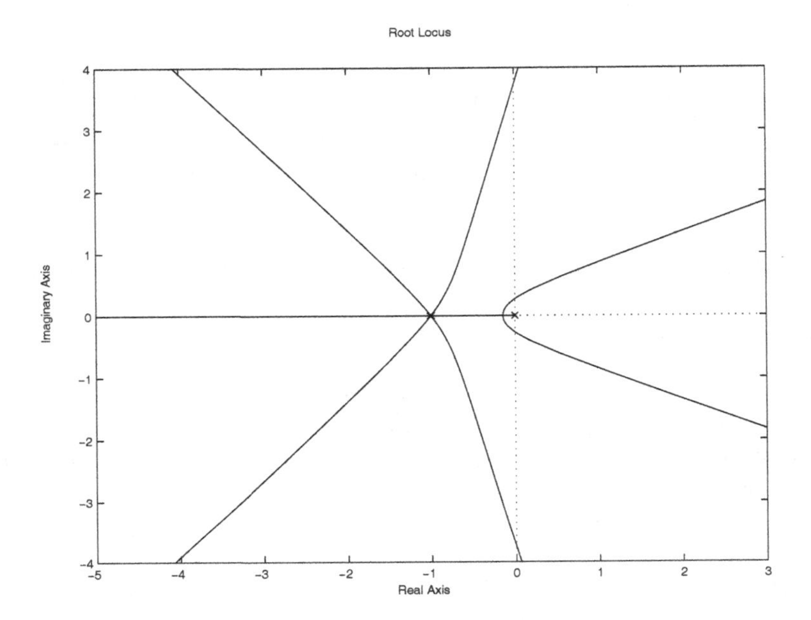

Fig. 11.28 The Root Locus Diagram of $G_p(s) = \frac{1}{s(s+1)^6}$

there are three more poles than zeros, the two branches must "balance" the branch heading off to $-\infty$ by heading towards the right half-plane and have real parts that tend to $+\infty$. See Figure 11.29.

15.b. The system's gain margin is finite.

15.c. To calculate the gain margin, consider the open-loop transfer function:

$$\frac{K}{s^3 + 2s^2 + 2s}.$$

The associated closed-loop transfer function is:

$$T(s) = \frac{K}{s^3 + 2s^2 + 2s + K}.$$

The Routh array associated with the denominator is:

$$\begin{array}{ll} 1 & 2 \\ 2 & K \\ 2 - K/2 & 0 \\ K & 0 \end{array}$$

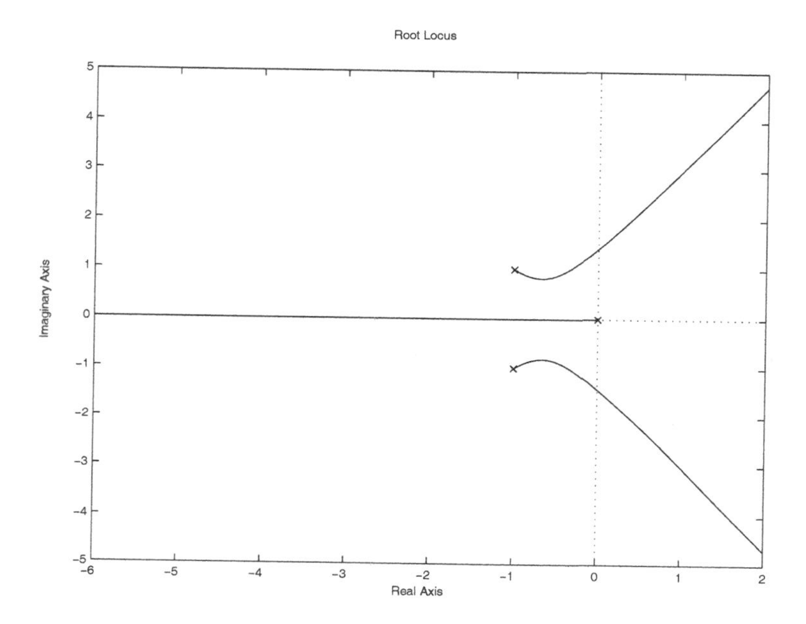

Fig. 11.29 The Root Locus Diagram of $G_p(s) = \frac{1}{s(s^2+2s+2)}$

We find that the system is stable as long as $0 < K < 4$. That is, the gain margin is finite—as we found in the previous section.

15.d. As the open-loop transfer function has one pole at the origin, the system tracks a step input but not a ramp input.

11.6.9 *Exercise 17*

17.a. The associated closed-loop transfer function is:

$$T(s) = \frac{Ks^2 + Ka^2}{s^3 + (3+K)s^2 + 2s + Ka^2}.$$

The Routh array associated with the denominator is:

$$\begin{array}{ll} 1 & 2 \\ 3+K & Ka^2 \\ 2-(K/(3+K))a^2 & 0 \\ Ka^2 & 0 \end{array}$$

In order for the system to be stable, $K > 0$ and:

$$\frac{K}{3+K}a^2 < 2.$$

As the fraction above is always less than one, if $a^2 \leq 2$, the system is stable for all positive K. If not, then $K < 6/(a^2-2)$.

17.b. Clearly, the gain margin is infinite for $0 \leq a \leq \sqrt{2}$.

17.c. It is clear that the root locus diagram has two branches that start at $s = 0$ and $s = -1$ meet and then head towards $s = \pm ja$ and one branch that goes from $s = -2$ and continues on to $-\infty$. The interesting point is the angle at which the branches approach the imaginary zeros.

Consider $s = ja + \epsilon$. In order for such a point to lie on the root locus, we find that:

$$\begin{aligned}\angle G_c(ja+\epsilon)G_p(ja+\epsilon) &= \angle\frac{(2ja+\epsilon)\epsilon}{(ja+\epsilon)(ja+1+\epsilon)(ja+2+\epsilon)}\\ &\approx \pi/2 + \angle\epsilon - \pi/2 - \tan^{-1}(a) - \tan^{-1}(a/2)\\ &= \angle\epsilon - \tan^{-1}(a) - \tan^{-1}(a/2)\\ &= -\pi.\end{aligned}$$

It is easy to check that when $a = \sqrt{2}$ the sum of the two arctangent terms is $\pi/2$ so that $\angle\epsilon = -\pi/2$ and the branch approaches the imaginary axis by becoming tangent to the axis. For smaller values of a, $\angle(\epsilon) < -\pi/2$ and the branch approaches the imaginary axis through the left half-plane. For larger values of a, the branch approaches the imaginary axis through the right half-plane and must, therefore, be unstable above some gain. (To see the root locus diagrams for three different values of a, see Figure 11.30.)

11.7 Chapter 7

11.7.1 *Exercise 1*

1. The transfer function of a phase-lag compensator is:

$$G_c(s) = \frac{s/\omega_o + 1}{s/\omega_p + 1}, \omega_p < \omega_o.$$

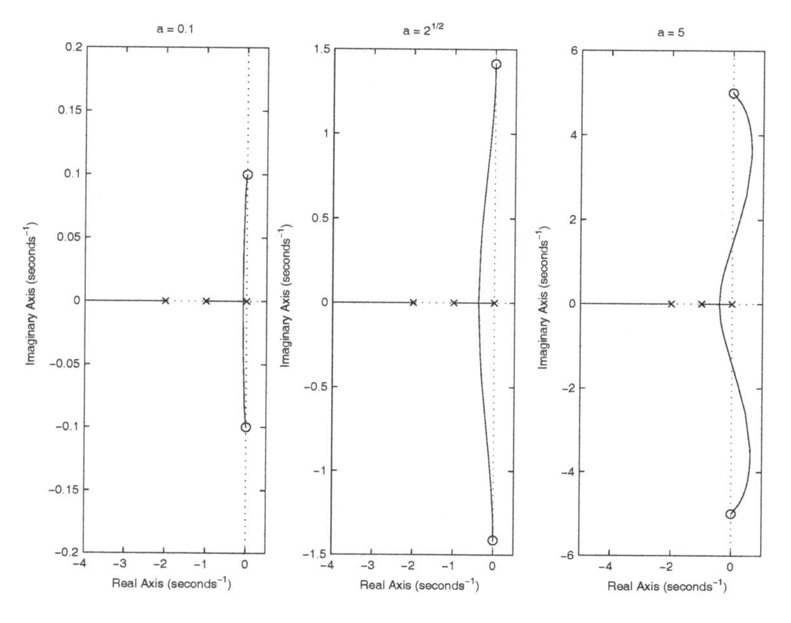

Fig. 11.30 The Root Locus Diagram of $G_p(s) = \frac{s^2+a^2}{s(s+1)(s+2)}$ for Three Values of a

As our plant is $G_p(s) = K/s^2$, we find that:

$$G_c(s)G_p(s) = \frac{K}{s^2}\frac{s/\omega_o + 1}{s/\omega_p + 1}.$$

Let us consider the root locus diagram that corresponds to such a system. As the degree of the denominator is two greater than the degree of the numerator, we know that the sum of the poles must be constant. Moreover we find that one branch of the root locus connects the pole at $-\omega_p$ to the zero at $-\omega_o$. As the pole on this branch moves to the left, there must be a compensatory net movement to the right on the part of the remaining poles. Since these poles start at the double pole at the origin, and since they cannot remain on the real axis (as the rules do not put any branches there), we find that these two poles leave the real axis. Thus, they have the same real part. In order for there to be a net movement to the right, both poles must move to the right—into the RHP. That is, the system is unstable. See Figure 11.31 for a plot of the root locus of such a system when $\omega_p = 1$ and $\omega_o = 2$.

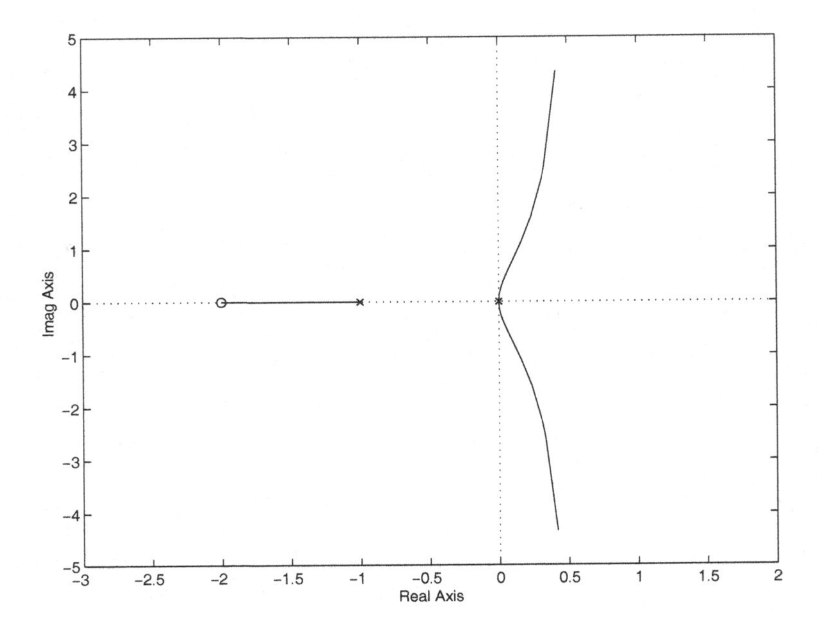

Fig. 11.31 The Root Locus Diagram of $G_c(s)G_p(s)$

11.7.2 *Exercise 3*

3. Here, we have:

$$G_p(s) = \frac{50}{s(2s+1)}, \text{ and } G_c(s) = K_P + K_D s.$$

Clearly:

$$G_c(s)G_p(s) = \frac{50(K_P + K_D s)}{s(2s+1)}$$

has one pole at the origin. Thus, the steady-state output of the system to a unit step input will be 1.

Now let us consider the root locus diagram that corresponds to such a system. Let us write the system in the form:

$$G_c(s)G_p(s) = 50K_P \frac{(1 + \frac{K_D}{K_P} s)}{s(2s+1)}.$$

Taking $50K_P$ as the gain parameter to be used in the root locus diagram, we find that we must plot the root locus diagram that corresponds to:

$$\frac{(1 + \frac{K_D}{K_P}s)}{s(2s+1)}.$$

This system has two poles—0 and $-1/2$—and one zero $-K_P/K_D$. If we choose $K_P/K_D > 1/2$, then we find that the root locus must:

- Have branches that connects the poles 0 and $-1/2$.
- Have branches that connects the zeros $-K_P/K_D$ and $-\infty$.
- Have branches that leave from some point between the two poles and reenter the real axis at some point between the two zeros.

An example of such a plot (with $K_P = K_D = 1$) is given in Figure 11.32.

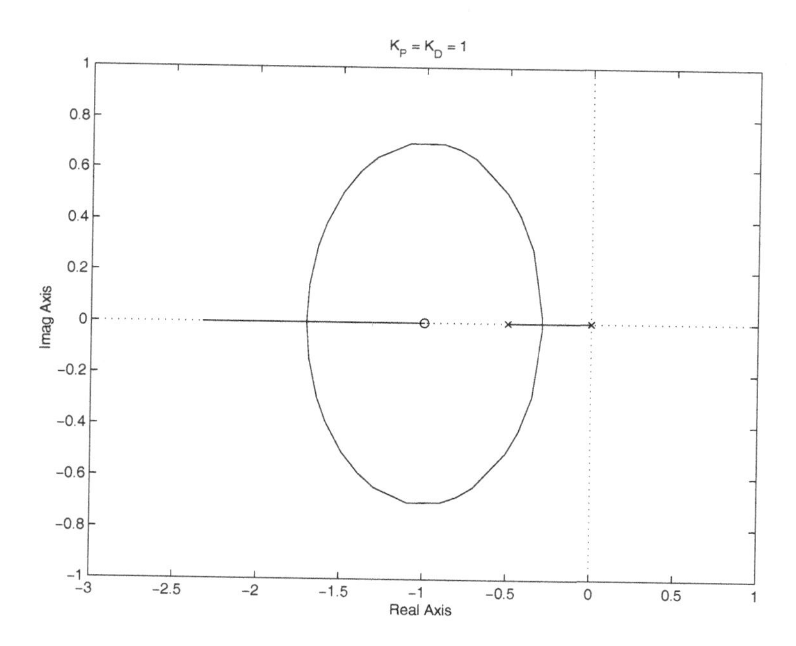

Fig. 11.32 The Root Locus Diagram of $G_c(s)G_P(s)$

We can fix K_P/K_D to be as large as we want. Then by taking K_P sufficiently large, we can force both of the poles of our system to be smaller than $-K_P/K_D$. That is, we can make the system as fast as we would like.

11.7.3 *Exercise 5*

5.a. The plant is:

$$G_p(s) = \frac{50}{s(2s+1)}.$$

We are to design a phase-lag compensator that will give us a phase margin of 45°. The first step is to find the value ω_1 for which:

$$\angle G_p(j\omega_1) = -180^\circ + 45^\circ + 5^\circ = -130^\circ.$$

Clearly:

$$\begin{aligned}\angle G_p(j\omega) &= \angle \frac{50}{j\omega(2j\omega+1)} \\ &= \angle 50 - \angle j\omega - \angle(2j\omega+1) \\ &= 0^\circ - 90^\circ - \tan^{-1}(2\omega).\end{aligned}$$

We find that:

$$\omega_1 = \frac{\tan(40^\circ)}{2} = 0.4195.$$

Our compensator has the form:

$$G_c(s) = \frac{s/\omega_o + 1}{s/\omega_p + 1}, \quad \omega_p < \omega_o.$$

From the standard rules for the design of a phase-lag compensator, we find that

$$w_o = \omega_1/10 = 0.04195.$$

We also know that:

$$\omega_p = \omega_o/|G_p(j\omega_1)|.$$

We find that:

$$|G_p(j\omega_1)| = 91.3, \text{ and } \omega_p = 0.00045956.$$

We plot the Bode plots in Figure 11.33. Note that we have achieved the phase margin we desired.

5.b. The step response of the system is given in Figure 11.34.

5.c. Because $G_p(s)$ has a pole at zero, the steady state response to a unit step input is 1. Because the compensator has a pole very near to zero the combined system takes a long time to reach its steady state.

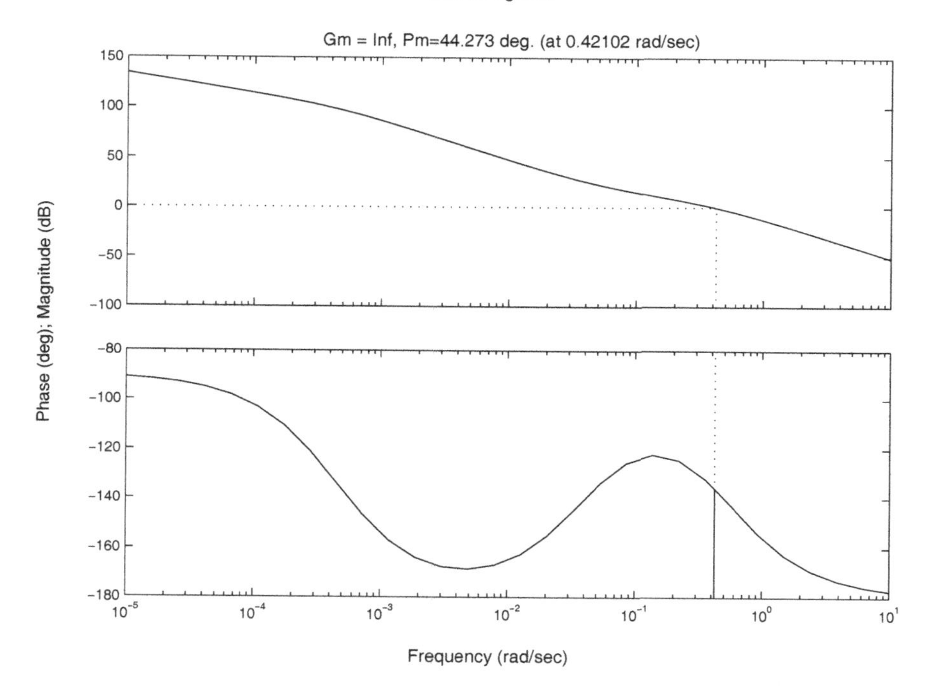

Fig. 11.33 The Bode Plots of the System

Another way to see that the system should take a long time to settle down is to use the formula:

$$T_s \approx \frac{8}{\tan(\varphi_M)\omega_1} = \frac{8}{\tan(45^{\circ})\omega_1} = \frac{8}{0.4195} = 19.07.$$

This is in reasonable accord with Figure 11.34.

11.7.4 *Exercise 7*

7.a. To find the phase margin, we must find those points for which $|G_p(j\omega)| = 1, \omega > 0$. In our case, it is pretty clear that only such point is $\omega = 1$. (This can, of course, be checked using MATLAB.) Plugging in $\omega = 1$, we find that:

$$G_p(j) = \frac{2}{j(j+1)^2} = -1 = 1\angle - 180^{\circ}.$$

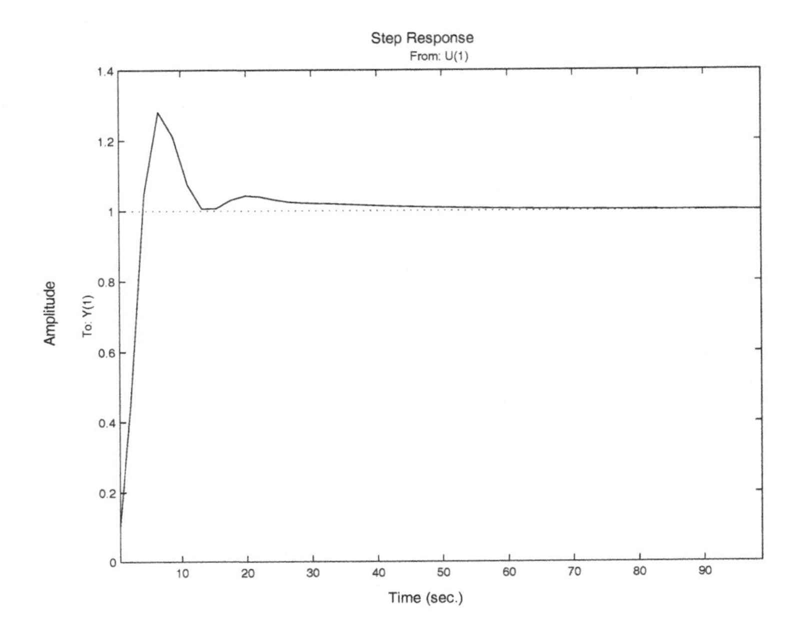

Fig. 11.34 The Step Response of the System

We find that the phase margin is zero degrees (and the gain margin is zero dB).

7.b. We must now design a lag-lead compensator to increase the phase margin to 55° while leaving the point at which it occurs—$\omega = 1$—alone. We do this by first designing a phase-lead compensator to increase the phase at $\omega = 1$ by 60° (remember the $-5°$ that the phase-lag compensator may add) and then adding a phase-lag compensator to make the gain at $\omega = 1$ equal to one. Our phase-lead compensator has the form:

$$G_{lead} = \frac{s/\omega_O + 1}{s/\omega_P + 1}, \quad \omega_O < \omega_P.$$

We would like:

$$\omega_{max} = \sqrt{\omega_O \omega_P} = 1 \tag{11.2}$$

and we want the phase at that point to be:

$$60^\circ = \tan^{-1}\left\{\frac{1}{2}\left(\sqrt{\frac{\omega_P}{\omega_O}} - \sqrt{\frac{\omega_O}{\omega_P}}\right)\right\}. \tag{11.3}$$

Rearranging (11.3), we find that:

$$\frac{1}{2}\left(\sqrt{\frac{\omega_P}{\omega_O}} - \sqrt{\frac{\omega_O}{\omega_P}}\right) = \sqrt{3}.$$

Multiplying this through by $\omega_{max} = \sqrt{\omega_O \omega_P}$ and rearranging the result slightly, we find that:

$$w_P - w_O = 2\sqrt{3}. \tag{11.4}$$

As:

$$\omega_{max} = \sqrt{\omega_O \omega_P} = 1,$$

we find that:

$$\omega_O = \frac{1}{\omega_P}.$$

Substituting this in (11.4), we find that:

$$\omega_P^2 - 2\sqrt{3}\omega_P - 1 = 0.$$

This equation has two solutions:

$$\omega_P = \sqrt{3} \pm 2.$$

As ω_P must be greater than its reciprocal, ω_O, we find that:

$$\omega_P = 2 + \sqrt{3}, \quad \omega_O = \frac{1}{2+\sqrt{3}} = 2 - \sqrt{3}.$$

As the gain of the phase-lead filter at ω_{max} is just:

$$|G_{lead}(j\omega_{max})| = \sqrt{\omega_P/\omega_O} = 2 + \sqrt{3},$$

we need our phase-lag compensator to reduce the gain by $2+\sqrt{3}$ at $\omega = 1$. The transfer function of the phase-lag compensator is :

$$G_{lag}(s) = \frac{s/\omega_o + 1}{s/\omega_p + 1}.$$

Following the regular procedure for designing a phase-lag compensator, we find that:

$$\omega_o = 0.1 \cdot 1 = 0.1, \text{ and } \omega_p = \frac{0.1}{2+\sqrt{3}} = 0.02679.$$

We find that the lag-lead compensator's transfer function is:

$$G_{lag-lead}(s) = \frac{(2+\sqrt{3})s+1}{(2-\sqrt{3})s+1} \frac{10s+1}{10(2+\sqrt{3})s+1}.$$

7.c. From the Bode plots of Figure 11.35, we find that the phase margin of system is 55.6° and the gain margin is 13.4dB.

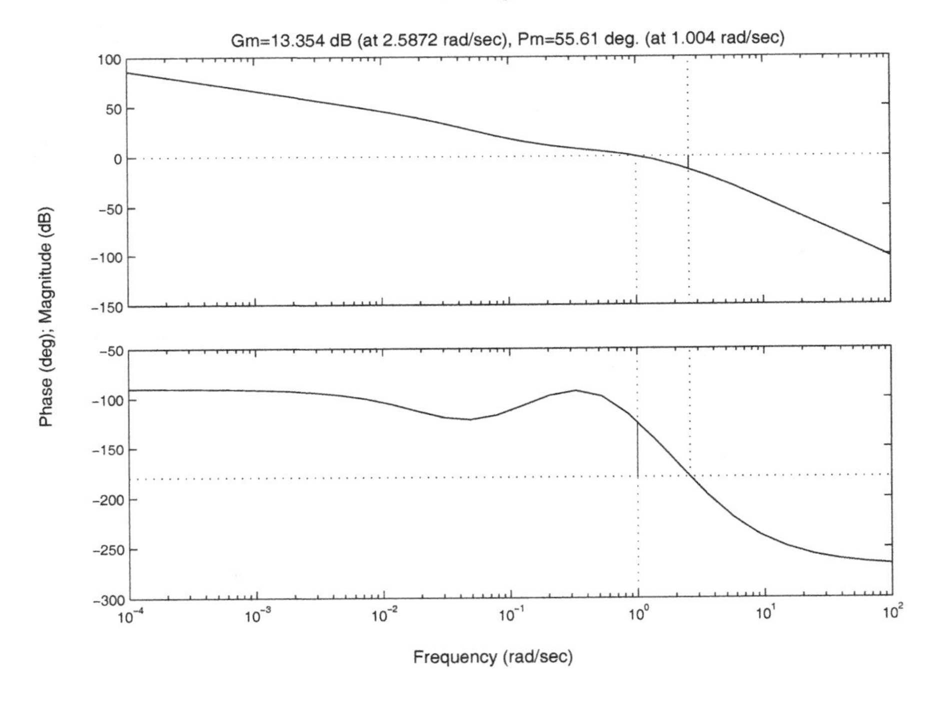

Fig. 11.35 The Bode Plots of the System

11.7.5 *Exercise 9*

9. If:

$$G_c(s) = K_P + \frac{K_I}{s} = \frac{sK_P + K_I}{s},$$

and:

$$G_p(s) = \frac{1}{s},$$

then:

$$G_c(s)G_P(s) = \frac{K_P s + K_I}{s^2}.$$

As this system has two poles at the origin, if the system is stable, it is able to track a ramp input.

Let us use the Routh-Hurwitz criteria to evaluate the stability of the system. We must check the zeros of:

$$\begin{aligned} 1 + G_c(s)G_p(s) &= 1 + \frac{K_P s + K_I}{s^2} \\ &= \frac{s^2 + K_P s + K_I}{s^2} \end{aligned}$$

The Routh array that corresponds to the numerator is:

$$\begin{array}{cc} 1 & K_I \\ K_P & 0. \\ K_I & 0 \end{array}$$

As long as $K_I, K_P > 0$, the system is stable.

11.7.6 *Exercise 11*

11.a.

$$T(s) = \frac{K}{s+K}, \qquad S^{T(s)}_{G_p(s)} = \frac{s}{s+K}.$$

11.b. Evaluating the integral (making use of the indefinite integral calculated in the text), we find that:

$$\begin{aligned} (1/2)\int_0^\infty & \left(\ln(\omega^2) - \ln(\omega^2 + K^2)\right) d\omega \\ &= (1/2) \lim_{R\to\infty} (\omega \ln(\omega^2) - 2\omega \\ &\quad - \left(\omega \ln(\omega^2 + K^2) - 2\omega + 2K\tan^{-1}(\omega/K)\right) |_0^R \\ &= -K\pi/2 \end{aligned}$$

where we have made use of the fact that $\lim_{x\to 0+} x\ln(x) = 0$. This agrees with the value given by the Bode sensitivity integral.

11.c. As K increases, the system is less and less sensitive to changes in the plant.

11.d. It is of relative degree one.

11.7.7 *Exercise 13*

13.a. Near the origin, $G_p(s) \approx d/s^n$. In order to draw the Nyquist plot, we need to determine the sign of d and the value of n. Considering Figure 7.34, we see that at low frequencies, the magnitude is decreasing at 40 db/dec. Thus, $n = 2$. This implies that at low frequencies, $G_p(j\omega) = d/(-\omega^2)$. As the phase at low frequencies is approaching -180°, d must be positive—its phase must be zero.

13.b. The results of 13.a. show that the system has a double pole at zero and, as $d > 0$, this results in a full or almost full circle in the counter-clockwise direction. One can use the Bode plots to find the behavior of the Nyquist plot for positive frequencies, and the negative frequencies can be taken care of by symmetry. The Nyquist plot of the system is given in Figure 11.36.

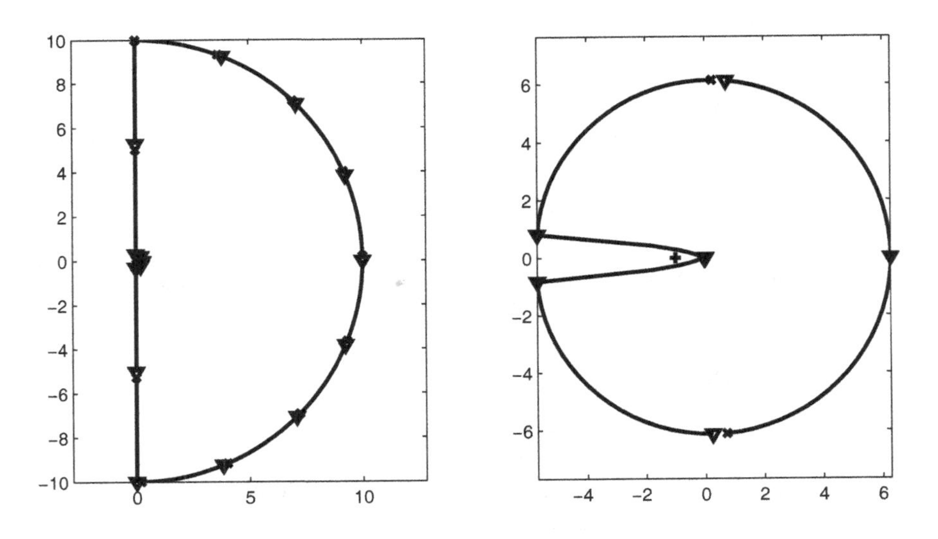

Fig. 11.36 The Nyquist Plot of the System

13.c. As the system to which $G_p(s)$ corresponds is marginally stable and the only frequency at which the frequency response tends to infinity is zero, the only pole of $G_p(s)$ on the imaginary axis is located at $s = 0$. Thus, $P = 0$. As the Nyquist plot does not encircle -1, $E = 0$. Thus, $N = 0$, and the system is closed-loop stable.

13.d. We make use of the approximate relationship:

$$0.2 \approx \frac{8}{\omega_1 \tan(60^\circ)}.$$

We find that $\omega_1 \approx 40/\sqrt{3} \approx 23\,\text{rad/s}$. Considering the Bode plots, the gain at this point is about $-50\,\text{dB} \approx 1/320$, and the phase is about -177°. Thus, $\Theta \approx 57^\circ$, and:

$$K_p \approx 320\cos(57^\circ) \approx 174, \qquad K_d \approx 320\sin(57^\circ)/\omega_1 \approx 11.7.$$

The transfer function of the compensator is $G_c(s) = 11.7s + 174$. (Note that at this point one should check that the system with the compensator is, in fact, stable.)

11.7.8 *Exercise 15*

15.a. The transfer function of the closed-loop system is:

$$\frac{K_D s + K_P}{s^3 + K_D s + K_P}.$$

Using the Routh Array or the results of Exercise 11, it is easy to see that there is no choice of K_P, K_D that stabilizes this system.

15.b. Making use of the approximate relation:

$$\omega_1 \approx \frac{8}{T_s \tan(\text{PM})},$$

we find that $\omega_1 = 0.92\,\text{rad/s}$. At that frequency, $G_c(j\omega_1)G_p(j\omega_1) = 1\angle -120^\circ$. Clearly, $G_c(j0.92) = 1/0.92^3\angle 150^\circ$. That is,

$$\begin{aligned} G_c(j0.92) &= -K_2 0.92^2 + K_0 + jK_1 0.92 \\ &= (0.92)^3\cos(150^\circ) + j(0.92)^3\sin(150^\circ). \end{aligned}$$

This fixes the value of K_1 but leaves us with one degree of freedom in the choice of K_0 and K_2. As the system is to track quadratic inputs, $G_c(s)G_p(s)$ must have a third-order pole at zero. Thus, $K_0 \neq 0$. In order to use the remaining degree of freedom wisely, we check the condition for the stability of the closed-loop system.

We consider the Routh array and find that it is:

$$\begin{array}{ll} 1 & K_1 \\ K_2 & K_0 \\ K_1 - K_0/K_2 & 0 \\ K_0 & 0 \end{array}$$

This system is stable when $K_2, K_0 > 0$ and $K_1 > K_0/K_2$. The values $K_0 = 0.1, K_1 = 0.42$, and $K_2 = 0.91$ provide us with a stable system.

15.c. Adding the additional K changes the system's Routh array to

$$\begin{array}{ll} 1 & KK_1 \\ KK_2 & KK_0 \\ KK_1 - K_0/K_2 & 0 \\ KK_0 & 0 \end{array}.$$

For this system to be stable, $K > K_0/(K_1K_2)$. That is, the system is only stable when K is *large* enough.

15.d. It is not really reasonable to speak of this system's gain margin. The assumption behind the gain margin is that too much gain is bad for the system. Here the reverse is true—too little gain is a problem.

11.7.9 *Exercise 17*

Considering the step responses, it seems that the settling times of both systems is about four seconds. Noting that the phase margin of the system with the lead compensator is 57.1°and its gain crossover frequency is $\omega_1 = 1.4\,\text{rad/s}$ and that the phase margin of the lag-lead compensated system is 65.6° and its gain crossover frequency is $\omega_1 = 1.0\,\text{rad/s}$, we find that the settling time of the lead compensated system should be approximately:

$$T_{s,lead} \approx \frac{8}{1.4\tan(57.1^\circ)} = 3.7\,\text{s},$$

and the settling time of the lag-lead compensated system should be

$$T_{s,lag-lead} \approx \frac{8}{1.0\tan(65.6^\circ)} = 3.6\,\text{s}.$$

That is, the settling times should both be approximately four seconds.

11.8 Chapter 8

11.8.1 *Exercise 1*

As always, when calculating the describing function, we consider the effect of our nonlinear element on $M\sin(\omega t)$. If we use such a signal as the input to our limiter, the output will have one of two possible forms:

- If $M \leq C$, then the signal passes through the limiter unchanged.
- If $M > C$, then the the sine-wave is clipped by the limiter. It will look like the signal in Figure 11.37.

We find that if $M \leq C$, then $D(\omega, M) = 1$—the sine wave is unchanged. If $M > C$, then we must find A and B by hand.

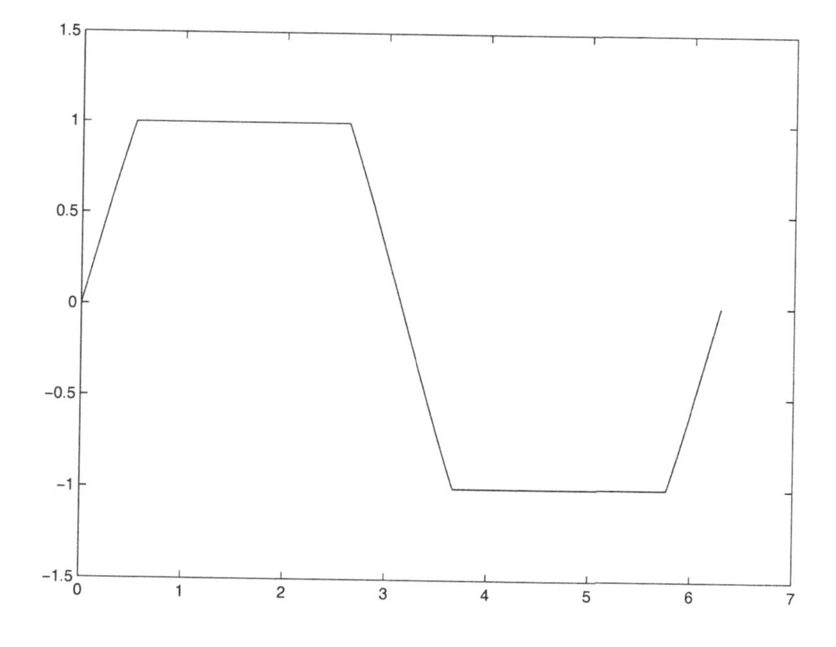

Fig. 11.37 A Clipped Sine Wave

As we have seen up until now, because of the symmetries of the clipped sine wave $B = 0$ and:

$$A = \frac{8}{T}\int_0^{T/4} \sin(\omega t) f(M\sin(\omega t))\, dt.$$

This integral must be broken into two pieces in order to calculate it. Until t reaches the value t_0 for which:

$$M\sin(\omega t_0) = C$$

it is clear that $f(M\sin(\omega t)) = M\sin(\omega t)$. After this point, $f(M\sin(\omega t)) = C$. Thus, we find that:

$$A = \frac{8}{T}\left(\int_0^{t_0} \sin(\omega t)M\sin(\omega t)\,dt. + \int_{t_0}^{T/4} \sin(\omega t)C\,dt\right).$$

To calculate the first integral, recall that:

$$\sin^2(\theta) = \frac{1-\cos(2\theta)}{2}.$$

We find that:

$$\begin{aligned}\int_0^{t_0} \sin(\omega t)M\sin(\omega t)\,dt &= M\int_0^{t_0} \sin^2(\omega t)\,dt\\ &= M\int_0^{t_0} \frac{1-\cos(2\omega t)}{2}\,dt\\ &= \frac{M}{2}\left(t_0 - \frac{\sin(2\omega t_0)}{2\omega}\right).\end{aligned}$$

As:

$$\sin(2\theta) = 2\sin(\theta)\cos(\theta),$$

we find that:

$$\int_0^{t_0} \sin(\omega t)M\sin(\omega t)\,dt = \frac{M}{2}\left(t_0 - \frac{\sin(\omega t_0)\cos(\omega t_0)}{\omega}\right).$$

We find that the second integral is equal to:

$$\begin{aligned}\int_{t_0}^{T/4} \sin(\omega t)C\,dt &= C\left.\frac{-\cos(\omega t)}{\omega}\right|_{t_0}^{T/4}\\ &\overset{\omega T=2\pi}{=} C\frac{-\cos(\pi/2)+\cos(\omega t_0)}{\omega}\\ &= \frac{C}{\omega}\cos(\omega t_0).\end{aligned}$$

We find that:

$$A = \frac{8}{T}\left\{\frac{M}{2}\left(t_0 - \frac{\sin(\omega t_0)\cos(\omega t_0)}{\omega}\right) + \frac{C}{\omega}\cos(\omega t_0).\right\}$$

As t_0 is defined by:

$$M\sin(\omega t_0) = C,$$

we find that:

$$\begin{aligned} t_0 &= \frac{\sin^{-1}(C/M)}{\omega} \\ \sin(\omega t_0) &= C/M \\ \cos(\omega t_0) &= \sqrt{1-\sin^2(\omega t_0)} \\ &= \sqrt{1-(C/M)^2}. \end{aligned}$$

Making use of the fact that $\omega T = 2\pi$, we find that:

$$\begin{aligned} A &= \frac{8}{T}\left\{\frac{M}{2}\left(\frac{\sin^{-1}(C/M)}{\omega} - \frac{(C/M)\sqrt{1-(C/M)^2}}{\omega}\right) + \frac{C}{\omega}\sqrt{1-(C/M)^2}\right\} \\ &= \frac{2}{\pi}\left\{M\left(\sin^{-1}(C/M) - (C/M)\sqrt{1-(C/M)^2}\right) + 2C\sqrt{1-(C/M)^2}\right\} \\ &= \frac{2}{\pi}\left(M\sin^{-1}(C/M) + C\sqrt{1-(C/M)^2}\right) \end{aligned}$$

Finally, we find that:

$$\begin{aligned} D(\omega, M) &= \frac{A}{M} \\ &= \begin{cases} 1 & M \le C \\ \frac{2}{M\pi}\left(M\sin^{-1}(C/M) + C\sqrt{1-(C/M)^2}\right) & M > C \end{cases} \end{aligned}$$

1.b. As $C \to 0$, we find that:

$$\begin{aligned} \sin^{-1}(C/M) &\to C/M \\ \sqrt{1-(C/M)^2} &\to 1. \end{aligned}$$

Thus, we find that as $C \to 0$, we have:

$$D(\omega, M) \to \frac{4C}{M\pi}.$$

That is, as the value at which the limiter starts to limit the signal tends to zero, the limiter's describing function tends to the describing function of a comparator of amplitude C. A little bit of reflection shows that this is as it ought to be.

11.8.2 *Exercise 3*

3.a. To determine the smallest value of K for which a limit cycle exists, we need to know the maximum value of the describing function as a function of M and C. We claim that the describing function is a non-increasing function of M. As for all $M < C$ the describing function equals one and as the describing function decreases towards zero as M increases without bound, we find that the describing function takes all values between 0 and 1 (including 1 but not including 0).

To see that the describing function is indeed a non-increasing function consider its derivative. For $M < C$ the describing function is constant and its derivative is zero. For $M > C$, we find that:

$$\begin{aligned}
\frac{d}{dM}D(\omega, M) &= \frac{d}{dM}\left(\frac{2}{\pi}\left(\sin^{-1}(C/M) + \frac{C}{M}\sqrt{1-(C/M)^2}\right)\right) \\
&= \frac{2}{\pi}\left(\frac{1}{\sqrt{1-(C/M)^2}}\frac{-C}{M^2} - \frac{C}{M^2}\sqrt{1-(C/M)^2}\right. \\
&\quad \left. + \frac{C}{M}\frac{1}{2\sqrt{1-(C/M)^2}}\frac{2C^2}{M^3}\right) \\
&= \frac{2}{\pi\sqrt{1-(C/M)^2}}\left(-\frac{C}{M^2} - \frac{C}{M^2}\left(1-(C/M)^2\right) + \frac{C}{M}\frac{C^2}{M^3}\right) \\
&= \frac{4C}{M^2\pi\sqrt{1-(C/M)^2}}\left(\frac{C^2}{M^2} - 1\right) \\
&< 0.
\end{aligned}$$

In order for the product of the describing function—which is never greater than one—and the frequency response to be equal to -1, the frequency response must be negative and have absolute value at least 1. The frequency response is negative when $\omega = 1$. At that point:

$$G_p(j) = -\mathrm{K}/2.$$

Thus, $\mathrm{K} \geq 2$. The smallest possible value of K for which the describing function technique predicts limit cycles is $\mathrm{K} = 2$.

3.b. If $\mathrm{K} = 2$, then the condition for a limit cycle is:

$$D(\omega, M) = 1.$$

Thus, $M \leq C$. In this region the derivative of the describing function is zero and there is no reason to expect stable limit cycles.

If K > 2, then the condition for a limit cycle is:

$$D(\omega, M) = \frac{2}{\mathrm{K}} < 1.$$

The M which solves this is less than C (as only for such M is the describing function less than one). In this region, as we saw above, the derivative of the describing function is negative. Thus, we expect the limit cycle predicted to be stable.

11.8.3 *Exercise 5*

5.a. As (8.6) can never be negative, and as the describing function of a comparator is positive, there can be no limit cycles.

5.b. Using the more exact transfer function one predicts limit cycles at the point at which the transfer function is negative—when $s = j\sqrt{1/\epsilon}$. (Note that this corresponds to a very high frequency.)

5.c. The crucial difference is that the more exact transfer function is negative at some frequency while the less exact transfer function is never negative.

11.8.4 *Exercise 7*

Note that:

$$\begin{aligned}\angle G_p(j\omega) &= 5\angle\frac{1}{j\omega} + 3\angle\frac{jw/\sqrt{3}+1}{jw\sqrt{3}+1} \\ &= -450^\circ + 3\angle\frac{jw/\sqrt{3}+1}{jw\sqrt{3}+1}.\end{aligned}$$

As:

$$\frac{s/\sqrt{3}+1}{s\sqrt{3}+1}$$

is a phase-lag compensator, we know that its phase is at a minimum for:

$$\omega_{min} = \sqrt{3}\frac{1}{\sqrt{3}} = 1.$$

It is easy to see that at $\omega = 1$ the phase of the filter is just -30°. We find that:

$$-450^\circ < \angle G_p(j\omega) \leq -540^\circ.$$

The phase is equal to -540° at one point—when $\omega = 1$. We find that:

$$\mathcal{I}(G_p(j\omega)) \leq 0$$

with equality when $\omega = 1$. Clearly then $\lambda(\omega)$ cannot be zero for any value of ω. I.e. this system does not support any limit cycles.

11.9 Chapter 9

11.9.1 *Exercise 1*

To find the eigenvalues, we must solve the equation:

$$\det(\mathbf{A} - \lambda I) = (-\lambda)(-1 - \lambda) = \lambda^2 + \lambda.$$

Clearly $\lambda = 0, -1$.

Using the now known values of the eigenvalues, we search for the eigenvectors. We start by looking for a vector $\vec{x} = [\, x_1 \; x_2 \,]^T$ that solves $\mathbf{A}\vec{x} = -\vec{x}$. This leads to the two conditions:

$$\begin{aligned} x_2 &= -x_1 \\ -x_2 &= -x_2. \end{aligned}$$

The second condition always holds. Thus, any vector of the form $[\, a \; -a \,]^T$ is an eigenvector whose eigenvalue is -1.

Now we search for the second eigenvector—the one that corresponds to $\lambda = 0$. We are now searching for vectors $\vec{x}$ that satisfy $\mathbf{A}\vec{x} = \vec{0}$. This leads to the two conditions:

$$\begin{aligned} x_2 &= 0 \\ -x_2 &= 0. \end{aligned}$$

Thus, vectors of the form $[\, a \; 0 \,]$ are eigenvectors with eigenvalue zero.

11.9.2 *Exercise 3*

3.a $\mathbf{A}$'s characteristic equation is

$$\det(\mathbf{A} - \lambda I) = (1 - \lambda)^2 = \lambda^2 - 2\lambda + 1 = 0.$$

Thus, we must check that $\mathbf{A}$ solves $\mathbf{A}^2 - 2\mathbf{A} + I = \mathbf{0}$. Plugging in $\mathbf{A}$ we find that:

$$\begin{pmatrix} 1 & 0 \\ 2 & 1 \end{pmatrix} - \begin{pmatrix} 2 & 0 \\ 2 & 2 \end{pmatrix} + \begin{pmatrix} 1 & 0 \\ 0 & 1 \end{pmatrix} = \begin{pmatrix} 0 & 0 \\ 0 & 0 \end{pmatrix}$$

which is just what we expected to see.

3.b. As $\mathbf{A}^2 = 2\mathbf{A} - \mathbf{I}$, $\mathbf{A}^3 = 2\mathbf{A}^2 - \mathbf{A} = 3\mathbf{A} - 2\mathbf{I}$.

11.9.3 *Exercise 4*

4.a. $y''(t) + 3y'(t) + 2y(t) = 3u(t)$.

4.b. Let $\vec{x}(t) = [y(t)\ y'(t)]^T$. We find that the state equations are:

$$\vec{x}'(t) = \begin{pmatrix} 0 & 1 \\ -2 & -3 \end{pmatrix} \vec{x}(t) + \begin{pmatrix} 0 \\ 3 \end{pmatrix} u(t), \qquad y(t) = [1\ 0]\vec{x}(t).$$

4.c. The controllability matrix is:

$$\text{Cont} = \begin{pmatrix} 0 & 3 \\ 3 & -9 \end{pmatrix}.$$

Its determinant is -9, so the system is controllable.

4.d. The observability matrix is:

$$\text{Obs} = \begin{pmatrix} 1 & 0 \\ 0 & 1 \end{pmatrix}.$$

As its determinant is 1, the system is observable.

11.9.4 *Exercise 5*

We must place the eigenvalues of the matrix $\mathbf{A}_f = \mathbf{A} - \vec{b}[\,k_1\ k_2\,]$ at -2 and -4.

We find that

$$\mathbf{A}_f = \begin{pmatrix} 0 & 1 \\ -2 - 3k_1 & -3 - 3k_2 \end{pmatrix}.$$

Thus:

$$\det\left(\mathbf{A}_f - \lambda\mathbf{I}\right)) = \det\begin{pmatrix} -\lambda & 1 \\ -2 - 3k_1 & -3 - 3k_2 - \lambda \end{pmatrix} = \lambda^2 + (3+3k_2)\lambda + 2 + 3k_1.$$

In order for the eigenvalues to be at -2 and -4, this polynomial must be equal to $(\lambda+2)(\lambda+4) = \lambda^2 + 6\lambda + 8$. Thus, $k_2 = 1$ and $k_1 = 2$. A sketch of the complete system is given in Fig. 11.38.

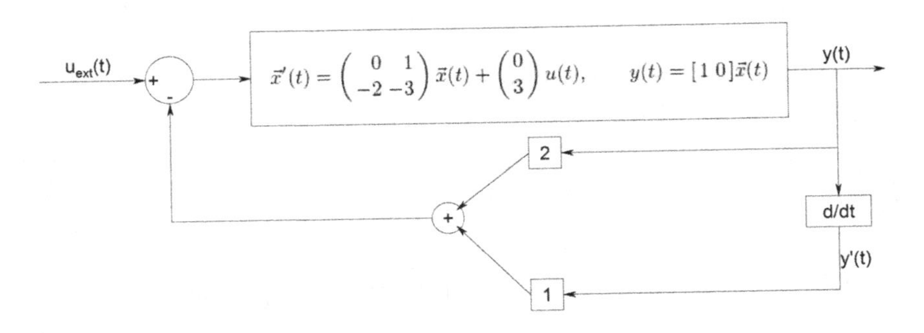

Fig. 11.38 The Complete System after Pole Placement

11.9.5 *Exercise 6*

(1) We find that:

$$(s^2 + 3s + 2)X(s) = U(s), \qquad Y(s) = (s+1)X(s).$$

Converting these equations to differential equations we find that:

$$x''(t) + 3x'(t) + 2x(t) = u(t), \qquad y(t) = x'(t) + x(t).$$

(2) If we let:

$$\vec{x}(t) = \begin{bmatrix} x(t) \\ x'(t) \end{bmatrix},$$

we find that:

$$\vec{x}'(t) = \begin{bmatrix} 0 & 1 \\ -2 & -3 \end{bmatrix} \vec{x}(t) + \begin{bmatrix} 0 \\ 1 \end{bmatrix} u(t), \qquad y(t) = \begin{bmatrix} 1\ 1 \end{bmatrix} \vec{x}(t).$$

(3) We find that:

$$\text{Con} = \begin{bmatrix} 0 & 1 \\ 1 & -3 \end{bmatrix}.$$

As $\det(\text{Con}) = -1$, we find that the system is controllable.

(4) We find that:

$$\text{Obs} = \begin{bmatrix} 1 & 1 \\ -2 & -2 \end{bmatrix}.$$

As det(Obs) = 0, we find that the system is not observable. Note that had we chosen to, we could have rewritten the transfer function as:

$$T(s) = \frac{s+1}{s^2+3s+2} = \frac{s+1}{(s+2)(s+1)} = \frac{1}{s+2}.$$

Thus, one does not really see the (internal) pole at -1 in the output of the system. It is not surprising that the system is not observable.

11.9.6 *Exercise 7*

If we let:

$$u = -\begin{bmatrix} k_1 & k_2 \end{bmatrix} \vec{x}(t) + u_{ext}(t),$$

then we find that the state equations can be rewritten as:

$$\vec{x}'(t) = \begin{bmatrix} 0 & 1 \\ -2-k_1 & -3-k_2 \end{bmatrix} \vec{x}(t) + \begin{bmatrix} 0 \\ 1 \end{bmatrix} u_{ext}(t).$$

The characteristic equation of this system is:

$$\det\left(\begin{bmatrix} -\lambda & 1 \\ -2-k_1 & -3-k_2-\lambda \end{bmatrix}\right) = \lambda^2 + (3+k_2)\lambda + 2 + k_1 = 0.$$

Comparing this with the equation with leading term λ^2 and with roots and -2 and -4—$\lambda^2+6\lambda+8=0$—we find that $k_1 = 6$ and $k_2 = 3$. This system is shown in Figure 11.39.

11.9.7 *Exercise 11*

11.a. $y'(t) + ay(t) = u(t)$.

11.b. $[y(t)]' = [-a][y(t)] + [1]u(t), \quad y(t) = [1]^T[y(t)]$.

11.c. Our very simple system's controllability matrix is just $[\vec{b}] = [1]$. This matrix clearly has a non-zero determinant.

11.d. Clearly, the observability matrix is $[1]^T$, its determinant is non-zero, and the system is observable.

11.e. As the only state is $y(t)$, state feedback involves feeding back $y(t)$—which is, essentially, proportional control.

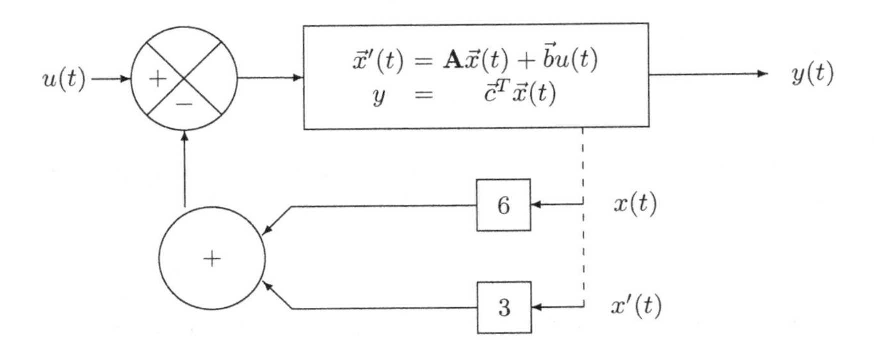

Fig. 11.39 The System with State Feedback Added

11.9.8 *Exercise 13*

If one changes the numerator to $s+3$, then taking $x(t)$ and $x'(t)$ to be the states of the system and letting $x''(t)+3x'(t) = u(t)$ and $y(t) = x'(t)+3x(t)$ one finds that:

$$\mathbf{A} = \begin{pmatrix} 0 & 1 \\ 0 & -3 \end{pmatrix}, \qquad \vec{c}^T = (\,3\ 1\,).$$

Thus,

$$\text{Obs} = \begin{pmatrix} 3 & 1 \\ 0 & 0 \end{pmatrix}.$$

Clearly $\det(\text{Obs}) = 0$, and the system is not observable.

11.9.9 *Exercise 15*

A simple calculation shows that:

$$\text{Cont} = \begin{pmatrix} 0 & 0 & 1 \\ 0 & 1 & -2 \\ 1 & -2 & 2 \end{pmatrix}.$$

We find that:

$$\det(\text{Cont}) = 1(0 \cdot (-2) - 1 \cdot 1) = -1.$$

Thus, the system is controllable.

11.10 Chapter 10

11.10.1 *Exercise 4*

As we know that the z-transform of p^k is:

$$\frac{z}{z-p},$$

and as we know that multiplication by z^{-1} is the z-domain is the same as shifting by one step in the time domain, we find that the inverse z-transform of:

$$z^{-1}\frac{z}{z-p}$$

is just the sequence $\{0, 1, p, p^2, \ldots\}$. This explains the results of exercise 3.

11.10.2 *Exercise 10*

10.a With $G_p(s) = 1/s^2$ and $H(s) = 1$, we find that:

$$w(t) = v(t) = \mathcal{L}^{-1}(1/s^3)(t) = \frac{t^2}{2}u(t).$$

We know that the z-transform of $Tku(k)$ is $Tz/(z-1)^2$. We need to find the z-transform of $T^2(k^2/2)u(k)$. Clearly this is:

$$V(z) = W(z) = \frac{T^2}{2}(-z)\frac{d}{dz}\left(\frac{z}{(z-1)^2}\right) = \frac{T^2}{2}\frac{z(z+1)}{(z-1)^3}.$$

Thus:

$$G_p(z) = \frac{T^2}{2}\frac{z+1}{(z-1)^2}.$$

10.b Let:

$$z = \frac{2+Tw}{2-Tw}.$$

We find that:

$$z+1 = \frac{4}{2-Tw}, \qquad z-1 = \frac{2Tw}{2-Tw}.$$

Thus, we find that:

$$G_p(w) = G_p(z)|_{z=\frac{2+Tw}{2-Tw}} = \frac{T^2}{2}\frac{4(2-Tw)}{(2Tw)^2} = 2T^2\frac{2-Tw}{(2Tw)^2}.$$

10.c With the above, we find that:

$$1 + G_p(w) = \frac{4T^2w^2 - 2T^3w + 4T^2}{(2Tw)^2}.$$

Clearly, the poles of the system are the zeros of the numerator. According to the Routh-Hurwitz criterion, these will all be in the left half-plane if and only if all the coefficients of the quadratic have the same sign. If $T > 0$, this cannot be. Thus, the system is unstable for all $T > 0$.

11.10.3 *Exercise 13*

As this signal is unchanged by the sample-and-hold circuit, in order to find the output as a function of time, we need only consider the inverse Laplace transform of:

$$\frac{1 - e^{-Ts}}{s}\frac{1}{s+1} = \frac{1 - e^{-Ts}}{s} - \frac{1 - e^{-Ts}}{s+1}.$$

Clearly this is:

$$v_o(t) = (1 - e^{-t})u(t) - (1 - e^{-(t-T)})u(t - T).$$

With $T = 0.1$, we find that:

$$v_o(k0.1 - m0.1) = \begin{cases} 0 & k = 0 \\ 1 - e^{-(k0.1 - m0.1)} & k = 1 \\ e^{-((k-1)0.l - m0.1)} - e^{-(k0.1 - m0.1)} & \\ \quad = (e^T - 1)e^{-(k0.1 - m0.1)} & k \geq 2 \end{cases}.$$

Let us solve the problem a second way. It is easy to see that we must find:

$$V(z) = \mathcal{Z}\left(\frac{e^{-mTs}}{s(s+1)}\right).$$

The inverse Laplace transform of $e^{-mTs}/(s(s+1)), 0 < m \leq 1$ is:

$$\begin{aligned} u(t - mT) - e^{-(t-mT)}u(t - mT) &= u(t - T) - e^{-(t-T-(m-1)T)}u(t - T) \\ &= u(t - T) - e^{(m-1)T}e^{-(t-T)}u(t - T). \end{aligned}$$

The z-transform of the samples of this function is:

$$V(z) = \frac{1}{z-1} - e^{(m-1)T}\frac{1}{z - e^{-T}}.$$

Thus, we find that:

$$\begin{aligned}\frac{z-1}{z}V(z) &= \frac{1}{z} - e^{(m-1)T}\frac{z-1}{z(z-e^{-T})}\\ &= \frac{1}{z} - e^{(m-1)T}\left(\frac{e^T}{z} + \frac{1-e^T}{z-e^{-T}}\right)\\ &= \frac{1-e^{mT}}{z} + \frac{(1-e^{-T})e^{mT}}{z-e^{-T}}.\end{aligned}$$

As our input has as its z-transform the function 1, we find that the above is the z-transform of the output of the system with the delay. Thus, we find that:

$$v_o(kT-mT) = \begin{cases} 0 & k=0\\ 1-e^{-(T-mT)} & k=1\\ (1-e^{-T})e^{mT}e^{-(k-1)T} = (e^T-1)e^{-(kT-mT)} & k\geq 2\end{cases}.$$

Plugging in $T = 0.1$ we find that the two expressions for $v_o(kT-mT)$ are identical.

11.10.4 *Exercise 16*

16.a We let $G_p(s) = s/(s+1)$ and $H(s) = 1$ in Figure 10.5. As $H(s) = 1$, we only need to calculate $V(z)$. We find that $V(s) = 1/(s+1)$. Thus, $v(t) = e^{-t}$ and $V(z) = z/(z-e^{-T})$. This shows that the transfer function is:

$$T(z) = \frac{\frac{z-1}{z-e^{-T}}}{1+\frac{z-1}{z-e^{-T}}} = \frac{z-1}{2z-1-e^{-T}}$$

16.b If the input to the system is a unit step function, then the z-transform of the output of the system is:

$$\frac{z-1}{2z-1-e^{-T}}\frac{z}{z-1} = \frac{z}{2z-1-e^{-T}} = \frac{1}{2}\frac{z}{z-(1+e^{-T})/2}.$$

We find that the samples of the output are:

$$\frac{1}{2}\{1, (1+e^{-T})/2, ((1+e^{-T})/2)^2, \ldots\}.$$

Note that when $T << 1$,

$$\frac{1+e^{-T}}{2} \approx \frac{1+1-T}{2} = 1 - T/2 \approx e^{-T/2}.$$

We see that we $T << 1$, the samples of the output are approximately:

$$\frac{1}{2}\{1, e^{-T/2}, e^{-T}, \ldots, e^{-kT/2}, \ldots\}.$$

Note that if one considers the system of Figure 10.5 *without* the sample and hold element, then the transfer function of the system is:

$$T(s) = \frac{\frac{s}{s+1}}{1 + \frac{s}{s+1}} = \frac{s}{2s+1}.$$

When one inputs a step function to the system, one finds that the output is:

$$\mathcal{L}^{-1}\left(\frac{s}{2s+1}\frac{1}{s}\right)(t) = \mathcal{L}^{-1}\left(\frac{1}{2s+1}\right)(t) = \frac{1}{2}e^{-t/2}.$$

Sampling this signal at $t = kT$, we find that the samples of the output are $e^{-kT/2}/2$. This is just what we found for the system *with* the sample and hold element in the limit as $T \to 0^+$.

11.10.5 *Exercise 17*

17.a We let $G_p(s) = 1/s$, $H(s) = 1$, and $D(z) = (1 - z^{-1})/T$ in Figure 10.7. As $H(s) = 1$, we only need to calculate $V(z)$. We find that:

$$V(z) = \mathcal{Z}(1/s^2)(z) = \mathcal{Z}(tu(t))(z) = \frac{Tz}{(1-z)^2}.$$

Thus, we find that:

$$T(z) = \frac{\frac{z-1}{Tz}\frac{z-1}{z}\frac{Tz}{(1-z)^2}}{1 + \frac{z-1}{Tz}\frac{z-1}{z}\frac{Tz}{(1-z)^2}} = \frac{1}{z+1}.$$

17.b If the input to the system is a unit step function, then the output is:

$$V_o(z) = \frac{1}{z+1}\frac{z}{z-1} = \frac{Az}{z+1} + \frac{Bz}{z-1} = \frac{1}{2}\left(\frac{z}{z-1} - \frac{z}{z+1}\right).$$

Thus, we find that:

$$v_o(kT) = \frac{1}{2}\left(1 - (-1)^k\right)u(k).$$

11.10.6 *Exercise 19*

19.a From the solution of Exercise 10 (see above) we know that:

$$V(z) = \mathcal{Z}(1/s^3)(z) = \frac{T^2}{2}\frac{z(z+1)}{(z-1)^3}, \qquad G_p(w) = \frac{2 - Tw}{2w^2}.$$

This system is unstable—as shown above. To stabilize it, we add the compensator $D(w) = K_P + K_D w$. We find that:

$$1 + G_p(w)D(w) = \frac{2w^2 + (2 - Tw)(K_P + K_D w)}{2w^2}.$$

The numerator of this expression—whose zeros are the poles of the final system—is:

$$(2 - TK_D)w^2 + (2K_D - TK_P)w + 2K_P.$$

Assuming that $K_P > 0$, we find that:

$$\begin{aligned} K_D &< 2/T \\ K_P &< 2K_D/T. \end{aligned}$$

19.b Converting from $D(w)$ to $D(z)$, we find that:

$$\begin{aligned} & D(w) = K_P + K_D w \\ \Rightarrow\ & D(z) = K_P + K_D \frac{2}{T}\frac{z-1}{z+1} \\ & = \frac{TK_P(z+1) + K_D 2(z-1)}{T(z+1)} \\ \Rightarrow\ & D(z) = \frac{TK_P(1+z^{-1}) + K_D 2(1-z^{-1})}{T(1+z^{-1})}. \end{aligned}$$

With $x(k)$ as the input and $y(k)$ as the output, we find that:

$$y(k) + y(k-1) = K_P(x(k) + x(k-1)) + 2K_D \frac{x(k) - x(k-1)}{T}.$$

Bibliography

K. J. Âström and R. M. Murray. *Feedback Systems: An Introduction for Scientists and Engineers.* Princeton University Press, Princeton, NJ, 2008.

Anonymous. In memoriam—Walter Richard Evans. `http://www.bently.com/articles/400inmem.asp`, 2000.

G. J. Borse. *Numerical Methods with MATLAB.* PWS Publishing Company, Boston, MA, 1997.

W. L. Brogan. *Modern Control Theory.* Quantum Publishers, Inc., New York, NY, 1974.

K. Y. Billah and R. H. Scanlan. Resonance, Tacoma Narrows bridge failure, and undergraduate physics textbooks. *American Journal of Physics*, 59(2), 1991.

R. V. Churchill and J. W. Brown. *Complex variables and applications.* McGraw-Hill, New York, NY, fifth edition, 1990.

M. A. B. Deakin. Taking mathematics to ultima thule: Horatio scott carslaw—his life and mathematics. *Autstralian Mathematical Society Gazette*, 24(1), 1997.

J. C. Doyle, B. A. Francis, and A. R. Tannenbaum. *Feedback Control Theory.* Macmillan, New York, NY, 1992.

S. Engelberg. Limitations of the describing function for limit cycle prediction. *IEEE Transactions on Automatic Control*, 47(11), 2002.

S. Engelberg. Tutorial 17: Control theory, part 3. *IEEE Instrumentation and Measurement Magazine*, 11(6), 2008.

G. C. Goodwin, S. F. Graebe, and M. E. Salgado. *Control System Design.* Prentice Hall, Upper Saddle River, NJ, 2000.

E. Gluskin. Let us teach this generalization of the final-value theorem. *European Journal of Physics*, 24:591–597, 2003.

B. Hamel. *C.E.M.V. no 17*, chapter Contribution á l'étude mathématique des systèmes de règale par tout-on-rien. Service Technique Aéronautique, Paris, 1949.

J. B. Hoagg and D. S. Bernstein. Nonminimum-phase zeros: Much to do about nothing. *IEEE Control Systems Magazine*, 27(3), 2007.

E. I. Jury. *Sampled-Data Control Systems.* John Wiley & Sons, Hoboken, NJ, 1958.

S. Kahne. Obituary: Nathaniel b. nichols 1914-1997. *Automatica*, 33(12):2101–2102, 1997.

J. Koughan. Undergraduate engineering review. `http://www.me.utexas.edu/~uer/papers/paper\_jk.html`, 1996.

F. L. Lewis. *Applied Optimal Control and Estimation.* Prentice-Hall, Englewood Cliffs, NJ, 1992.

G. Meinsma. Elementary proof of the Routh-Hurwitz test. *System and Control Letters*, 25:237–242, 1995.

J. J. O'Connor and E. F. Robertson. The Mactutor history of mathematics archive. `http://turnbull.dcs.st-and.ac.uk/~history/Mathematicians`.

G. Stein. Respect the unstable. *IEEE Control Systems Magazine*, 23(4):12–25, 2003.

I. Stewart. *Galois Theory.* CRC Press, Boca Raton, FL, third edition, 2003.

M. R. Stojic. In memoriam: Yakov Zalmanovich Tsypkin 1919-1997. `http://factaee.elfak.ni.ac.yu/facta9801/inmemoriam.html`.

G. B. Thomas Jr. *Calculus and Analytic Geometry.* Addison-Wesley Publishing Co., Reading, MA, 1968.

Ya. Z. Tsypkin. *Relay Control Systems.* Cambridge University Press, Cambridge, UK, 1984.

Wikipedia, the free encyclopedia. `http://www.wikipedia.com/wiki/Alexander_graham_bell`.

J. G. Ziegler and N. B. Nichols. Optimum settings for automatic controllers. *Transactions of the ASME*, 64:759–768, 1942.

Index

Abel, N. H., 88
aliasing, 330
alternating harmonic series, 309
available bandwidth, 239

Babylonians, 87
bandwidth
  available, 239
Bell, Alexander Graham, 42
BIBO stability, 26, 323
bilinear transform, 331
  behavior as $T \to 0$, 336
block diagram, 55
  feedback connection, 58
Bode plots, 42
  and stability, 132
Bode's sensitivity integral, 230, 233, 234, 236
  general form, 237
Bode, Hendrik, 42, 106

capacitor, 20, 25, 38
Cayley-Hamilton theorem, 278
chain rule, 7
characteristic equation
  of a matrix, 278
comparator, 247
compensation, 67, 169, 193
  attenuators, 193
  integral, 75, 241
  lag-lead compensation, 206
  open-loop, 69
  PD controller, 207
  phase-lag compensation, 194
    design, 196
  phase-lead compensation, 201
  PI controller, 207
  PID controller, 207
    design equations, 210
    tuning, 214, 224
    Ziegler-Nichols rules, 222
  proportional control, 225
conditional stability, 130
  example of, 169
controllability, 279
controllability matrix, 280
critically damped systems, 47

damping factor, 46
DC, 45
DC Motor, 51–52
  transfer function of, 51
degree
  of a polynomial, 31
delta function
  definition of, 312
describing function, 248
  comparator, 249
  comparator with dead zone, 256
  definition of, 249
  for prediction of limit cycles, 251
    graphical method, 258
  simple quantizer, 258
digital compensators, 321, 337

discrete-time systems
  stability of, 323

eigenvalues, 274
  relation to poles of transfer
      function, 277
eigenvectors, 274
encirclement, 106
Enestrom's theorem, 326
Evans, Walter R., 151

feedback, 55
  block diagram, 58
  disturbance rejection properties,
      80, 82, 83
  internal stability of systems with,
      78
  position, 193
  unity, 68
filter
  high-pass, 38
    frequency response, 41
  low-pass, 99
final value theorem, *see* Laplace
  transform, properties
  generalized, *see* Laplace transform,
      properties
frequency response, 328
  definition, 41
  high-pass filter, 41

gain crossover frequency, 136
gain margin, 128
  calculation of, 133
  infinite, 128
  of systems with zeros in the RHP,
      169

Heaviside function, 1
Heaviside, Oliver, 1
Hurwitz, A., 88
hybrid systems
  control of, 303
  definition, 303

ideal sampler, 312, 313
inductor, 20
integrator, 75
internal stability, 78, 183
inverse Laplace transform, 15

Kirchoff's voltage law, 21

Lagrange, J. L., 87
Laplace transform, 1
  of functions
    cosine, 5
    delta function, 312
    exponential, 2
    hyperbolic sine, 20
    sine, 3, 5
    unit step, 1
  of integro-differential equations, 19
  properties
    delay, 8
    differentiation, 4
    dilation, 7
    final value theorem, 8
    generalized final value
      theorem, 10, 34
    initial value theorem, 11
    integration, 5
    linearity, 2
    multiplication by e^{-at}, 6
    multiplication by t, 5
    second derivative, 4
    uniqueness, 12
  whose denominators have repeated
      roots, 6
left half-plane, 26
limit cycles, 248
  prediction, 251
  stability, 252
  Tsypkin's method, *see* Tsypkin's
      method
linear
  definition, 247
low-pass systems, 115

Maclaurin series, 282
MATLAB
  commands

bode, 43, 148
c2d, 319
d2c, 335
margin, 133, 148
nyquist, 120
residue, 31
rlocus, 152
roots, 30
ss(A,B,C,D), 292
step, 52
tf, 43, 318
assignments, 29
introduction, 29–32
numerical artifact of, 320
matrix
controllability, 280
norm of, 297
observability, 282
matrix differential equations, 273
calculation of, 276
inhomogeneous, 277
solutions
uniqueness of, 274, 278
matrix exponential
calculation of, 276
definition, 273
Meinsma, G., 91
Minorsky, Nichlas, 207
modern control, 271
integral compensation with, 290

natural frequency, 46
Newton's second law, 23
Nichols, N. B., 223
Nyquist plot, 114
delays, 121
fourth-order system
example of, 140
low-pass systems, 115
Nyquist, Harry, 106

observability, 281
observability matrix, 282
observer, 283
design of, 283
example, 287, 296
ODE, *see* ordinary differential equation
Ohm's law, 20, 38
operational amplifier, 48–51
configured as a buffer, 81
configured as an (inverting) amplifier, 49
gain-bandwidth product, 51
order
of a polynomial, 31
of system, 74
ordinary differential equation, 3
solution of
by means of the Laplace transform, 12
over-damped systems, 47
overshoot
definition, 45
in under-damped systems, 48

partial fraction expansion, 16
phase margin, 128
calculation of, 133
in second order systems, 135
relation to settling time, 136, 212
phase-lock loop, 179
piecewise continuous, 12
PLL, *see* phase-lock loop
pole, 15
pole placement, 280
connection to observability, 284
example, 289
pole-zero cancellation, 182
position feedback, 193
principle of the argument, 106
proof of, 107
pulse transfer function, 316
lack of, 340

rational functions, 88
relative degree, 231
relative stability, *see* stability, relative
resistor, 20, 25
resonance, 22, 64
resonant frequency, 24
rise time

definition, 45
in first order systems, 46
in under-damped systems, 48
root locus, 151
asymptotic behavior, 155
departure from real axis, 158
for systems that are not low-pass, 174
grouping near the origin, 166
plotting conventions, 152
symmetry, 154
treatment of delays, 176–179
Routh array, 94
number of rows in, 96
premature termination, 96
sign changes in, 95
Routh, E. J., 88
Routh-Hurwitz criterion, 91
delays, 125
proof of, 91
Ruffini, P., 88

sample-and-hold, 310
sampled-data systems, 309
second order system
phase margin, 135
settling time of, 136
sensitivity, 69
sensitivity integral, 230
settling time
second order system, 136
simple satellite, 52–53, 64, 101, 241
transfer function of, 53
sinusoidal steady state, 40
spring-mass system, 22
stability
and the location of poles, 324
BIBO, 26, 33, 38, 64
conditional, *see* conditional stability
definition, 26
marginal, 27, 38, 152, 324
relative, 127
state equation, 272
state variables, 271
for a system characterized by an ODE, 286
state equation, 272
steady state output
to a unit ramp, 75
to a unit step, 72–75
to sinusoids, *see* frequency response, definition
steady-state
amplification, 329
behavior, 330
superposition, 247

Tacoma Narrows Bridge, 29
time constant
definition of, 241
tracking inputs
of the form $t^n u(t)$, 76
conditions for, 77
transfer function, 38
relative degree of, 231
from input to error, 77, 191, 230
transient response
definition, 46
triangle inequality, 8, 325
generalized, 9, 11
Tsypkin's Method, 260

undamped systems, 64
under-damped systems, 47
overshoot, 48
rise time, 48
undershoot, 54, 87
unit pulse
definition of, 321
unity feedback, 68
unstable systems, 37, 38

variation of parameters, 277

waterbed effect, 231

z-transform, 303
modified, 303, 341
of a sine, 305
of a unit pulse, 321

of a unit step, 304
of an exponential, 304
properties
final value, 305
initial value, 307
linearity, 305
multiplication by k, 308
translation, 307
zero-order hold, 312, 313
zeros
left half-plane
necessary condition, 103
right half-plane, 54
undershoot, 54
Ziegler, J. G., 223
Ziegler-Nichols rules, 222
gain margin, 228
phase margin, 228
settling time, 229